AF322979

Climate Change Impacts on Urban Pests

CABI CLIMATE CHANGE SERIES

Climate change is a major environmental challenge to the world today, with significant threats to ecosystems, food security, water resources and economic stability overall. In order to understand and research ways to alleviate the effects of climate change, scientists need access to information that not only provides an overview of and background to the field, but also keeps them up to date with the latest research findings.

This series addresses many topics relating to climate change, including strategies to develop sustainable systems that minimize impact on climate and/or mitigate the effects of human activity on climate change. Coverage will encompass all areas of environmental and agricultural sciences. Aimed at researchers, upper level students and policy makers, titles in the series provide international coverage of topics related to climate change, including both a synthesis of facts and discussions of future research perspectives and possible solutions.

Titles Available

1. Climate Change and Crop Production
   Edited by Matthew P. Reynolds

2. Crop Stress Management and Climate Change
   Edited by José L. Araus and Gustavo A. Slafer

3. Temperature Adaptation in a Changing Climate: Nature at Risk
   Edited by Kenneth Storey and Karen Tanino

4. Plant Genetic Resources and Climate Change
   Edited by Michael Jackson, Brian Ford-Lloyd and Martin Parry

5. Climate Change Impact and Adaptation in Agricultural Systems
   Edited by Jürg Fuhrer and Peter J. Gregory

6. Livestock Production and Climate Change
   Edited by Pradeep K. Malik, Raghavendra Bhatta, Junichi Takahashi,
   Richard A. Kohn and Cadaba S. Prasad

7. Climate Change and Insect Pests
   Edited by Christer Björkman and Pekka Niemelä

8. Climate Change and Agricultural Water Management in Developing Countries
   Edited by Chu Thai Hoanh, Vladimir Smakhtin and Robyn Johnston

9. Climate Change Challenges and Adaptations at Farm-level
   Edited by Naveen Prakash Singh, Cynthia Bantilan, Kattarkandi Byjesh and
   Swamikannu Nedumaran

10. Climate Change Impacts on Urban Pests
    Edited by Partho Dhang

Climate Change Impacts on Urban Pests

Edited by

Partho Dhang

Independent Consultant, Manila, Philippines

CABI is a trading name of CAB International

CABI	CABI
Nosworthy Way	745 Atlantic Avenue
Wallingford	8th Floor
Oxfordshire OX10 8DE	Boston, MA 02111
UK	USA
Tel: +44 (0)1491 832111	Tel: +1 (617)682-9015
Fax: +44 (0)1491 833508	E-mail: cabi-nao@cabi.org
E-mail: info@cabi.org	
Website: www.cabi.org	

A catalogue record for this book is available from the British Library, London, UK.

Library of Congress Cataloging-in-Publication Data

Title: Climate change impacts on urban pests / edited by Partho Dhang.
Other titles: CABI climate change series ; 10.
Description: Boston, MA : CABI, [2016] | Series: CABI climate change series ; 10 |
    Includes bibliographical references and index.
Identifiers: LCCN 2016022770 (print) | LCCN 2016023630 (ebook) |
    ISBN 9781780645377 (hbk : alk. paper) | ISBN 9781780645384 (ePDF) |
    ISBN 9781786391162 (ePub)
Subjects: LCSH: Urban pests--Climatic factors. | Climatic changes.
Classification: LCC SB603.3 .C55 2016 (print) | LCC SB603.3 (ebook) |
    DDC 628.9/6091732--dc23
LC record available at https://lccn.loc.gov/2016022770

ISBN-13: 978 1 78064 537 7

Commissioning editor: Rachael Russell
Editorial assistant: Emma McCann
Production editor: Tim Kapp

Typeset by AMA DataSet Ltd, Preston, UK.
Printed and bound in the UK by Antony Rowe, CPI Group (UK) Ltd.

Contents

Contributors

Arthur G. Appel, Department of Entomology and Plant Pathology, 301 Funchess Hall, Auburn University, Auburn, Alabama 36849, USA. E-mail: appelag@auburn.edu

Martha Macedo de Lima Barata, Av. Epitacio Pessoa, 4560 apto 601, Lagoa – CEP 22471-003, Rio de Janeiro, Brazil. E-mail: baratamml@gmail.com

Richard F. Comont, Bumblebee Conservation Trust, Centre for Ecology & Hydrology Wallingford, Benson Lane, Crowmarsh Gifford, Oxfordshire OX10 8BB, UK. E-mail: richardcomont@gmail.com

Carrie Cottone, City of New Orleans Mosquito, Termite & Rodent Control Board, New Orleans, Louisiana, USA. E-mail: cbowens@nola.gov

Teresa de Troya, Head of Wood Protection Laboratory, Forest Research Centre (CIFOR), National Institute for Agricultural and Food Research and Technology (INIA) Ctra. Coruña km 7, 28040-Madrid, Spain.

Partho Dhang, 2410 Belarmino Street, Bangkal, Makati City, 1233, Philippines. E-mail: partho@urbanentomology.com

Ramesh C. Dhiman, National Institute of Malaria Research (ICMR), Sector-8, Dwarka, Delhi-110077, India. E-mail: r.c.dhiman@gmail.com

Donald Ewart, PO Box 1044 Research 3095, Australia. E-mail: don_ewart@drdons.net

Aleksandra Gliniewicz, National Institute of Public Health – National Institute of Hygiene, Chocimska 24, 00-791 Warsaw, Poland. E-mail: agliniewicz@pzh.gov.pl

Yousif E. Himeidan, Africa Technical Research Centre, Vector Health International, P.O. Box 15500, Arusha, Tanzania

Nildimar Alves Honório, Laboratório de Mosquitos Transmissores de Hematozoários, Instituto Oswaldo Cruz, Fundação Oswaldo Cruz, Avenida Brasil, 4365, Manguinhos, Rio de Janeiro, Brazil, CEP 21040-360

Grzegorz Karbowiak, W. Stefański Institute of Parasitology of Polish Academy of Sciences, Twarda 51/55, 00-818 Warsaw, Poland.

Philip G. Koehler, Department of Entomology and Nematology, UF/IFAS Extension, University of Florida, Gainesville, Florida 32611, USA. E-mail: pgk@ufl.edu

Agnieszka Królasik, National Institute of Public Health – National Institute of Hygiene, Chocimska 24, 00-791 Warsaw, Poland.

Magdalena Kutnik, Head of Biology Laboratory, Chair of CEN/TC 38, Technological Institute FCBA, Allées de Boutaut – BP 227, 33028 Bordeaux Cedex, France

Eliningaya J. Kweka, Division of Livestock and Human Diseases Vector Control, Tropical Pesticides Research Institute, PO Box 3024, Arusha, Tanzania and Department of Medical Parasitology and Entomology, Catholic University of Health and Allied Sciences, PO Box 1464, Mwanza, Tanzania. E-mail: pat.kweka@gmail.com

David Liszka, ICB Pharma, ul. Mozdzierzowcow 6a, Jaworzno, Poland. E-mail: david.liszka@icbpharma.pl

Humphrey D. Mazigo, Department of Medical Parasitology and Entomology, Catholic University of Health and Allied Sciences, PO Box 1464, Mwanza, Tanzania

Ewa Mikulak, National Institute of Public Health – National Institute of Hygiene, Chocimska 24, 00-791 Warsaw, Poland

Chester G. Moore, Department of Microbiology, Immunology & Pathology, Arthropodborne & Infectious Diseases Laboratory (AIDL), 1690 Campus Delivery, Colorado State University, Fort Collins, Colorado 80523-1690, USA. E-mail: chester.moore@colostate.edu

Domenica Morona, Department of Medical Parasitology and Entomology, Catholic University of Health and Allied Sciences, PO Box 1464, Mwanza, Tanzania

Stephen Munga, Centre for Global Health Research, Kenya Medical Research Institute, Kisumu, Kenya

Joanna Myślewicz, Warsaw University of Life Sciences, Nowoursynowska 166, 02-787 Warsaw, Poland

Lina Nunes, LNEC, National Laboratory for Civil Engineering, Structures Department, Lisbon, Portugal

Tamara Nunes de Lima-Camara, Laboratório de Entomologia em Saúde Pública/Culicídeos, Departamento de Epidemiologia, Faculdade de Saúde Pública da USP, Avenida Doutor Arnaldo, 715, Cerqueira César, São Paulo, Brazil, CEP 01246-904. E-mail: limacamara@usp.br

Roberto M. Pereira, Department of Entomology and Nematology, UF/IFAS Extension, University of Florida, Gainesville, Florida 32611, USA. E-mail: rpereira@ufl.edu

Mohamed F. Sallam, Department of Entomology and Nematology, UF/IFAS Extension, University of Florida, Gainesville, Florida 32611, USA and Department of Entomology, College of Science, Ain Shams University, Cairo 11566, Egypt. E-mail: mdsallam@ufl.edu

Steven R. Sims, Blue Imago LLC, 1973 Rule Ave, Maryland Heights, Missouri 63043, USA. E-mail: steve.sims@blueimago.com

Poonam Singh, National Institute of Malaria Research (ICMR), Sector-8, Dwarka, Delhi-110077, India. E-mail: punamsingh10@gmail.com

Nan-Yao Su, Fort Lauderdale Research and Education Center, University of Florida, Fort Lauderdale, Florida, USA. E-mail: nysu@ufl.edu

Marta Supergan-Marwicz, Department of General Biology and Parasitology, Medical University of Warsaw, Zwirki and Wigury 61, 02-091 Warsaw, Poland

Pawel Swietoslawski, ICB Pharma, ul. Mozdzierzowcow 6a, Jaworzno, Poland

Acknowledgements

I am thankful to all individual contributors for their immensely valuable insights. I would like to thank my parents for their continued inspiration, which helped me complete this book. I also acknowledge my special appreciation to the publisher, CAB International, for their acceptance of this book for publication.

Partho Dhang
15 April 2016
Manila, Philippines

1 Climate Change Effects on Urban Pest Insects

Richard F. Comont*

Bumblebee Conservation Trust, Centre for Ecology & Hydrology Wallingford, Oxfordshire, UK

1.1 Introduction

In recent decades, climate change has become a major issue worldwide. During the last century, the global temperature has increased by 0.8°C as the atmospheric CO_2 concentration has risen from 280 to 370 ppm: this is expected to double by 2100, with a concomitant temperature increase of 1.1–5.4°C above the 1900 level (IPCC, 2007). This also causes changes in existing weather patterns, including increased frequency and magnitude of droughts, floods and other extreme weather events (Easterling *et al.*, 2000; Greenough *et al.*, 2001; IPCC, 2007; Pall *et al.*, 2011). These changes are mainly attributed to increased human exploitation of natural resources, especially fossil fuels (releasing greenhouse gases to the atmosphere), but also industrialization, deforestation and urbanization affecting the character of the landscape (IPCC, 2007). Within the UK, the Central England Temperature record shows a *c.*1°C rise since the 1970s (Tubby and Webber, 2010). Future changes are difficult to predict reliably, as they depend upon changes within human society, but the UK Climate Impacts Programme (UKCIP) project (Murphy *et al.*, 2009) illustrates both 'high' and 'low' emissions scenarios based on an ongoing high dependence on fossil fuels or a rapid uptake of sustainable policies and technologies, respectively. While the degree of change varies, the British climate is expected to change to hotter, drier summers and milder, wetter winters than currently occur (Murphy *et al.*, 2009).

Urbanization itself has been defined as the replacement of nature by culture (Rolston, 1994). More specifically, the urban environment is a complex of wholly or partially anthropogenically influenced habitats created from natural areas, sometimes via the intermediate state of agricultural land (Robinson, 2005). In the context of the urban environment, 'urban' comprises not just the city centre, but also the surrounding suburbs, urban greenspace, industrial areas, and the rural–urban fringe (Robinson, 2005). This is an increasing, and inevitable, phenomenon: in 1960, 60% of the global population lived a rural or semi-rural existence in villages and small towns, but by 2030 it is predicted that 60% of the world's people will live in metropolitan areas, and the quality of life for future generations will be inexorably linked to the urban environment (Robinson, 2005).

This environment is very different to the original natural environment, with many habitats and their associated assemblages destroyed during the building process, but even dense modern conurbations contain a wide range of habitats suitable for wildlife, across a range of soil types, above- and below-ground areas, disturbed and

*E-mail: richardcomont@gmail.com

undisturbed regions. These may have been created intentionally for wildlife, such as areas of wildlife-friendly gardening in parks or private gardens; they may be highly altered versions, or equivalents, of the original natural environment, for instance, canalized rivers and street trees; or they may be entirely human-created and human-oriented habitats which are nevertheless suitable for colonization, such as food stores, landfill sites, or even roadside puddles (Davis, 1976; Gemmel and Connell, 1984; Robinson, 2005; Jones and Leather, 2013).

It is this last category where colonizers are particularly likely to have reputations as 'pests'. Pests have been defined as 'a plant or animal detrimental to humans or human concerns' (Merriam-Webster, 2015) or even, more broadly, as 'a competitor of humanity' (Encyclopaedia Britannica, 2015), but perhaps the best definition, following the well-known definition of a weed as 'a plant in the wrong place', is 'an animal in the wrong place'.

Pest status overall is generally associated with an economic loss, a potential medical threat, or a persistent nuisance. Pest status specifically in the agricultural ecosystem, and the development and deployment of control measures, is based largely on economics and the financial value of crops lost to the potential pest species. While the financial or human cost of some types of pest can be assessed in urban areas (human- or pet-disease vectoring, structural damage to buildings, etc.), control decisions are at least as likely to be based on the emotional response to the presence of a species. The extensively human-created nature of much of the urban environment – particularly indoors – means that animals are perhaps more likely to become pests than in the natural or agricultural environments, as the density of humans, and their increasing disconnection from nature, means that the tolerance threshold for wildlife may be as low as one individual. These are likely to be regarded as non-statutory nuisances – annoyances, perhaps – and generally a level of detriment necessary for a species to be called a true pest requires an abundance which most species never reach.

Urban environments generally impoverish biodiversity, and the original wildlife of the natural area can be removed entirely (Robinson, 2005). However, for some species with particular traits (Table 1.1), urbanization can hugely increase both the area of occupancy and the extent of occurrence at a local or global scale. This can happen either directly, by providing a thermal niche which would otherwise not be present, e.g. species which are synanthropic at high latitudes but which are found in the wild in more tropical regions, such as the leaf-mining fly *Liriomyza huidobrensis* (Blanchard) (IPPC, 2009); or indirectly, by providing large quantities of food, for example, *Anthrenus verbasci* (L.) or *Plodia interpunctella* (Hübner), attacking dried organic material inside buildings. These latter species are likely to be found in the wild in the same geographic area.

The processes driving these colonizers to reach pest proportions, however these are defined, work at several spatial scales. At the extremely local scale, provision of breeding habitat or attractants is important; for instance, mosquitoes will breed in ephemeral pools, e.g. inside discarded vehicle tyres, or wasps will nest in sheltered spaces such as sheds. But these are linked to landscape-scale factors (for instance, river floodplain management or wetland draining), and in turn to regional or even global factors such as species' distributional changes in response to climate change. There is considerable interlinkage between the scales, and effects can be chaotic and difficult to predict: for instance, malaria is thought to be indigenous to Britain, particularly in the Kent marshes (Dobson, 1989), but local- and landscape-scale changes such as marsh drainage and improvements in housing are thought to have eradicated it, although many mosquito species able to carry the plasmodium remain (Vaile and Miles, 1980; Dobson, 1989; Chin and Welsby, 2004). It is now thought that, as south-east Britain warms by 2–3.5°C under the effects of global climate change, malaria may return to the country (Chin and Welsby, 2004). There is a 16°C lower limit to plasmodium development which generates a clear isotherm of threat, which will move northwards as the

Table 1.1. Traits exhibited by the archetypal pest insect species.

Trait	Definition and reason
Synanthropy	A life-history strategy which brings the species into constant/repetitive contact with humans, either directly, e.g. blood-feeding parasites, or indirectly as a result of other life-history strategy parameters
Overwintering ability	A relatively low thermal minimum, allowing a good level of survival during outdoor overwintering, in turn allowing a population reservoir to build up, which is capable of infesting buildings. Organisms may not need this capability if they can instead overwinter on an alternative host (parasitic species) or indoors.
Resilience	Resilient populations capable of surviving unfavourable conditions, e.g. cold, application of pesticides, lack of food, etc. This can be either the result of a resistant life stage (e.g. flea pupae, which, until triggered by specific mechanical or chemical stimuli, can remain dormant for several months), or by having a large amount of redundancy built into the population structure (e.g. the potential for parthenogenesis in cockroaches, or the polygynous, ephemeral-nesting nature of several tramp ant species).
Diet	A generalist diet, particularly on human-created detritus, or alternatively a need to feed on humans, or on items precious to humans.
r-selection	A short generation time and high reproductive potential, allowing the species to take advantage of favourable conditions, particularly indoors, by producing large numbers of offspring quickly.
Anthropogenic dispersal	A high anthropogenic dispersive potential, giving the species a good chance of being transported between human settlements at a large scale, and between individual houses at a local scale.
Natural dispersal	A relatively low natural dispersal tendency. Species with a high natural dispersive ability will tend not to aggregate to nuisance numbers, but those with a low ability to disperse and a high reproductive potential will reach nuisance levels in a small area relatively quickly, while still allowing some element of dispersal, minimizing the chances of the entire population being eradicated.

climate warms, but the predicted changes in rainfall (up to 30% more during the winters, but up to 50% less during the summers) are likely to affect distribution and development of the vector species in a less predictable way (Chin and Welsby, 2004).

Insects are poikilothermic, and as such, temperature is generally the single most important environmental factor governing distribution, behaviour, survival, reproduction and development (Stewart and Dixon, 1989; Yamamura and Kiritani, 1998; Bale *et al.*, 2002; Dixon *et al.*, 2009). Other environmental factors, and species traits and interactions, play roles which can be significant, but these generally affect the realized niche within the fundamental niche of climate suitability (Roy *et al.*, 2009b; Jarošík *et al.*, 2015). The increased temperatures associated with climate change, coupled with anticipated extreme weather conditions (more and longer droughts, more frequent storms and increased rainfall) (IPCC, 2007), are predicted to impact on insect population dynamics. It has been recognized for many years that climate affects biochemical, physiological and behavioural processes in insects (Thomas and Blanford, 2003). Even modest changes to the climate are expected to have a rapid impact on the distribution and abundance of pest insects because of their physiological sensitivity to temperature, short life cycles, high mobility and high reproductive potential (Ayres and Lombardero, 2000; Roy *et al.*, 2009a). Many non-pest insects are already responding rapidly to climate change (Parmesan and Yohe, 2003) and expanding northwards (Asher *et al.*, 2001; Hickling *et al.*, 2006). Milder and shorter winters will lengthen the breeding period of some insects.

Several key traits are likely to govern the sensitivity of insect pests to climate change: these are outlined below.

- Persistence outside housing: The higher the proportion of a species' life cycle spent outside, the more likely it is that the species will be impacted by climate change. Species which have a dispersive phase outside buildings (e.g. clothes moths) are more likely to be impacted by climate changes than species which live entirely inside houses and premises.
- Overwintering ability: In temperate regions such as the UK, species which overwinter outside and which have a relatively low thermal minimum are more likely to increase in abundance as the climate warms. Species which do not diapause, where adults are produced and are active year-round, may be most likely to benefit from warmer environments.
- *r*-selection: Provided that the species persists in the wider environment, short generation times and a high reproductive potential are likely to allow the species to take advantage of increasingly warm conditions by producing large numbers of offspring quickly.
- Breeding sites: Species which breed outside are more likely to be impacted by climate change than species which breed exclusively indoors. The species which have aquatic or semi-aquatic breeding sites are likely to be negatively impacted by predicted summer droughts, and benefit from increased winter precipitation and flooding events.
- Resource specialism: Species which are highly specialized in resource use (e.g. diet, hosts or habitat use) are less likely to increase due to climate change than generalists, as the preferred host/prey must also increase with climate change.
- Dispersal: Species which are dispersed by humans or on hosts which undergo human-mediated transport may be better able to take advantage of climate change-induced increases in habitat availability than species that undergo active dispersal.

This opens up several similar ways in which the pest burden in an area may change.

Existing synanthropic species such as cockroaches may become able to survive outside, increasing the potential to spread without being transported by humans: though this may also mean that these species become less dependent on domestic areas. In Sweden, the scale insect *Pulvinaria floccifera* (Westwood), traditionally a pest within greenhouses, has recently become established outside (Gertsson, 2005). This is in line with the rapid spread of previously thermally limited (albeit non-synanthropic) species such as Roesel's bush-cricket (*Metrioptera roeselii* (Hagenbach)) across the UK (Gardiner, 2009). Wholly synanthropic species which can only survive in the human-inhabited environment are vulnerable to control measures, and warming which allows them to persist in the urban environment but outside dwellings is likely to increase their pest status: for instance, the American cockroach (*Periplaneta americana* (L.)), which re-infests buildings from a network of reservoir populations in sewage pipes (Robinson, 2005).

A similar, though less dramatic, degree of change in the reaction to increased temperatures could allow species currently occurring in the area at low, non-pest levels to greatly increase their abundance as they cease to be limited by temperature (directly, e.g. increased survival and fecundity, or indirectly, e.g. where dependent on a temperature-limited food species) and achieve pest proportions. In the Czech Republic, the European corn-borer, *Ostrinia nubialis* Hubner, is likely to become bivoltine under the warming conditions expected over the period 2025–2050, and thus pose a greater threat to crops (Trnka *et al.*, 2007).

Urban areas are already known to be warmer than surrounding countryside, a phenomenon known as the urban heat island effect, produced from the cumulative effect of increased heat from vehicles and machines, heat retention from man-made surfaces, the windbreak effect of buildings reducing heat dissipation, and reduced evaporative cooling (Oke, 1973). This means that they are likely to reach developmental threshold temperatures first and allow species or behaviours which will only gradually be seen outside of urban areas. For instance, the buff-tailed bumblebee (*Bombus terrestris* (L.)) is now winter-active across many British cities, at least partially because of climatic effects (Stelzer *et al.*, 2010; Owen *et al.*, 2013; Holland and Bourke, 2015).

A variant of the same process could see new species arrive and establish where previously they have been thermally limited or excluded, either from warmer areas of the same country or species which are currently non-native, arriving either naturally or via human-mediated dispersal. The scarce bordered straw, *Helicoverpa armigera* Hübner, is native to the Mediterranean, tropics and sub-tropics, but has spread naturally northwards with climate change, including a 'phenomenal' increase in Britain during 1969–2007 (Parsons and Davey, 2007), and has been recorded as a pest outdoors in Germany (FAO, 2008). The scale insect, *Icerya purchasi* Maskell, has spread naturally northwards with climate change, but is also spreading from introduction points well in advance of the natural spread, including outbreaks in Paris and London (Watson and Malumphy, 2004; Smith *et al.*, 2007).

Other climatic changes can have effects as well. The plant and human health pest *Thaumetopoea processionea* L. is thought to be spreading northwards partially as a result of a reduction in late frosts, while a major factor in the northwards spread of the mountain pine beetle, *Dendroctonus ponderosae* Hopkins, in the Pacific Northwest USA is increased drought stress on the food plant trees (FAO, 2008).

Lastly, and most difficult to predict, human behavioural and engineering changes in response to the changing climate may create new colonizable niches, or destroy existing ones. The UK has been predicted to have warmer, drier summers, potentially leading to more people-hours spent outside during summer evenings and a greater susceptibility to vector or nuisance-biting mosquito species (Chin and Welsby, 2004), but water shortages exacerbated by decreased summer rainfall (Thomsen, 1993) may limit the availability of water for these mosquitoes to breed, in turn reducing the risk. In European aphid species, climatic models predicted aphids across Europe to be recorded 8 days earlier in the year as a result of increased temperatures, but species traits, including food-plant and life-cycle type, had major effects on the changes (Harrington *et al.*, 2007; Bell *et al.*, 2015).

1.2 Non-native Species

The Convention on Biological Diversity (CBD: http://www.cbd.int) defines non-native species as 'a species, subspecies or lower taxon, introduced outside its natural past or present distribution; includes any part, gametes, seeds, eggs, or propagules of such species that might survive and subsequently reproduce' (CBD, 2002) (COP 6, decision VI/23). The term has many synonyms in the literature, including alien, introduced, exotic, foreign and non-indigenous. The use of the term 'introduced' in the definition is important, as it makes explicit the fact that a species can only be called non-native if it has arrived via human-mediated dispersal, or natural spread from a human-mediated introduction: simply benefiting from human-mediated environmental changes, such as the establishment of suitable habitat or an increase in temperature from climate change in order to arrive naturally, simply makes the species a new native. There is little disagreement that, as species continue to move northwards with climate change, more species are likely to arrive and establish in temperate areas such as the UK, but there is considerable disagreement over which species these will be, and the effects each of them is likely to have (Walther *et al.*, 2002, 2009; Parmesan, 2006; Blackburn and Jeschke, 2009; Blackburn *et al.*, 2009, 2015; Roy *et al.*, 2014).

Only a small subset of the non-native species introduced to a new area will actually become established, and a yet smaller subset of these will go on to become invasive (Lodge, 1993). This is often referred to as the 'tens rule', whereby one species establishes from every ten introduced, and of every ten established species, one will become invasive, although the exact proportions are often variable (Williamson, 1996;

Vander Zanden, 2005). For a species to become a pest, it must generally become invasive: indeed, the legal definition of an invasive species in the USA is 'an alien species whose introduction does or is likely to cause environmental or economic harm or harm to human health' (EO, 1999). Climate suitability is just one of many factors influencing the arrival, establishment and spread of non-native invertebrates (Smith *et al.*, 2007), and very few non-native species are likely to arrive and arise as pests solely because of climate change, at least in the near future.

Far more likely is a scenario whereby a species which is currently adventive or a casual non-spreading introduction is able to establish and spread out (Hardwick *et al.*, 1996; Thuiller *et al.*, 2006; Callaway *et al.*, 2012). Bio-climate models have been used to estimate the potential UK distribution of insects (Baker *et al.*, 1996; Gevrey and Worner, 2006; Poutsma *et al.*, 2008; Comont *et al.*, 2013; Purse *et al.*, 2014), marine crustaceans (Gallardo *et al.*, 2012) and many others: without fail, these analyses find that species are likely to at least increase the area of their fundamental niche, even if their actual realized niche remains constrained by local factors.

There is no shortage of potential organisms: a recent systematic assessment of non-native species established within Great Britain found there were at least 3758 non-native species in the country, although around one-third of those were somewhat ambiguous and status could not be allocated with complete confidence (Roy *et al.*, 2012b). Established non-native species numbered 1795, and although the vast majority of these are plants (74%), some 269 insect species fall within that category and could potentially be in the lag phase of an invasion.

The report lists 282 non-native species as currently invasive in Great Britain (Roy *et al.*, 2012b), and these are estimated to have a direct cost of £1.7 billion per year (Williams *et al.*, 2010). As this cost is largely related to control measures (Williams *et al.*, 2010), an increase in pest species could see a considerable rise in the financial burden of

non-natives. Most analyses have focused on the biodiversity impacts of non-native species (Evans *et al.*, 2011; Roy *et al.*, 2012a; Comont *et al.*, 2013), but for a species to become known as a pest, the biodiversity impacts are generally less important than when considered from an ecological point of view. To be known as a pest, a species (native or non-native) is generally a threat in at least one of three main ways: threats, or apparent threats, to health of humans or domestic animals; structural or aesthetic damage to property or amenity plantings; and nuisance impacts.

1.3 Medical Threats

Species can be medical pests in two main ways: causing direct harm, e.g. by biting, or by inducing illness, mainly by vectoring diseases. Most insect orders contain species which at least have potential to be pests but perhaps the most significant are Diptera (particularly the disease-vectoring mosquitoes) and Hymenoptera (stinging and biting bees, wasps and ants). Very few medically important pest species are entirely synanthropic (the most frequently encountered of these are probably head lice, *Pediculus humanus capitis* (de Geer), and bed bugs, *Cimex lectularius* L.): the majority are peridomestic, living and breeding in semi-natural areas and other urban habitats outside dwellings (Robinson, 2005). None of these is exclusively urban in distribution: indeed, many occur more widely in rural environments, but their importance as pests is exacerbated by their proximity to people in urban areas.

Largely synanthropic species, such as bed bugs, and human-parasitic species, such as head lice, or pubic lice, *Pthirus pubis* L., are unlikely to experience major, if any, changes in population size or areal extent from climate change, unless they drive a major change in human behaviour. It has been posited that a behavioural change (increased levels of pubic hair removal) may be reducing the area of habitat available for the pubic louse, with a corresponding drop in abundance of the species (Armstrong and Wilson, 2006), but more recent papers dispute both the existence of an excessive level of hair removal (Tiggemann and Hodgson, 2008; Herbenick *et al.*, 2010) and the decreased incidence of lice infestation (Dholakia *et al.*, 2014), which appears to be steady at around 2% (Uribe-Salas *et al.*, 1996; Anderson and Chaney, 2009). As the time period covered by these studies was the hottest decade yet recorded (Hansen *et al.*, 2010), some pattern would probably be becoming evident if considerable change was likely in the foreseeable future.

Worldwide, the major groups of disease-vectoring arthropods in the urban environment are mosquitoes, ticks and assassin bugs (Hemiptera: Reduviidae) (Robinson, 2005). In particular, several species of *Culex*, *Aedes* and *Anopheles* mosquitoes occur widely in urban environments, sometimes at high abundance, and are drawn to lights, carbon dioxide, and olfactory plumes given off by mammals, including humans. Many of these species ancestrally breed in tree holes and have switched easily to utilizing the plethora of ephemeral pools found in urban areas. The worldwide spread of species such as the Asian tiger mosquito, *Aedes albopictus* (Skuse), demonstrates the potential for human-aided dispersal and introduction, and forthcoming climate change is predicted to allow establishment at higher latitudes than is currently possible for the species (Roy *et al.*, 2009a). Both species have desiccation-resistant eggs, which allow them to survive long-distance travel to new areas, and an ability to breed in small containers, particularly old vehicle tyres, which has allowed the species to establish worldwide in warm climates (Hawley, 1988; Romi *et al.*, 2006). As this species is a major vector of arboviruses, including dengue fever (Hawley, 1988; Gratz, 2004; Messina *et al.*, 2015), and other pathogens and parasites (Cancrini *et al.*, 2003; Nunes *et al.*, 2015), climate change is likely to be the root cause of major public-health threats and potential regional epidemics.

Several other species are not as well adapted to living within urban areas, but have dispersal ranges long enough to allow them to breed outside cities and feed within them. Several other *Aedes* species, including (in North America) *Ae. dorsalis* (Meigen), *Ae. vexans* (Meigen), *Ae. squamiger* (Coquillett), *Ae. sollicitans* (Walker), and *Ae. taeniorhynchus* (Wiedemann), develop and breed in saltmarshes and floodplains, but have flight ranges of 6.4–64 km, enough to forage within huge areas of nearby cities (Robinson, 2005). As sea-level rise is a predicted major outcome of climate change (IPCC, 2007), abundance and areas of occupancy of these and similar species are likely to change significantly, although the direction of this change will result, at least in part, from human decisions on defending the coastline. Hard defences such as sea walls are likely to be used in some areas, and the rising sea levels are predicted to wash away mudflats and marshes in front of the walls, removing the mosquito's breeding areas, but the prohibitive cost means that soft management policies, such as managed retreat, are likely to be commonly employed (Nicholls *et al.*, 1995; Galbraith *et al.*, 2002). The policy of managed retreat is likely to result in a net increase of saltmarsh suitable for breeding, but existing areas may well be lost (Pethick, 1993, 2001; King and Lester, 1995), making future impacts difficult to predict.

Other species, such as *Culex tritaeniorhynchus* and *Culex tarsalis*, vectors for the Japanese and Western equine encephalitis viruses respectively, are floodwater species and thus may increase in abundance in the urban environment during and after flooding events, which are predicted to increase with climate change (Easterling *et al.*, 2000; Greenough *et al.*, 2001; Pall *et al.*, 2011).

The effects of peri-domestic species such as these, which are mainly (but not entirely) pests outside, are also likely to be mitigated by human behaviours (some of which are likely to change in response to climate change). In urban areas of Japan, a decrease in Japanese encephalitis has been attributed to a behavioural shift away from being outdoors in the evening, and towards staying indoors in air-conditioned houses watching the television (Robinson, 2005).

Insects can cause a severe reaction without vectoring a disease, however. Allergic disease is common, affecting around 40% of the world's population, and while most allergies cause minor reactions such as headaches, itching and rashes, some are severe, from difficulty breathing up to anaphylaxis requiring immediate hospitalization (Reisman, 1992; Van der Linden *et al.*, 1994; Robinson, 2005; Demain *et al.*, 2010). A wide range of insects can induce allergic reactions, either by their presence alone (e.g. cockroaches, fleas, etc.) or by stinging to defend themselves (bees, ants and wasps) (Robinson, 2005; Roy *et al.*, 2009a). As climate change is predicted to increase insect numbers, the encounter probability is likely to rise, although this should be balanced against the ongoing decline in abundance and distribution for many species (Conrad *et al.*, 2006; Potts *et al.*, 2010; Roy *et al.*, 2012a; Lebuhn *et al.*, 2013).

1.4 Nuisance Pests

The pest status of many species in urban areas, however, is based solely on an intolerance of the presence of species other than humans and companion animals within a living space, home or garden. The presence of 'unauthorized' insects is considered unacceptable, and control is based on an emotional or aesthetic threshold rather than a financial or cost–benefit-based analysis. Inside houses, the pests most commonly controlled at a low density are cockroaches (Insecta: Blattodea), silverfish (*Lepisma saccharina* L.), moth flies (Diptera: Psychodidae), and carpet beetles (Coleoptera: *Anthrenus* spp.) (Robinson, 2005).

The aesthetic threshold for control extends beyond the indoor environment. Gardens and municipal plantings, such as roadside amenity trees in urban environments, frequently suffer from pests which affect the appearance but do little to no real harm to the host (Raupp *et al.*, 2010; Zvereva *et al.*, 2010; Dale and Frank, 2014a). It has been found that urban trees grow quicker in warmer areas, but are also more water-stressed, which leaves them more vulnerable to pests (Coffelt and Schultz, 1993; Tubby and Webber, 2010; Dale and Frank, 2014a). As pest insects have been found to be more fecund in warmer areas and are likely to increase in abundance in such areas (Dale and Frank, 2014b), there is likely to be a multiplicative effect on the plantings with climate change, resulting in a poor outlook for the urban forest.

In the UK, section 79(1) (fa) of the Environmental Protection Act 1990 (as amended) states that 'any insects emanating from relevant industrial, trade or business premises and being prejudicial to health or a nuisance' shall constitute a statutory nuisance and thus be subject to controls. This wording indicates the two-limbed structure of legal nuisance: insects do not need to spread disease or provoke an allergic reaction to act as pests. Often, the mere presence of insects within dwellings is enough to provoke a response (particularly those species seen as dirty or threatening in some way), and numerous papers report the increase in prevalence of nuisance insects (Brenner *et al.*, 2003). For example, cockroaches are one of the most common pests found in apartments, homes, food handling establishments, hospitals and health care facilities worldwide (Bonnefoy *et al.*, 2008). Many people find cockroaches objectionable; in a London study, 80% of residents from uninfested apartments felt that cockroach infestations were worse than poor security, dampness, poor heating and poor repair (Majekodunmi *et al.*, 2002), while 90% of pesticides (which can themselves have human health effects) applied in apartments in the United States are directed at cockroaches (Whyatt *et al.*, 2002).

When abundant or long-lasting, insects can themselves constitute a nuisance in law, though this depends on the circumstances and the effects that the insects have on people or property. Nuisance has been defined as 'a condition or activity which unduly interferes with the use or enjoyment of land' (Dugdale and Jones, 2006: Paragraph 20-01), so, allowing insects to remain or cause an infestation may (in severe cases) comprise a legal nuisance. Generally, this is a private nuisance and a tort, or civil wrong (interfering with the right of property owners to use it free from unreasonable interferences from neighbouring property), but when the effect is widespread it may comprise a public nuisance, defined by Lord Justice Denning as a nuisance which is:

> so widespread in its range or so indiscriminate in its effect that it would not be reasonable to expect one person to take proceedings on his own responsibility to put a stop to it, but that it should be taken on the responsibility of the community at large (EWCA, 1957).

This means that if a class or group of people suffers to an unreasonable extent from insects emanating from a person's land, then a prosecution could be brought either by the local authority or by a private individual against the person responsible. As with private nuisance, an injunction could be sought to prevent recurrence of the nuisance in the High Court or the County Court (Roy *et al.*, 2009a).

Nuisance insects can emanate from a wide range of sources, but it is expected that most complaints of insect nuisance will be from the following sources: poultry and other animal houses; buildings on agricultural land including manure and silage storage areas; sewage treatment works; stagnant ditches and drains; landfill sites and refuse tips; waste transfer premises; commercial, trade or business premises; slaughterhouses; and used car tyre recycling businesses. Such areas (except for commercial, trade and business premises) are rare in city centres, and it is likely that most cases of purely nuisance insect infestation will be around the edges of urban areas, where these businesses are close to residential areas. As with other groups of outdoor-breeding insects, the increased temperatures from climate change are likely to decrease the generation time and increase the abundance, potentially increasing the severity of a current minor nuisance, or creating a nuisance at a greater distance from the source.

Most nuisances from insect pests are not severe or wide-ranging enough to be classed as statutory nuisances. The distinction between statutory nuisance and non-statutory nuisance will vary on a case-by-case basis; an insect will be a statutory nuisance in some scenarios but not others

(particularly in relation to the source and abundance of the insects).

1.5 Conclusions

Climate change is happening. Human behaviour may minimize the extent of changes, but it is too late to avoid the effects entirely; indeed, they are already being felt. These ongoing changes to climate are likely, at least in temperate areas, to increase the abundance and diversity of pest species, especially those with some element of their life cycle outside. The impacts of the pest species are likely to scale with abundance and diversity, and new pests are likely to arise as a result of introductions or better adaptations to the new climate regime.

With urban areas, changes to the built environment and to human behaviour (including adaptations to the changing climate) will affect the impacts of pest species in ways that are hard to predict. A further level of complexity is added by the potential for multi-species interactions: pest species do not live in isolation, but interact with, and are affected by, a multitude of other species, including parasites, predators and pathogens, which will themselves be affected by the changing environment.

References

Anderson, A.L. and Chaney, E. (2009) Pubic lice (*Pthirus pubis*): history, biology and treatment vs. knowledge and beliefs of US college students. *International Journal of Environmental Research and Public Health* 6, 592–600.

Armstrong, N.R. and Wilson, J.D. (2006) Did the 'Brazilian' kill the pubic louse? *Sexually Transmitted Infections* 82, 265–266.

Asher, J., Warren, M.S., Fox, R., Harding, P., Jeffcoate, G. and Jeffcoate, S. (2001) *The Millennium Atlas of Butterflies in Britain and Ireland.* Oxford University Press, Oxford, UK.

Ayres, M.P. and Lombardero, M.J. (2000) Assessing the consequences of global change for forest disturbance from herbivores and pathogens. *Science of the Total Environment* 262, 263–286.

Baker, R.H.A., Cannon, R.J.C. and Walters, K.F.A. (1996) An assessment of the risks posed by selected non-indigenous pests to UK crops under climate change. *Aspects of Applied Biology* 45, 323–330.

Bale, J.S., Masters, G.J., Hodkinson, I.D., Awmack, C., Bezemer, T.M., Brown, V.K., Butterfield, J., Buse, A., Coulson, J.C., Farrar, J., Good, J.E.G., Harrington, R., Hartley, S., Jones, T.H., Lindroth, R.L., Press, M.C., Symrnioudis, I., Watt, A.D. and Whittaker, J.B. (2002) Herbivory in global climate change research: direct effects of rising temperature on insect herbivores. *Global Change Biology* 8(1), 1–16.

Bell, J.R., Alderson, L., Izera, D., Kruger, T., Parker, S., Pickup, J., Shortall, C.R., Taylor, M.S., Verrier, P. and Harrington, R. (2015) Long-term phenological trends, species accumulation rates, aphid traits and climate: five decades of change in migrating aphids. *Journal of Animal Ecology* 84(1), 21–34.

Blackburn, T.M. and Jeschke, J.M. (2009) Invasion success and threat status: two sides of a different coin? *Ecography* 32, 83–88.

Blackburn, T.M., Cassey, P. and Lockwood, J.L. (2009) The role of species traits in the establishment success of exotic birds. *Global Change Biology* 15, 2852–2860.

Blackburn, T.M., Lockwood, J.L. and Cassey, P. (2015) The influence of numbers on invasion success. *Molecular Ecology* 24(9), 1942–1953.

Bonnefoy, X., Kampen, H. and Sweeney, K. (2008) *Public Health Significance of Urban Pests.* World Health Organization, Copenhagen, Denmark.

Brenner, B.L., Markowitz, S., Rivera, M., Romero, H., Weeks, M., Sanchez, E., Deych, E., Garg, A., Godbold, J., Wolff, M.S., Landrigan, P.J. and Berkowitz, G. (2003) Integrated pest management in an urban community: a successful partnership for prevention. *Environmental Health Perspectives* 111, 1649–1653.

Callaway, R., Shinn, A.P., Grenfell, S.E., Bron, J.E., Burnell, G., Cook, E.J., Crumlish, M., Culloty, S., Davidson, K., Ellis, R.P., Flynn, K.J., Fox, C., Green, D.M., Hays, G.C., Hughes, A.D., Johnston, E., Lowe, C.D., Lupatsch, I., Malham, S., Mendzil, A.F., Nickell, T., Pickerell, T., Rowley, A.F., Stanley, M.S., Tocher, D.R., Turnbull, J.F., Webb, G., Wootton, E. and Shields, R.J. (2012) Review of climate change impacts on marine aquaculture in the UK and Ireland. *Aquatic Conservation: Marine and Freshwater Ecosystems* 22, 389–421.

Cancrini, G., Frangipane di Regalbono, A., Ricci, I., Tessarin, C., Gabrielli, S. and Pietrobelli, M. (2003) *Aedes albopictus* is a natural vector of *Dirofilaria immitis* in Italy. *Veterinary Parasitology* 118, 195–202.

CBD (2002) *Decision VI/23* of the Conference of the Parties to the CBD*, Annex, footnote to the Introduction. Convention on Biological Diversity, Montreal, Canada.

Chin, T. and Welsby, P.D. (2004) Malaria in the UK: past, present, and future. *Postgraduate Medicine Journal* 80, 663–666.

Coffelt, M.A. and Schultz, P.B. (1993) Larval parasitism of orange striped oakworm (Lepidoptera: Saturniidae) in the urban shade tree environment. *Biological Control* 3, 127–134.

Comont, R.F., Roy, H.E., Harrington, R., Shortall, C.R. and Purse, B.V. (2013) Ecological correlates of local extinction and colonisation in the British ladybird beetles (Coleoptera: Coccinellidae). *Biological Invasions* 16(9), 1805–1817.

Conrad, K.F., Warren, M.S., Fox, R., Parsons, M.S. and Woiwod, I.P. (2006) Rapid declines of common, widespread British moths provide evidence of an insect biodiversity crisis. *Biological Conservation* 132(3), 279–291.

Dale, A. and Frank, S. (2014a) The effects of urban warming on herbivore abundance and street tree condition. *PLoS ONE* 9(7), article ID e102996. DOI: 10.1371/journal.pone.0102996.

Dale, A. and Frank, S. (2014b) Urban warming trumps natural enemy regulation of herbivorous pests. *Ecological Applications* 24, 1596–1607.

Davis, B.N.K. (1976) Wildlife, urbanisation and industry. *Biological Conservation* 10, 249–291.

Demain, J.G., Minaei, A.A. and Tracy, J.M. (2010) Anaphylaxis and insect allergy. *Current Opinion in Allergy and Clinical Immunology* 10, 318–322.

Dholakia, S., Buckler, J., Jeans, J.P., Pillai, A., Eagles, N. and Dholakia, S. (2014) Pubic lice: an endangered species? *Sexually Transmitted Diseases* 41(6), 388–391.

Dixon, A.F.G., Honek, A., Keil, P., Kotela, M.A.A., Šizling, A.L. and Jarošík, V. (2009) Relationship between the minimum and maximum temperature thresholds for development in insects. *Functional Ecology* 23(2), 257–264.

Dobson, M.J. (1989) History of malaria in England. *Journal of the Royal Society of Medicine* 82, 3–7.

Dugdale, A.M. and Jones, M.A. (eds) (2006) *Clerk & Lindsell on Torts*, 19th edn. Sweet & Maxwell, London, UK, chapter 20.

Easterling, D.R., Meehl, G.A., Parmesan, C., Changnon, S.A., Karl, T.R. and Mearns, L.O. (2000) Climate extremes: observations, modeling, and impacts. *Science* 289, 2068–2074.

Encyclopaedia Britannica (2015) Pest, vermin. Available at: http://www.britannica.com/science/pest-vermin (accessed 23 June 2015).

EO (1999) Invasive species. *Federal Register* 64(25), 6183–6186. Executive Order 13112.

Available at: https://www.gpo.gov/fdsys/pkg/FR-1999-02-08/pdf/99-3184.pdf.

Evans, E.W., Comont, R.F. and Rabitsch, W. (2011) Alien arthropod predators and parasitoids: interactions with the environment. *BioControl* 56, 395–407.

EWCA (1957) Lord Justice Denning in *Attorney-General vs PYA Quarries Ltd* [1957] 2 QB 169. England and Wales Court of Appeal (Civil Division) Decisions. Available at: http://www.bailii.org/ew/cases/EWCA/Civ/1958/1.html (accessed 12 October 2016).

FAO (2008) HLC/08/BAK/4 *Climate-related Transboundary Pests and Diseases*. Food and Agriculture Organization, Rome, Italy.

Galbraith, H., Jones, R., Park, R., Clough, J., Herrod-Julius, S., Harrington, B. and Page, G. (2002) Global climate change and sea level rise: potential losses of intertidal habitat for shorebirds. *Waterbirds* 25(2), 173–183.

Gallardo, B., Errea, M.P. and Aldridge, D.C. (2012) Application of bioclimatic models coupled with network analysis for risk assessment of the killer shrimp, *Dikerogammarus villosus*, in Great Britain. *Biological Invasions* 14(6), 1265–1278.

Gardiner, T. (2009) Macropterism of Roesel's bush-cricket *Metrioptera roeselii* in relation to climate change and landscape structure in eastern England. *Journal of Orthoptera Research* 18(1), 95–102.

Gemmel, R.P. and Connell, R.K. (1984) Conservation and creation of wildlife habitats on industrial land in Greater Manchester. *Landscape and Urban Planning* 11, 175–186.

Gertsson, C.-A. (2005) New species and new province-records of scale insects from Sweden (Hemiptera: Coccoidea) up to the year 2004. *Entomologisk Tidskrift* 126, 35–42.

Gevrey, M. and Worner, S.P. (2006) Prediction of global distribution of insect pest species in relation to climate by using an ecological informatics method. *Journal of Economic Entomology* 99, 979–986.

Gratz, N.G. (2004) Critical review of the vector status of *Aedes albopictus*. *Medical and Veterinary Entomology* 18, 215–227.

Greenough, G., McGeehin, M., Bernard, S.M., Trtanj, J., Riad, J. and Engelberg, D. (2001) The potential impacts of climate variability and change on health impacts of extreme weather events in the United States. *Environmental Health Perspectives* 109 Suppl. 2, 191–198.

Hansen, J., Ruedy, R., Sato, M. and Lo, K. (2010) Global surface temperature change. *Reviews of Geophysics* 48(4), RG4004. DOI: 10.1029/2010RG000345.

Hardwick, N.V., Armstrong, A.C., Ellis, S.A., Gladders, P., Yarham, D.J. and Parker, W.E. (1996) The impact of climate change on crop pests and diseases. *Aspects of Applied Biology* 45, 261–268.

Harrington, R., Clark, S.J., Welham, S.J., Verrier, P.J., Denholm, C.H., Hulle, M., Maurice, D., Rounsevell, M.D. and Cocu, N. (2007) Environmental change and the phenology of European aphids. *Global Change Biology* 13(8), 1550–1564.

Hawley, W.A. (1988) The biology of *Aedes albopictus*. *Journal of the American Mosquito Control Association* 4(Suppl. 1), 1–39.

Herbenick, D., Schick, V., Reece, M., Sanders, S. and Fortenberry, J.D. (2010) Pubic hair removal among women in the United States: prevalence, methods, and characteristics. *Journal of Sexual Medicine* 7, 3322–3330.

Hickling, R., Roy, D.B., Hill, J.K., Fox, R. and Thomas, C.D. (2006) The distributions of a wide range of taxonomic groups are expanding polewards. *Global Change Biology* 12, 450–455.

Holland, J.G. and Bourke, A.F.G. (2015) Colony and individual life-history responses to temperature in a social insect pollinator. *Functional Ecology* 29(9), 1209–1217.

IPCC (2007) *Contribution of Working Groups I, II and III to the Fourth Assessment Report of the Intergovernmental Panel on Climate Change*. Intergovernmental Panel on Climate Change, Geneva, Switzerland.

IPPC (2009) IPPC Official Pest Report, GBR-20/1. UK freedom from *Liriomyza huidobrensis* and *L. trifolii*. Intergovernmental Panel on Climate Change, Rome, Italy.

Jarošík, V., Kenis, M., Honěk, A., Skuhrovec, J. and Pyšek, P. (2015) Invasive insects differ from non-invasive in their thermal requirements. *PLoS ONE* 10(6), article ID e0131072. DOI: 10.1371/journal.pone.0131072.

Jones, E.L. and Leather, S.R. (2013) Invertebrates in urban areas: a review. *European Journal of Entomology* 109, 463–478.

King, S.E. and Lester, J.N. (1995) The value of salt marsh as a sea defence. *Marine Pollution Bulletin* 30(3), 180–189.

Lebuhn, G., Droege, S., Connor, E.F., Gemmill-Herren, B., Potts, S.G., Minckley, R.L., Griswold, T., Jean, R., Kula, E., Roubik, D.W., Cane, J., Wright, K.W., Frankie, G. and Parker, F. (2013) Detecting insect pollinator declines on regional and global scales. *Conservation Biology* 27(1), 113–120.

Lodge, D.M. (1993) Biological invasions: lessons for ecology. *Trends in Ecology and Evolution* 8, 133–137.

Majekodunmi, A., Howard, M.T. and Shah, V. (2002) The perceived importance of cockroach [*Blatta orientalis* (L.) and *Blattella germanica* (L.)] infestation to social housing residents. *International Journal of Environmental Health Research* 1, 80–87.

Merriam-Webster (2015) Pest. Available at: http://www.merriam-webster.com/dictionary/pest (accessed 23 June 2015).

Messina, J.P., Brady, O.J., Pigott, D.M., Golding, N., Kraemer, M.U.G., Scott, T.W., Wint, G.R., Smith, D.L. and Hay, S.I. (2015) The many projected futures of dengue. *Nature Reviews Microbiology* 13, 230–239.

Murphy, J., Sexton, D.M., Jenkins, G., Boorman, P., Booth, B.B. , Brown, C., Clark, R., Collins, M., Harris, G. and Kendon, E. (2009) *UK Climate Projections Science Report: Climate Change Projections*. Met Office Hadley Centre, Exeter, UK.

Nicholls, R.J., Leatherman, S.P., Dennis, K.C. and Volonte, C.R. (1995) Impacts and responses to sea-level rise: qualitative and quantitative assessments. *Journal of Coastal Research* 14, 26–43.

Nunes, M.R.T., Faria, N.R., de Vasconcelos, J.M., Golding, N. and Kraemer, M.U.G. (2015) Emergence and potential for spread of Chikungunya virus in Brazil. *BMC Medicine* 13. DOI: 10.1186/s12916-015-0348-x.

Oke, T.R. (1973) City size and the urban heat island. *Atmospheric Environment* 7(8), 769–779.

Owen, E.L., Bale, J.S. and Hayward, S.A.L. (2013) Can winter-active bumblebees survive the cold? Assessing the cold tolerance of *Bombus terrestris audax* and the effects of pollen feeding. *PLoS ONE.* 8(11), e80061 DOI: 10.1371/journal.pone.0080061.

Pall, P., Aina, T., Stone, D.A., Stott, P.A. and Nozawa, T. (2011) Anthropogenic greenhouse gas contribution to flood risk in England and Wales in autumn 2000. *Nature* 470, 382–385.

Parmesan, C. (2006) Ecological and evolutionary responses to recent climate change. *Annual Review of Ecology, Evolution, and Systematics* 37, 637–669.

Parmesan, C. and Yohe, G. (2003) A globally coherent fingerprint of climate change impacts across natural systems. *Nature* 421, 37–42.

Parsons, M.S. and Davey, P. (2007) The rise of the scarce bordered straw *Helicoverpa armigera* (Hübner) (Lep.: Noctuidae). *Entomologists Record and Journal of Variation* 119, 185–192.

Pethick, J. (1993) Shoreline adjustments and coastal management: physical and biological processes under accelerated sea-level rise. *The Geographical Journal* 159(2), 162–168.

Pethick, J. (2001) Coastal management and sea-level rise. *Catena* 42, 307–322.

Potts, S.G., Biesmeijer, J.C., Kremen, C., Neumann, P., Schweiger, O. and Kunin, W.E. (2010) Global pollinator declines: trends, impacts and drivers. *Trends in Ecology and Evolution* 25, 345–353.

Poutsma, J., Loomans, A.J.M., Aukema, B. and Heijerman, T. (2008) Predicting the potential geographical distribution of the harlequin ladybird, *Harmonia axyridis*, using the CLIMEX model. *BioControl* 53, 103–125.

Purse, B.V., Comont, R., Butler, A., Brown, P.M.J., Kessel, C. and Roy, H.E. (2014) Landscape and climate determine patterns of spread for all colour morphs of the alien ladybird *Harmonia axyridis*. *Journal of Biogeography* 42(3), 575–588.

Raupp, M.J., Shrewsbury, P.M. and Herms, D.A. (2010) Ecology of herbivorous arthropods in urban landscapes. *Annual Review of Entomology* 55, 19–38.

Reisman, R.E. (1992) Natural history of insect sting allergy: relationship of severity of symptoms of initial sting anaphylaxis to re-sting reactions. *Journal of Allergy and Clinical Immunology* 90, 335–339.

Robinson, W.H. (2005) *Urban Insects and Arachnids: a Handbook of Urban Entomology*. Cambridge University Press, Cambridge, UK.

Rolston, H. (1994) *Conserving Natural Value*. Columbia University Press, New York, USA.

Romi, R., Severini, F. and Toma, L. (2006) Cold acclimation and overwintering of female *Aedes albopictus* in Roma. *Journal of the American Mosquito Control Association* 22, 149–151.

Roy, H.E., Hails, R.S., Hesketh, H., Roy, D.B. and Pell, J.K. (2009a) Beyond biological control: non-pest insects and their pathogens in a changing world. *Insect Conservation and Diversity* 2, 65–72.

Roy, H.E., Beckmann, B.C., Comont, R.F., Hails, R.S., Harrington, R., Medlock, J., Purse, B. and Shortall, C.R. (2009b) *An Investigation into the Potential for New and Existing Species of Insect with the Potential to Cause Statutory Nuisance to Occur in the UK as a Result of Current and Predicted Climate Change*. DEFRA report NANR 274. Department for Environment, Food and Rural Affairs, London, UK.

Roy, H.E., Adriaens, T., Isaac, N.J.B., Kenis, M., Onkelinx, T., San Martin, G., Brown, P.M.J., Hautier, L., Poland, R., Roy, D.B., Comont, R.F., Eschen, R., Frost, R., Zindel, R., Van Vlaenderen, J., Nedved, O., Ravn, H.P., Gregoire, J., de Biseau, J. and Maes, D. (2012a) Invasive alien predator causes rapid declines of native European ladybirds. *Diversity and Distributions* 18(7), 717–725.

Roy, H.E., Bacon, J., Beckmann, B., Harrower, C.A., Hill, M.O., Isaac, N.J.B., Preston, C.D., Rathod, B., Rorke, S.L., Marchant, J.H., Musgrove, A.J., Noble, D.G., Sewell, J., Seeley, B., Sweet, N., Adams, L., Bishop, J., Jukes, A.R., Walker, K.J. and Pearman, D.A. (2012b) *Non-native Species in Great Britain: Establishment, Detection and Reporting to Inform Effective Decision Making.* DEFRA, London, UK.

Roy, H.E., Peyton, J., Aldridge, D.C., Bantock, T., Blackburn, T.M., Britton, R., Clark, P., Cook, E., Dehnen-Schmutz, K., Dines, T., Dobson, M., Edwards, F., Harrower, C., Harvey, M.C., Minchin, D., Noble, D.G., Parrott, D., Pocock, M.J.O., Preston, C.D., Roy, S., Salisbury, A., Schonrogge, K., Sewell, J., Shaw, R.H., Stebbing, P., Stewart, A.J. and Walker, K.J. (2014) Horizon scanning for invasive alien species with the potential to threaten biodiversity in Great Britain. *Global Change Biology* 20, 3859–3871.

Smith, R.M., Baker, R.H., Malumphy, C.P., Hockland, S., Hammon, R.P., Ostoja-Starzewski, J.C. and Collins, D.W. (2007) Recent non-native invertebrate plant pest establishments in Great Britain: origins, pathways, and trends. *Agricultural and Forest Entomology* 9(4), 307–326.

Stelzer, R.J., Chittka, L., Carlton, M. and Ings, T.C. (2010) Winter active bumblebees (*Bombus terrestris*) achieve high foraging rates in urban Britain. *PLoS ONE* 5(3), e9559. DOI: 10.1371/journal.pone.0009559.

Stewart, L.A. and Dixon, A.F.G. (1989) Why big species of ladybird beetles are not melanic. *Functional Ecology* 3, 165–177.

Thomas, M.B. and Blanford, S. (2003) Thermal biology in insect–parasite interactions. *Trends in Ecology and Evolution* 18(7), 344–350.

Thomsen, R. (1993) Future droughts, water shortages in parts of western Europe. *Eos, Transactions of the American Geophysical Union* 74(14), 161–165.

Thuiller, W., Richardson, D.M., Rouget, M., Proche, S. and Wilson, J.R.U. (2006) Interactions between environment, species traits, and human uses describe patterns of plant invasions. *Ecology* 87, 1755–1769.

Tiggemann, M. and Hodgson, S. (2008) The hairlessness norm extended: reasons for and predictors of women's body hair removal at different body sites. *Sex Roles* 59(11), 889–897.

Trnka, M., Muška, F., Semerádová, D., Dubrovský, M., Kocmánková, E. and Žalud, Z. (2007) European Corn Borer life stage model: Regional estimates of pest development and spatial distribution under present and future climate. *Ecological Modelling* 207, 61–84.

Tubby, K.V. and Webber, J.F. (2010) Pests and diseases threatening urban trees under a changing climate. *Forestry* 83, 451–459.

Uribe-Salas, F., Del Río-Chiriboga, C., Conde-Glez, C.J., Juárez-Figueroa, L., Uribe-Zúñaga, P., Calderón-Jaimes, E. and Hernández-Avila, M. (1996) Prevalence, incidence, and determinants of syphilis in female commercial sex workers in Mexico City. *Sexually Transmitted Diseases* 23, 120–126.

Vaile, M. and Miles, S.J. (1980) Mosquitoes and malaria in the North Kent marshes. *Community Medicine* 2, 298–301.

Van der Linden, P.W., Hack, C.E., Struyvenberg, A. and van der Zwan, J.K. (1994) Insect-sting challenge in 324 subjects with a previous anaphylactic reaction: current criteria for insect-venom hypersensitivity do not predict the occurrence and the severity of anaphylaxis. *Journal of Allergy and Clinical Immunology* 94, 151–159.

Vander Zanden, M.J. (2005) The success of animal invaders. *Proceedings of the National Academy of Sciences of the United States of America* 102, 7055–7056.

Walther, G.-R., Post, E., Convey, P., Menzel, A., Parmesan, C., Beebee, T.J.C., Fromentin, J., Hoegh-Guldberg, O. and Bairlein, F. (2002) Ecological responses to recent climate change. *Nature* 416, 389–395.

Walther, G.-R., Roques, A., Hulme, P.E., Sykes, M.T., Pyšek, P., Kuhn, I., Zobel, M., Bacher, S., Botta-Dukat, Z., Bugmann, H., Czucz, B., Dauber, J., Hickler, T., Jarosik, V., Kenis, M., Klotz, S., Minchin, D., Moora, M., Nentwig, W., Ott, J., Panov, V.E., Reineking, B., Robinet, C., Semenchenko, V., Solarz, W., Thuiller, W., Vila, M., Vohland, K. and Settele, J. (2009) Alien species in a warmer world: risks and opportunities. *Trends in Ecology and Evolution* 24, 686–693.

Watson, G.W. and Malumphy, C.P. (2004) *Icerya purchasi* Maskell, cottony cushion scale (Hemiptera: Margarodidae), causing damage to ornamental plants growing outdoors in London. *British Journal of Entomology and Natural History* 17, 105–109.

Whyatt, R.M., Camann, D.E., Kinney, P.L., Reyes, A., Ramirez, J., Dietrich, J., Diaz, D., Holmes, D. and Perera, F.P. (2002) Residential pesticide use during pregnancy among a cohort of urban minority women. *Environmental Health Perspectives* 110, 507–514.

Williams, F., Eschen, R., Harris, A., Djeddour, D., Pratt, C., Shaw, R.S., Varia, S., Lamontagne-Godwin, J., Thomas, S.E. and Murphy, S.T. (2010) *The Economic Cost of Invasive*

Non-Native Species on Great Britain. CABI, Wallingford, UK. 198 pp.

Williamson, M. (1996) *Biological Invasions*. Chapman & Hall, London, UK.

Yamamura, K. and Kiritani, K. (1998) A simple method to estimate the potential increase in the number of generations under global warming in temperate zones. *Applied Entomology and Zoology* 33(2), 289–298.

Zvereva, E.L., Lanta, V. and Kozlov, M.V. (2010) Effects of sap-feeding insect herbivores on growth and reproduction of woody plants: a meta-analysis of experimental studies. *Oecologia* 163, 949–960.

2 Climate Change and Urban Pest Management

Partho Dhang*

Bangkal, Makati City, Philippines

2.1 Introduction

Climate change may have far-reaching consequences affecting human lives. Among them could be challenges posed purely by urban pests. Climate change can modify pest life history, resulting in increased diversity and density of pests, consequently posing significant risk to humans. It is also notable that the impact of climate change on various parameters could be amplified in urban centres differently from natural areas, possibly due to the creation of urban heat islands, making most urban areas warmer than surrounding areas (Douglas and James, 2015). This phenomenon in certain latitudes could influence pests, either by increasing their activity or by causing non-pest species to move into these warmer centres and become pests, as evident from examples described by Douglas and James (2015). A number of comprehensive reports are available which discuss the effects of climate change on public pests in both indoor and outdoor conditions (Patz *et al.*, 2003; Epstein and Mills, 2005; Chartered Institute of Environmental Health, 2008; USEPA, 2010).

Evidence of climate change and its impact are now visible in many parts of the world. There are reports of insect vectors and vector-borne diseases in previously unrecorded elevations in eastern and central Africa, Latin America, and Asia. The incidence of malaria is on the rise in highland urban cities, such as Nairobi, and rural highland areas, such as those of Papua New Guinea (Reiter, 2008). The mosquito *Aedes aegypti*, once limited by temperature to approximately 1000 m in elevation, has recently been found at 1700 m in Mexico and 2200 m in the Colombian Andes (Epstein, 2004). In addition to mosquitoes, other pest species such as flies, lice, ticks, fleas, bats, rodents, snails and termites could significantly increase their distribution range and contribute to the spread of disease and damage to humans, influenced by changes in climate. This is apparent in one WHO report, which highlights the evidence linking climatic factors such as temperature, precipitation and sea-level rise to the life cycles of infectious disease vectors, including both direct and indirect associations via ecological processes (Patz *et al.*, 2003).

Climate change, defined by Epstein (2004, p. 381) as 'the heating of the inner atmosphere, oceans, and land surfaces of the Earth, resulting in more intense and extreme weather events and the altered timing, intensity, and distribution of precipitation', has generated enormous interest in the scientific world in the last two decades. Global mean surface temperatures have risen by 0.74°C±0.18°C when estimated by a linear trend during the period 1906–2005. The rate of warming in the last 50 of these years is almost double that over the century as a whole (IPCC, 2007a).

A report by USEPA (2010) notes the impact that climate change is expected to have on the outdoor environment, and focuses on those aspects most likely to have

**E-mail: partho@urbanentomology.com*

significant impact. The following points summarize what is expected to result from climate change in a region such as the USA:

1. The average temperature of the Earth has risen about 0.8°C (1.5°F) since 1900. By 2100, it is projected to rise by between 1°C (2°F) and 6.3°C (11.5°F). By the end of this century, the average temperature in the United States is projected to increase by between about 4°C (7°F) and 6.1°C (11°F) under high emissions scenarios and about 2°C (4°F) and 3.5°C (6.5°F) under low-emissions scenarios.

2. Excess heat events that now occur once every 20 years are projected to occur about every other year in much of the country by the end of this century, and these very hot days are projected to be about 5.5°C hotter than they are today.

3. Heavy downpours that are now one-in-20-year occurrences are projected to occur about every 4–15 years by the end of this century, depending on the location.

4. The destructive energy of Atlantic hurricanes has increased in recent decades. The intensity of these storms is likely to increase in this century.

5. Cold-season storm tracks will continue to shift northwards, and the strongest storms are likely to become stronger and more frequent, with greater wind speeds and more extreme wave heights in northern areas.

6. Assuming historical, geological forces continue, a 2-foot (0.6 m) rise in global sea level (within the range of recent estimates) by the end of this century would result in a relative sea-level rise of 2.3 feet (0.7 m) in New York City.

7. Unforeseen ecological changes could result in massive dislocations of species or in pest outbreak. With increased global trade and shorter travel time, disease flare-ups brought about by climate change in any part of the world, particularly in poorer nations, have the potential to reach the United States, where extreme weather events could undermine the public-health infrastructure and make people more vulnerable, as disease transmission from food, water, and insects are likely to increase.

8. The projected rapid rate and large degree of climate change over this century will challenge the ability of society and natural systems to adapt. Adaptation will be particularly challenging because society will not be responding to a new steady state, but rather to a rapidly moving target. Climate will be changing continually and rapidly at a rate outside the range to which society has adapted in the past.

It is beyond the scope of this chapter to evaluate the reasons for climate change; however, it is important to mention that overwhelming evidence shows human activity is partially responsible. Gore (2006) in his well-publicized book, *An Inconvenient Truth*, mentions 928 peer-reviewed articles dealing with climate change that agree to this fact. Greenhouse gases are held up as the primary reason for the increase in global mean surface temperature (IPCC, 2007b). This has notably happened only in the second half of the 20th century and the beginning of the 21st century (Bonnefoy *et al.*, 2008); coinciding with industrialization, further confirmation of human involvement.

In a scenario of limited evidence, complicated by numerous interactive biotic and abiotic factors, climate change is often a challenging subject to discuss. This chapter is an effort to present a general review on how climate change could affect urban pest dynamics and their management.

2.2 Climate Change and Indoor Environment

Humans live in man-made structures, and climate change is expected to bring stress to human lives through alterations in the indoor environment. These structures, in spite of modern materials and design, are subjected to stress owing to changes in outdoor conditions. Changes are likely to affect the rate of degradation of building materials, thus affecting the service life of structures, which would have huge implications on the indoor environment (IOM, 2011). In addition to the materials used in constructing them, the location of these structures also plays a vital role in the effects of climate change. It is estimated that half the world's population is concentrated in cities (WHO, 1997), and most of these cities are coastal or located near rivers, considered most vulnerable to the elements of climatological changes (IPCC, 2007a).

Structures which protect humans from the outdoor climate are also extremely well suited for a group of organisms or pests, which have associated their lives with humans and their activities. The USEPA (2010) report serves as a good example of how climate change could mitigate or influence indoor environment and pests. It states that a rapidly changing climate, such as extremes of heat and precipitation, will create changes in the indoor environment in many ways, influencing these pests. It mentions the prevalence of dampness and mould in US buildings due to lack of proper building defences against excess water, leading to rampant mould problems caused by flooding during Hurricane Katrina. This is evidence of how climate can make structures vulnerable, allowing pest invasion or in other cases trigger an outbreak of the resident pests.

2.3 Climate Change and Urban Pests

The essential life elements for urban pests, such as air, water and food, are indispensable, and are often found in excess in their immediate environment. However, one element which has significant influence in the life of these pests is temperature. Insects are cold-blooded organisms and cannot regulate their body temperature: their bodies are approximately the same temperature as the immediate environment. Therefore, temperature is probably the single most important environmental factor influencing pest behaviour, distribution, development, survival and reproduction (Petzoldt and Seaman, 2010). It could be safely assumed that the influence of temperature on insects largely overwhelms all other environmental factors (Bale *et al.*, 2002). It has been estimated that with a 2°C rise in temperature,

insects might experience one to five additional life cycles per season (Yamamura and Kiritani, 1998).

The effect of climate on urban pests has generated interest, mostly through research on insect vectors such as mosquitoes. It is, however, expected that changes in the environment will also affect a number of other pests inhabiting and frequenting urban areas. Roy *et al.* (2009) undertook an interesting study to determine the effect of climate change on nuisance insect species in the United Kingdom. The report highlights that the nuisance insect species unlikely to be affected by a warm climate are *Blattella germanica* (German cockroach), *Cimex lectularius* (bed bug), *Monomorium pharaonis* (Pharaoh ant), *Anobium punctatum* (woodworm), *Ctenocephalides felis* (cat flea), *Lyctus brunneus* (powderpost beetle), *Hylotrupes bajulus* (house longhorn beetle), *Tineola bisselliella* (common clothes moth), *Dolichovespula media* (median wasp) and *Vespa crabro* (European hornet). The ten species most likely to increase with warming are *Tinearia alternata* (moth fly), *Lasius neglectus* (invasive garden ant), *Thaumetopoea processionea* (oak processionary moth), *Linepithema humile* (Argentine ant), *Reticulitermes grassei* (Mediterranean termite), *Culex pipiens molestus* (urban mosquito), *Culex pipiens pipiens* (mosquito), *Aedes vexans* (wetland mosquito), *Ochlerotatus cantans* (woodland mosquito) and *Musca domestica* (housefly). The ten species most likely to increase with changes in precipitation are the same as for increasing temperature with the exception of *Musca domestica* and inclusion of *Phlebotomus mascittii* (sand fly).

Examples from another region such as the St Louis area of the USA was used by Sims and Appel (2013) to illustrate weather-influenced effects on pests. This article reported termite swarming unusually early in the year, following the extremely mild winter of 2011–2012. The mild winter was followed by a very warm and dry summer and this was also associated with a significant increase in the number of houses invaded by brown recluse spiders that normally reside outdoors. Quarles (2007) suggested that temperature increases in the

USA favour warm-weather pests such as ants, termites, clothes moths, flies, mosquitoes, fleas, stored-product moths, wood-boring beetles, and even bed bugs. In addition, the increase in geographic range of mosquitoes following warm temperature gradients into previously unknown territories in the USA (Rochlin *et al.*, 2013), serves as another good example of how temperature can influence urban pest dynamics.

2.4 Climate Change Influence on Pest Distribution

The subject of climate change and pest distribution is complex as it is interdependent on a number of factors. Alteration in the environment as a result of flooding or drought in addition to changes in land use pattern can influence pest distribution in urban areas (Bonnefoy *et al.*, 2008). In addition, climate change also has the potential to modify the impact of exotic invasive species by affecting their source, pathway and destination (Hobbs, 2000).

Urban pests, however, will be differentially affected by climate change, as part of their life is dependent on activities related to humans. This dependence has made their distribution common throughout the world. Pests like German cockroach, housefly, bed bug, and stored-product pests, commensal rodents, etc. are present in all territories where humans have set habitats. However, certain pests like mosquitoes, houseflies and termites face geographical limitations, which restrict their distribution. These are tropical pests and prevalent in warm areas, but as tropical weather expands into temperate regions, their distribution pattern is expected to change.

The presence of mosquitoes and termites in previously unrecorded latitudes is being reported. Warmer temperatures and shorter winters allow pests to be active for longer periods, activating higher population build-up, and providing the opportunity to disperse. For example, a study by Culler *et al.* (2015) on the Arctic mosquito (*Aedes nigripes*) shows that warmer spring temperatures caused the young to emerge 2 weeks earlier. They had a shortened development time through the larval and pupal stages by about 10% for every 1°C increase in temperature. Although warming increased the number of mosquitoes being eaten by diving beetles, the mosquitoes' accelerated growth in their vulnerable juvenile stages lessened their time with aquatic predators. This ultimately increased their chance of surviving to adulthood. With a 2°C warming scenario, the model predicts that the mosquitoes' probability of survival will increase by 53%. The population increase described above would further cause the pest population to thrive and migrate northward into previously unoccupied territories. Pests that could benefit and expand their distribution would be mosquitoes, termites, rodents and flies (Epstein, 2001).

In addition to general variations in climate, socioeconomic conditions and human behaviour can influence pest distribution. The complex interlinking between pests and human habits further complicates distribution. For example, in the case of mosquitoes, water-storage practices and various intervention methods used to prevent the intrusion of mosquitoes into homes affect their distribution and dissemination. Such factors can mix up basic associations between climate parameters and mosquito abundance (Lozano-Fuentes *et al.*, 2012) and affect distribution models. The close association of *Aedes* mosquitoes to humans is an example of where human action can be a determinant factor in deterring pest distribution and density, irrespective of climate.

The case of invasive termite species serves as another proven example of human-assisted pest distribution. Characteristics of these species to eat and live in their food makes them transportable along with goods, leading to the establishment of new distribution zones in similar climatic territories. Reports of tropical species, such as *Coptotermes gestroi*, becoming established in the subtropics (Grace, 2006), and subtropical species, such as *Coptotermes formosanus*, expanding their range northwards into more temperate areas (Jenkins *et al.*, 2002) substantiate the role of humans in pest distribution. A few examples of pest distribution relative to climate change are discussed further below.

2.4.1 Mosquitoes

A combination of various habits like anthropophilism, zoophilism, endophagy, exophagy and exophily, based on geography, location and seasons make mosquitoes a complex group to study and it is difficult to predict the effects of climate change on them. However, all anthropophilic species of mosquitoes either live or breed in urban areas or frequent urban dwellings for a blood meal. Among a few species, *Aedes* has become a truly urban mosquito, as it has adapted to breed in urban environments, particularly in artificial containers containing fresh water and (its preference) human blood. This makes studies on associations between climate parameters and *Ae. aegypti* complicated (Lozano-Fuentes *et al.*, 2012). However, temperature is an important driver for population growth of *Ae. aegypti*, particularly during rainy periods of the year when water-filled containers are most abundant (de Almeida *et al.*, 2010; Farnesi *et al.*, 2009).

Mosquitoes, as major public pests and disease vectors, have attracted the highest number of studies with respect to climate change. A literature review shows that increases in temperature and rainfall will in general favour mosquito breeding. There are a large number of mosquito species breeding alongside urban communities around the world and it is predicted that each one will respond slightly differently to climate change, according to the Chartered Institute of Environmental Health (2008). This report predicts that warmer summers and milder winter temperatures will favour mosquito development and extend the biting season of some species. Additionally, that wetter winters will provide more temporary and underground aquatic sites for some species during winter and spring. Drier summers could, however, reduce possible aquatic sites for other species.

Coastal areas would have different effects: an increase in sea levels and increased storm surges could eliminate some mosquito sites, for example species breeding in salt marshes, but create new sites where there is an inundation of salt water (Chartered Institute of Environmental Health, 2008). In a survey after the tsunami of 2004, the Vector Control Research Institute (VCRC) in Pondicherry in South India found a breeding population of *Anopheles stephensi* in the tsunami water-inundated habitats that had not previously been recorded. Other vector species of mosquito, such as *Culex quinquefasciatus* and *Cx. tritaeniorhynchus*, were also recorded in tidal water with measurable salinity between 2541 to 17,468 ppm. Low-lying paddy fields and fallow lands with salinity ranging from 3000 to 42,505 ppm were also found to support high breeding populations of *Anopheles sundaicus* and *Anopheles subpictus* in the Andaman Islands (VCRC, 2005). This report shows that sea-level rise and sea storms could increase vector distribution through adaptation.

Climate change can also initiate a number of behavioural changes in mosquitoes, such as the capacity to breed in new habitats, including slurry pits, rainwater pools and used car tyres (Chartered Institute of Environmental Health, 2008), which could influence distribution patterns in a localized area. Behavioural changes and adaptation to the changing environment could play a role in species distribution. This is evident from the behaviour of the Asian tiger mosquito *Aedes albopictus*, which outcompeted the local population of *Ae. aegypti* in the United States. *Ae. albopictus* is adapted to breed in large numbers in nutrient-depleted water and shows tolerance to higher temperatures, which helped it to displace *Ae. aegypti* (Juliano, 1998).

2.4.2 Flies

Distribution of flies is weather-dependent but can also be influenced by the availability of food. Urban centres produce considerable quantities of waste, with a large organic component suitable as breeding sites for many types of calyptrate fly species (Goulson *et al.*, 2005). Species that are particularly likely to be human health hazards include the housefly, *Musca domestica* Linnaeus (Muscidae), and bluebottle flies, *Calliphora* spp. (Calliphoridae). The larval development of most flies, and especially the housefly, is influenced by temperature. The optimal temperature for larval development is 35–38°C, though larval survival is greatest at 17–32°C.

Models were produced for the housefly, *Musca domestica*, and blowflies, *Calliphora* spp., which showed that predictions based on climatic factors were strongly correlated, suggesting fly population changes are largely driven by the weather rather than biotic factors (Goulson *et al.*, 2005). The models predict that under likely scenarios of climate change in the UK, fly populations could increase substantially, with increases of up to 244% by 2080, compared to current levels (Goulson *et al.*, 2005). Prolonged warmth and the extension of warmer conditions to cooler zones will help fly populations to establish and expand their distribution.

2.4.3 Termites

Termites have shown distribution around the tropical countries. On a global scale, termite diversity is greatest in tropical areas and decreases with increasing latitude, and only a few termite genera are found beyond 40° latitude (Eggleton, 2000). It is observed that termite distribution and abundance is closely linked to distribution and abundance of rainfall (Buxton, 1981; Wood and Johnson, 1986), temperature and relative humidity (Cabrera and Rust, 1994). Also, the termite-foraging patterns are linked to key abiotic parameters (Haagsma and Rust, 1995; Dibog *et al.*, 1998; Haverty *et al.*, 1974; Evans and Gleason, 2001; Messenger and Su, 2005; Moura *et al.*, 2006). Foraging activities of *Coptotermes lacteus* (Froggatt) was found to be correlated with both soil and air temperature (Evans and Gleason, 2001). Preferred soil temperatures of the western subterranean termite, *Reticulitermes hesperus*, range from 29°C to 32°C (Smith and Rust, 1994), indicating the role

of temperature as a dominant factor in determining distribution.

In the context of global distribution, the latitudinal restriction of termites is a well-established fact which remains unexplained (Eggleton, 2000). One possibility is that the hindgut mutualistic system existing in termites limits the metabolism below 23°C (Eggleton, 2000). If this is the reason, then the termite population will move to higher latitudes with warmer temperatures. *Coptotermes formosanus*, the best-known invasive species studied to date, is a tropical termite species that has so far been limited to southern areas by cold winter temperatures in the USA (Potter, 1997); with global warming, its range is likely to expand northwards.

Interestingly, another factor linking climate change to termite diversity and population density is the presence of semi-natural and natural habitats. A decline in termite diversity has been reported with forest clearance (Collins, 1980; Decaens *et al.*, 1994). However, this may not be of significance when studying pest termite species in urban areas.

Drywood termites (*Incisitermes minor*), unlike their subterranean cousins, live and feed on wood and are mostly exposed to the elements of environmental conditions. This exposure to environmental change is made possible by having an impermeable cuticle (Minnick *et al.*, 1973) and highly specialized rectal glands for reabsorption of water from the faeces (Collins, 1969). Also, the nymphs of *Incisitermes minor* show clumping behaviour and seal themselves in a carton-like casing under dry conditions (Pence, 1956). In the USA, drywood termites are now found mostly at the southern edge and along the Pacific Coast (Potter, 1997). Their range will probably expand when more areas are able to consistently reach the preferred swarming temperature of about 27°C (Harvey, 1946).

2.4.4 Other pests

Distribution of a number of species will be influenced by changes in climate. Invasive soil-borne ant species, such as Argentine ants (*Linepithema humile*), could possibly benefit from climate change. These have already expanded throughout southern and central California, as they are better competitors than native ant species at higher temperatures (Dukes and Mooney, 1999). Red imported fire ants (*Solenopsis invicta*) have already spread beyond the south-east and are now found in southern California. Due to global warming, their range in the USA is expected to increase by 5–21% over the next century (Morrison *et al.*, 2005).

Reports of tick populations expanding their ranges as a result of global warming are being observed in higher latitudes. In Sweden, the disease-carrying tick, *Ixodes ricinus*, has become more abundant along its northernmost range. Numbers of ticks found on dogs and cats have increased from 22 to 44% between 1980 and 1994 (Parmesan, 2007).

2.5 Climate Change Influence on Pest Diversity and Density

Changes in climate have a significant influence on pest diversity and density. Studies have shown that broad-scale geographical variation in species diversity is strongly correlated with climate. At regional scales, across continents and globally, species richness of most taxa covaries strongly with climate (Currie, 1991). Richness–climate relationships are largely consistent among continents (Currie and Francis, 2004), suggesting a mechanistic link between climate and species richness that applies very broadly. This hypothesized causal effect of climate on species richness is highly debated in the literature because the high correlation between climate and richness could be due to covariance with other factors, particularly historical ones (Latham and Ricklefs, 1993; Fine and Ree, 2006)

However, climate is only one variable which determines abundance of a given species. Other non-climatic variables such as host, pathogen, predator and competitors can have a significant effect on attainment of a sizeable population (Sutherst and Maywald, 2005). In the absence of predators and competitors, as in the urban environment, pests can attain densities based purely on the availability of food and breeding opportunities. These two parameters, combined with favourable temperature, could lead to explosive populations, as evident during catastrophic events. In recent times, pests have been shown to attain unnaturally high densities following natural disasters such as tsunamis (Srinivasan *et al.*, 2006; Lee, 2012).

An important challenge facing ecologists is to understand how climate change may affect species performance and species interactions, both determining diversity and density. However, predicting changes in abiotic variables associated with climate change on species performance can also depend on a number of biotic factors, such as density, food availability and entomopathogenic diseases. A three-year study by Laws and Belovsky (2010) was conducted to determine how the herbivorous grasshopper *Camnula pellucida* (Scudder) responds to manipulations of temperature and population density. It was shown that grasshopper survival and fecundity decreased with density, indicating the importance of intraspecific competition. Female fecundity tended to increase with temperature, whereas survival exhibited a unimodal response to temperature, with highest survival at intermediate temperatures. Grasshopper performance responses to temperature also depended on density. Peak survival in the low-density treatment occurred in warmer conditions than for the high-density treatment, indicating that the intensity of intraspecific competition varies with temperature. Data clearly indicated the importance of a number of biotic components in mediating species responses to climatic factors associated with global change. It is, however, not known how various factors, both abiotic and biotic, would interact with and influence urban pest density and diversity.

2.6 Climate Change and Influence of Temperature on Pests

Changes in temperature due to climate change are expected to be one of the important driving forces for changes in both natural and managed systems (IPCC, 2007a). Climate change is linked to temperature, among other parameters, so it is imperative that the effect of temperature on insect pests is given much attention. Temperature is considered as a primary indicator used to determine changes in climate, and it influences all cold-blooded animals, including insects. Nearly every aspect of an insect's life is influenced by temperature, such as kinetics of enzymatic reactions, various physiological functions, behaviour (Lee, 1991) and metabolic rate (Hawkins, 1995; Angilletta *et al.*, 2002), and it controls nearly all physiological and biochemical processes (Huey and Berrigan, 2001). In addition, a large amount of experimental data on temperature and its influence are available, which is useful for interpretation.

There are minimum and maximum temperature ranges beyond which growth and development ceases and death occurs. The temperature range varies between species and also may vary within a population of a particular species. Generally, it is observed that temperatures below accepted human comfort levels of 20–25°C increasingly affect an insect's metabolism, slowing down movement, feeding and reproduction (Mullen and Arbogast, 1979). In many insect pests which are acclimatized to human comfort conditions within buildings, reproduction slows or stops below 15°C and activity ceases below 10°C (Child, 2007). At the other end of the scale, prolonged exposure to temperatures above 35°C can negatively affect or kill insects (Child, 2007).

In general, insects are known to live in a wide range of thermal climates, but there is very little variability in the maximum temperature (40–50°C) at which they can survive (Heinrich, 1981). For bed bugs *Cimex lectularius* and *C. hemipterus*, a truly indoor urban pest, the upper thermal limit for short-term exposure of eggs, nymphs and adults is 40–45°C. Long-term exposure to temperatures above 35°C results in a significant reduction in survival. Also, eggs for *C. lectularius* and *C. hemipterus* are no longer viable when held below 10°C or above 37°C throughout embryogenesis (Benoit, 2011).

Cockroaches, particularly the German cockroach, *Blatella germanica*, the most common indoor household pest in global terms, may not be exposed directly to climate change. However, under natural conditions cockroaches are influenced by temperature changes. *B. germanica*, shows almost double the growth rate with a 3°C (5.4°F) increase in temperature (Noland *et al.*, 1949). Similarly, storage pests sheltering indoors show increased growth and development, such as the Indian meal moth, *Plodia interpunctella*, at a 5°C increase (Cox and Bell, 1991). Child (2007) noticed that pests such as the common clothes moth, *Tineola bisselliella*, wood-boring furniture beetle, *Anobium punctatum* and death watch beetle, *Xestobium rufovillosum*, complete their larval stages in an appreciably shorter time than previously observed, possibly due to higher temperatures.

Outdoor pests such as mosquitoes and flies are also likely to respond to climate changes, especially to temperature. Mohammed (2011) showed a near-complete hatching in *Ae. aegypti* after 48 h at 24–25°C and 80% relative humidity, but the rate significantly declined as temperatures increased from 29°C to 35°C. Similarly, Hemme *et al.* (2009) reported the absence of *Ae. aegypti* in containers in which water temperatures exceeded 32°C. The development time for *Ae. aegypti* reared under the diurnal temperature regimen was 7–10 days at 25°C and 7–9 days at 30°C. However, at 35°C the development time (from first instar to adult stage) was 6–7 days, showing that as temperatures increase, the mosquito development time is reduced (Mohammed, 2011). Apart from development time, morphological characteristics such as size of wings in adults showed a strong negative correlation with temperature, suggesting a direct relationship between temperature and size of adults. So higher temperatures produced significantly smaller adults (Mohammed, 2011). The same study also found modification in the male/female sex ratios, with larvae reared at 33°C and 35°C producing M/F ratios of 0.9 and 0.79, respectively, which was significantly lower than the M/F ratios found at the lower temperatures.

There are a number of similar citations available on the effect of temperature on mosquito growth and development, the variation in the data is mostly in the strain of insect selected and the geography for the test (Lyimo *et al.*, 1992; Tun-Lin *et al.*, 2000; Carrington *et al.*, 2013). The overall results suggest that as the climate changes, the mosquitoes may become efficient vectors, with alteration in body sizes, change in sex ratio, increase in ability to adapt to temperature ranges thus shortening extrinsic incubation periods for arboviruses, eventually increasing their propensity to affect humans.

2.7 Effects of Climate and Temperature Change on Pest Management

Temperature could determine not only geographical distributions and population dynamics of pests, but also their management (Régnière *et al.*, 2012). Change in climate will trigger changes in pest dynamics in urban areas, and some, if not most, would require alterations in the existing pest management strategy. Pesticides are often the single most important intervention method in pest control. Urban pest control currently uses insecticides not only to keep human damages in check, but also to keep nuisance pests away. Pesticides are also incorporated in paints, furnishings, furniture, screens and other household items to control pests.

The efficacy of a pesticide is determined by its active ingredient. The active ingredient is usually formulated with other materials and this is the product as sold, but it may be further diluted in use. Formulation improves the properties of a chemical for handling, storage and application, and may substantially influence effectiveness and safety. However, various chemical and physical properties of insecticide, such as stability, vaporization, penetration, activity and degradation, are dependent on temperatures at the time of use. A review of the literature shows that the effect of insecticides is more rapid on insects at higher temperatures, although they do not always show a linear relationship with temperature (Uddin and Ara, 2006). Temperature has shown a positive effect on organochlorine, organophosphate and carbamates in general, but has shown a negative effect on synthetic pyrethroids (Uddin and Ara, 2006; Wang and Shen, 2007).

The impact of temperature on the efficacy of insecticides on various urban pest species is available in the literature, but the general consensus is that research in this subject lacks detailed analysis. A few studies have made an effort to evaluate the subject with relevance to climate change, particularly against vectors. It is thought, however, from the available knowledge, that climate change could significantly impact on the efficacy of insecticides and alter the result of pest-control activity through changes in temperature. Temperature will influence the storage, transportation, application and efficacy of most insecticides. The examples discussed below indicate that temperature has significant synergic influence on the efficacy of insecticides against urban pests.

2.7.1 Cockroaches

Cockroach control in the urban environment is the commonest pest-control activity in most parts of Europe and throughout the temperate world. Its proximity to food and cryptic nature often requires careful use of insecticides. It has been shown that temperature can affect the toxicity of most insecticides. The subject was reviewed by Rust (1995). Insecticide gel baits which have grown popular in managing cockroaches have shown variation in efficacy, depending on storage temperatures. Oz *et al.* (2010) found that cockroach mortality was greater when the gels were stored at 30°C than at 23°C. The authors concluded that this is possibly due to the increased concentration of the active ingredients in the formulations, resulting from increased evaporation of the gel moisture at the higher temperature.

Surface sprays using pyrethroid is a common method used for cockroach control and ambient temperature during the application could influence efficacy of the insecticide. Toxicity of DDT and pyrethrins, when applied topically, reduced with an increase in temperature (Guthrie, 1950). The toxicity of two pyrethroid insecticides, S-bioallethrin and cypermethrin, was investigated over time at 12, 25 and 31°C in susceptible and knockdown-resistant strains of *Blattella germanica* (L.) by Scott (1987). Both strains showed greater kill with decreasing temperature for S-bioallethrin. The susceptible strain had a negative temperature coefficient for knockdown, but a positive temperature coefficient for mortality towards cypermethrin. The resistant strain had a negative temperature coefficient towards cypermethrin at all times. Resistance to S-bioallethrin was generally greatest at 25°C initially, although the difference between temperatures and the level of resistance diminished with time. Resistance to cypermethrin was significantly less at 12°C than at 25 or 31°C. A similar negative temperature coefficient of toxicity (greater toxicity at lower temperature) toward λ-cyhalothrin was observed for the Orlando strain but not the knockdown-resistant cockroaches (Valles *et al.*, 1998).

Values of LC_{50} were negatively related to temperature when ten different pyrethroids were tested by topical application of male *Blattella germanica*. Temperature-toxicity responses of five out of seven alpha-cyano pyrethroids were parallel, possibly indicating qualitatively identical but quantitatively different levels of detoxification enzymes (Wadleigh *et al.*, 1991).

2.7.2 Flies

Worldwide, insecticides are used commonly to control houseflies. Outdoor areas such as breeding sites and resting surfaces are sprayed with insecticides to manage the fly population. Studies have shown that insecticides could be affected by the prevailing environmental conditions, particularly temperature. Khan and Akram (2014) showed that within a temperature range of 20–34°C, the toxicity of chlorpyrifos, profenofos, emamectin and fipronil increased with an increase in temperature, showing a positive temperature coefficient, whereas the toxicities of cypermethrin, deltamethrin and spinosad decreased, showing a negative temperature coefficient.

2.7.3 Termites

Termites are mostly controlled by soil treatment and performance of termiticide is dependent on a number of soil parameters, including moisture and temperature. In their reviews, Kamble and Saran (2005) and Wiltz (2010) clearly highlighted the importance of temperature, among other factors, on the efficacy of soil termiticides. Both reviews state that soil temperature affects termiticide bioavailability through its influence on solubility and adsorption. Temperature also has an effect on the physical and chemical properties of the pesticide and the rate of microbial degradation. Several studies have demonstrated that temperature affects adsorption of pesticides to soil, but it is notable that the nature of this effect varies among pesticides. In general, termiticides will remain more efficacious and persistent in soils with low temperatures and low moisture contents. Warm soil temperatures and moist conditions can enhance the activity of insecticide-degrading microorganisms, thereby increasing degradation of the compound (Kamble and Saran, 2005).

In addition to physical and chemical degradation of termiticides due to weathering and soil, certain biological functioning of the active ingredient may also be influenced by temperature. Studies suggest that temperature is one of the key factors affecting the rate of uptake and subsequent horizontal transfer of [14C]-fipronil in *Reticulitermes flavipes* (Spomer *et al.*, 2008). Non-repellent termiticides function by the principle of horizontal transfer and this study showed that the highest level of uptake occurred by termites held at 22–32°C, and decreased at lower temperatures.

2.7.4 Mosquitoes

Insecticides evaluated against mosquitoes have shown a dependency on temperature. Das and Needham (1961) studied the effects of a change in temperature ranging between 15 and 28°C on the toxicity of DDT to larvae of *Ae. aegypti*. An increase in temperature during exposure to DDT (0.02 ppm for about 1 h) increased the toxic action. When larvae were left in the suspensions for the duration of the test (3 h–4 days), an increase in temperature throughout the test decreased the toxic action of a very low concentration of DDT (0.002 ppm), but had no effect with higher concentrations (0.1–0.2 ppm). Toxic action was greater in larvae held at a low temperature than in larvae held at a high temperature after treatment (0.025 ppm for 3 h). However, such toxic action was reversible, a change from high to low temperature increased paralysis, and larvae, paralysed at a low temperature, recovered when the temperature was raised.

Temperature can also influence the resistance of mosquito larvae to insecticides. Karen *et al.* (2012) examined the effects of increasing larval-rearing temperatures on the resistance status of Trinidadian populations of *Aedes aegypti* to organophosphate (OP) insecticides. The study showed a positive association between resistance to OP insecticides and increased activities of α- and β-esterase in larval populations reared at 28±2°C. Although larval populations reared at higher temperatures showed variations in resistance to OPs, there was a general increase in susceptibility. However, increases or decreases in activity levels of enzymes did not always correspond to an increase or decrease in the proportion of resistant individuals reared at higher temperatures (Polson *et al.*, 2011).

It is thus evident that populations of mosquito could only be classified as susceptible or resistant to a given chemical, depending on the temperature at which the mosquitoes were exposed. Glunt *et al.* (2014) showed that lowering the exposure temperature from the laboratory standard 26°C strongly reduced the susceptibility of female *An. stephensi* to the WHO resistance-discriminating concentration of malathion. The susceptibility of these mosquitoes to the resistance-discriminating concentration of permethrin was not as strongly temperature-dependent. For permethrin, especially, the thermal history of the mosquito was important in determining the ultimate outcome of insecticide exposure for survival. This led the authors to conclude that investigations on the performance of insecticides under different temperature conditions is very important to better understand the epidemiological significance of insecticide resistance and select the most effective products.

Investigating the performance of vector control tools under different temperature conditions will augment the ability to better understand the epidemiological significance of insecticide resistance and select the most effective products for a given environment (Glunt *et al.*, 2014).

2.8 Conclusion

Currently, man-made climate change has become a defining moral and political issue, and speculations on its potential impacts often focus on infectious diseases (Reiter, 2008). Its influence on aspects such as building materials, quality of urban structures, non-vectors, nuisance pests and prevention are often marginalized and neglected. This chapter conclusively shows that climate change will affect various aspects of urban pests, most notably their distribution. Pests limited in the past by geography will now move to unoccupied regions of the world. All these new regions will require extensive preparations to counter and prevent damage by the new occupants.

Temperature, the primary indicator commonly used to determine climate change, has a significant influence on pest biology and behaviour. Temperature also influences the efficacy of all pest-control activities. The efficacy of all insecticides currently in use varies considerably, depending on method of usage, dosage, application device, level of training and environmental conditions. All critical components that are dependent on humans can be modified and improved, except environment. Although very little is known about the effect of environment on the quality of insecticides and their application, research is showing that it has a significant impact, which remains unaddressed (Glunt *et al.*, 2014).

Pyrethroid and organophosphate insecticides are the most commonly used insecticide classes in urban pest control. Both classes show sensitivity to temperature; pyrethroids have a negative and organophosphates a positive temperature coefficient, respectively (Musser and Shelton, 2005). However, some studies also revealed variation in the toxicity within a given insecticide class (Scott, 1995; Muturi *et al.*, 2011) between insect species and temperature range tested (Boina *et al.*, 2009; Muturi *et al.*, 2011). Therefore, generalization of the temperature-toxicity trend could be misleading within a given class, and for different insect species (Khan and Akram, 2014).

In addition to the direct impact of temperature on the efficacy of insecticides, temperature can also influence a number of tools and methodologies which make use of them, such as in insecticide-treated nets (ITN), long-lasting insecticidal nets (LLIN), insecticide residual treatment (IRS), and odour-baited traps. It can be safely concluded that climate change and resulting temperature regimens, in particular, will have a profound influence on urban pests and their management strategies.

Acknowledgements

I would like to acknowledge my sincere thanks to Dr Ravi Joshi and Dr K.P. Sanjayan for their valuable comments in reviewing the manuscript.

References

Angilletta, M.J., Niewiarowski, P.H. and Navas, C.A. (2002) The evolution of thermal physiology in ectotherms. *Journal of Thermal Biology* 27, 214–223.

Bale, J.S., Masters, G.J., Hodkinson, I.D., Awmack, C., Bezemer, T.M., Brown, V.K. and Butterfield, J. (2002) Herbivory in global climate change research: direct effects of rising temperature on insect herbivores. *Global Change Biology* 8, 1–16.

Benoit, J.B. (2011) Stress tolerance of bed bugs: a review of factors that cause trauma to *Cimex lectularius* and *C. hemipterus. Insects* 2, 151–172.

Boina, D.R., Onagbola, E.O., Salyani, M. and Stelinski, L.L. (2009) Influence of posttreatment temperature on the toxicity of insecticides against *Diaphorina citri* (Hemiptera: Psyllidae). *Journal of Economic Entomology* 102, 685–691.

Bonnefoy, X., Kampen, H. and Sweeney, K. (2008) Introduction. In: Bonnefoy, X., Kampen, H. and Sweeney, K. (eds) *Public Health Significance of Urban Pests*. World Health Organization Regional Office for Europe, Copenhagen, Denmark, pp. 1–6.

Buxton, R.D. (1981) Changes in the composition and activities of termite communities in relation to changing rainfall. *Oecologia* 51, 371–378.

Cabrera, B.J. and Rust, M.K. (1994) The effect of temperature and relative humidity on the survival and wood consumption of the western drywood termite, *Incisitermes minor* (Isoptera: Kalotermitidae). *Sociobiology* 24(2), 95–113.

Carrington, L.B., Armijos, M.V., Lambrechts, L., Barker, C.M. and Scott, T.W. (2013) Effects of fluctuating daily temperatures at critical thermal extremes on *Aedes aegypti* life-history traits. *PLoS ONE* 8(3), e58824. DOI: 10.1371/journal.pone.0058824.

Chartered Institute of Environmental Health (2008) *A CIEH Summary Based on Urban Pests and Their Public Health Significance*. WHO Regional Office for Europe, Copenhagen, Denmark.

Child, R.E. (2007) Insect damage as a function of climate. In Padfield, T. and Borchersen, K. (eds) *Museum Microclimates*. National Museum of Denmark, Copenhagen, Denmark, pp. 57–60.

Collins, M.S. (1969) Water relations in termites. In: Krishna, K. and Weesner, F.M. (eds), *Biology of termites, Vol. 1*. Academic Press, New York and London, pp. 433–458.

Collins, N.M. (1980) The distribution of soil macro fauna on the west ridge of Gunung (Mount) Mulu, Sarawak. *Oecologia (Berl.)* 44, 263–275.

Cox, P.D. and Bell. C.H. (1991) Biology and ecology of moth pests of stored foods. In: Gorham, J.R. (ed.) *Ecology and Management of Food Industry Pests*. US Food and Drug Administration Technical Bulletin No. 4, pp. 181–193.

Culler, L.E., Ayres, M.P. and Virginia, R.A. (2015) In a warmer Arctic, mosquitoes avoid increased mortality from predators by growing faster. *Proceedings of the Royal Society of London. Series B: Biological Sciences* 282(1815).

Currie, D.J. (1991) Energy and large-scale patterns of animal- and plant-species richness. *The American Naturalist* 137, 27 – 49.

Currie, D.J. and Francis, A.P. (2004) Taxon richness and climate in angiosperms: is there a globally consistent relationship that precludes region effects? *The American Naturalist* 163, 780–785.

Das, M. and Needham, P.H. (1961) Effect of time and temperature on toxicity of insecticides to insects. *Annals of Applied Biology* 49, 32–38.

de Almeida, C., de Mendonca, S., Correia, J.C. and de Albuquerque, C.M.R. (2010) Impact of small variations in temperature and humidity on the reproductive activity and survival of *Aedes aegypti* (Diptera, Culicidae). *Review of Brazil Entomology* 54, 488–493.

Decaens, T., Lavelle, P., Jimenez-Jaen, J.J., Escobar, G. and Rippstein, G. (1994) Impact of land management on soil macro fauna in the Oriental Llanos of Colombia. *European Journal of Soil Biology* 30, 157–168.

Dibog, I. Eggelton, P. and Forzi, F. (1998) Seasonality of soil termites in a humid tropical forest, Mbalmayo, Southern Cameroon. *Journal of Tropical Ecology* 14, 841–850.

Douglas, I. and James, P. (2015) Adapting to change. In: Douglas, I. and James, P. *Urban Ecology – An Introduction.* Routledge, Abingdon, Oxfordshire, UK; New York, USA, pp. 367–394.

Dukes, J.S. and Mooney, H.A. (1999) Does global change increase the success of biological invaders? *Trends in Ecology and Evolution* 14(9), 135–139.

Eggleton, P. (2000) Global patterns of termite diversity. In: Abe, T., Bignell, D.E. and Higashi, M. (eds) *Termites: Evolution, Sociality, Symbioses, Ecology.* Kluwer Academic Publishers, Dordrecht, The Netherlands, pp. 25–51.

Epstein, P.R. (2001) Climate change and emerging infectious diseases. *Microbes and Infection* 3, 747–754.

Epstein, P.R. (2004) Climate change and public health: emerging infectious diseases. In: Ayres, R.U., Costanza, R., Goldemberg, J., Ilic, M.D., Jochem, E., Kaufmann, R., Lovins, A.B., Munasinghe, M., Pachauri, R.K., Sheinbaum Pardo, G., Peterson, P., Schipper, L., Smil, V., Worrell, E. and Cleveland, C.J. (eds) *Encyclopedia of Energy*, Vol. 1, pp. 381–392. Elsevier Inc., San Diego, California, USA. Available at: http://www.bvsde.paho.org/bvsacd/cd35/8else.pdf (accessed September 2015).

Epstein, P.R. and Mills, E. (2005) *Climate Change Futures: Health, Ecological and Economic Dimensions.* Center for Health and the Global Environment, Harvard Medical School, Boston, Massachusetts, USA. Available at: http://coralreef.noaa.gov/aboutcrcp/strategy/reprioritization/wgroups/resources/climate/resources/cc_futures.pdf (accessed May 2016).

Evans, T.A. and Gleason, P.V. (2001) Seasonal and daily activity patterns of subterranean, wood eating termite foragers. *Australian Journal of Zoology* 49, 311–321.

Farnesi, L.C., Martins, A.J., Valle, D. and Rezende, G.L. (2009) Embryonic development of *Aedes aegypti* (Diptera: Culicidae): Influence of different constant temperatures. *Memórias do Instituto Oswaldo Cruz* 104, 124–126.

Fine, P.V.A. and Ree, R.H. (2006) Evidence for a time integrated species-area effect on the latitudinal gradient in tree diversity. *The American Naturalist* 168, 796–804.

Glunt, K.D., Paaijmans, K.P., Read, A.F., and Thomas, M.B. (2014) Environmental temperatures significantly change the impact of insecticides measured using WHOPES protocols. *Malaria Journal* 13, 350.

Gore, A. (2006) *An Inconvenient Truth: The Planetary Emergency of Global Warming and What We Can Do About It.* Rodale Books, Emmaus, Pennsylvania, USA.

Goulson, D., Derwent, L.C., Hanley, M.E., Dunn, D.W. and Abolins, S.R. (2005) Predicting calyptrate fly populations from the weather, and probable consequences of climate change. *Journal of Applied Ecology* 42, 795–804.

Grace, J.K. (2006) When invasives meet: *Coptotermes formosanus* and *Coptotermes vastator* in the Pacific. In: Forschler, B.T. (ed.) *Proceedings of the 10th Pacific-Rim Termite Research Group Conference.* Pacific-Rim Termite Research Group, Kyoto University, Kyoto, Japan. Available at: http://www.ctahr.hawaii.edu/gracek/pdfs/285.pdf (accessed May 2016).

Guthrie, F. (1950) Effect of temperature on toxicity of certain organic insecticides. *Journal of Economic Entomology* 43, 559–560.

Haagsma, K.A. and Rust, M.K. (1995) Colony size estimates, foraging trends, and physiological characteristics of western termite (Isoptera: Rhinotermitidae). *Environmental Entomology* 24, 1520–1528.

Harvey, P.A. (1946) Life history of *Kalotermes minor.* In: Kofoid, C.A. (ed.) *Termites and Termite Control.* University of California Press, Berkeley, California, USA, pp. 217–233.

Haverty, M.I., LaFage, J.P. and Nutting, W.L. (1974) Seasonal activity and environmental control of foraging of the subterranean termite, *Heterotermes aureus* (Snyder) in desert grassland. *Life Science* 15, 1091–1101.

Hawkins, A.J. (1995) Effects of temperature change on ectotherm metabolism and evolution: metabolic and physiological interrelations underlying the superiority of multi-locus heterozygotes in heterogeneous environments. *Journal of Thermal Biology* 20, 23–33.

Heinrich, B. (1981) Ecological and evolutionary perspectives. In: Heinrich, B. (ed.), *Insect Thermoregulation*. John Wiley, New York, pp. 235–302.

Hemme, R.R., Tank, J.L., Chadee, D.D. and Severson, D.W. (2009) Environmental conditions in water storage drums and influences on *Aedes aegypti* in Trinidad West Indies. *Acta Tropica* 112, 59–66.

Hobbs, R.J. (2000) Land-use changes and invasions. In: Mooney, H.A. and Hobbs, R. (eds) *Species in a Changing World*. Island Press, Washington, DC, USA.

Huey, R.B. and Berrigan, D. (2001) Temperature, demography, and ectotherm fitness. *The American Naturalist* 158, 204–210.

IOM (2011) *Climate Change, the Indoor Environment, and Health* (Institute of Medicine), National Academies Press, Washington, DC, USA. Available at: http://www.chgeharvard.org/sites/default/files/resources/ClimateChange Indoor%20EnviornmentAndHealth.pdf (accessed 20 May 2016).

IPCC (2007a) *Climate Change 2007: Working Group I: The Physical Science Basis* (Intergovernmental Panel on Climate Change). Available at: http://www.ipcc.ch/publications_and_data/ar4/wg1/en/ch3.html (accessed 15 October 2015).

IPCC (2007b) *Climate Change 2007: Working Group II: Impacts, Adaptation and Vulnerability* (Intergovernmental Panel on Climate Change). Available at: http://www.ipcc.ch/publications_and_data/ar4/wg2/en/ch6s6-es.html (accessed 15 October 2015).

Jenkins, T.M., Dean, R.E. and Forschler, B.T. (2002) DNA technology, interstate commerce, and the likely origin of Formosan subterranean termite (Isoptera: Rhinotermitidae) infestations in Atlanta, Georgia. *Journal of Economic Entomology* 95, 381–389.

Juliano, S.A. (1998) Species introduction and replacement among mosquitoes: interspecific resource competition or apparent competition? *Ecology* 79, 255–268.

Kamble, S.T. and Saran, R.K. (2005) Effect of concentration on the adsorption of three termiticides in soil. *Bulletin of Environmental Contamination and Toxicology* 75, 1077–1085.

Karen, A.P., Brogdon, W.G., Rawlins, S.C. and Chadee, D.D. (2012) Impact of environmental temperatures on resistance to organophosphate insecticides in *Aedes aegypti* from Trinidad. *Review of Panam Salud Publica* 32, 212–214.

Khan, H.A.A. and Akram, W. (2014) The effect of temperature on the toxicity of insecticides against *Musca domestica* L.: implications for the effective management of diarrhea. *PLoS ONE* 9(4), e95636. DOI: 10.1371/journal.pone.0095636.

Latham, R.E. and Ricklefs, R.E. (1993) Global patterns of tree species richness in moist forests: energy-diversity theory does not account for variation in species richness. *Oikos* 67, 325–333.

Laws, A.N. and Belovsky, G. E. (2010) How will species respond to climate change? Examining the effects of temperature and population density on an herbivorous insect. *Environmental Entomology* 39, 312–319.

Lee Jr, R.E. (1991) Principles of insect low temperature tolerance. In Lee, R. (ed.) *Insects at Low Temperature*. Springer, Berlin, Germany, pp. 17–46.

Lee, C.-Y. (2012) *Best Practices in IPM*. Available at: http://www.ipmcenters.org/ipmsymposium12/Plenary_Lee.pdf (accessed 15 September 2015).

Lozano-Fuentes, S., Hayden, M.H., Welsh-Rodriguez, C., Ochoa-Martinez, C., Tapia-Santos, B., Kobylinski, K.C., Uejio, C.K., Zielinski-Gutierrez, E., Delle-Monache, L., Monaghan, A.J., Steinhoff, D.F. and Eisen, L. (2012) The dengue virus mosquito vector *Aedes aegypti* at high elevation in México. *American Journal of Tropical Medicine and Hygiene* 87(5), 902–909. Available at: http://www.ncbi.nlm.nih.gov/pmc/articles/PMC3516267 (accessed 15 October 2015).

Lyimo, E.O., Takken, W. and Koella, J.C. (1992) Effect of rearing temperature and larval density on larval survival, age at pupation and adult size of *Anopheles gambiae*. *Entomologia Experimentalis et Applicata* 63, 265–271.

Messenger, M.T. and Su, N.Y. (2005) Colony characteristics and seasonal activity of the Formosan subterranean termite (Isoptera: Rhinotermitidae) in Louis Armstrong Park, New Orleans, Louisiana. *Journal of Entomological Science* 40, 268–279.

Minnick, D.R., Kerr, S.H. and Wilkinson, R.C. (1973) Humidity behavior of the drywood termite (*Cryptotermes brevis*). *Environmental Entomology* 2, 597–601.

Mohammed, A. (2011) Effects of different temperature regimens on the development of *Aedes aegypti* (L.) (Diptera: Culicidae) mosquitoes. *Acta Tropica* 119, 38–43.

Morrison, L.W., Korzukin, M.D. and Porter, S.D. (2005) Predicted range expansion of the invasive fire ant, *Solenopsis invicta*, in the Eastern United States based on the VEMAP global warming scenario. *Diversity and Distributions* 11, 199–204.

Moura, F.M., Vasconcellos, S.A., Araujo, V.F.P. and Bandiera, A.G. (2006) Seasonality in foraging behavior of *Constrictotermes cyphergaster* (Termitidae, Nasutitermitinae) in the Caatinga of northeastern Brazil. *Sociobiology* 53, 453–479.

Mullen, M.A. and Arbogast, R.T. (1979) Time temperature–mortality relationships for various stored-product insect eggs and chilling times for selected commodities. *Journal of Economic Entomology* 72, 476–478.

Musser, F.R. and Shelton, A.M. (2005) The influence of post-exposure temperature on the toxicity of insecticides to *Ostrinia nubilalis* (Lepidoptera: Crambidae). *Pest Management Science* 61, 508–510.

Muturi, E.J., Lampman, R., Costanzo, K. and Alto, B.W. (2011) Effect of temperature and insecticide stress on life-history traits of *Culex restuans* and *Aedes albopictus* (Diptera: Culicidae). *Journal of Medical Entomology* 48, 243–250.

Noland, J.E., Lilly, J.H. and Bauman. C.A. (1949) A laboratory method for rearing cockroaches and its application for dietary studies on the German cockroach. *Annals of Entomological Society of America* 42, 63-70.

Oz, E., Cetin, H., Cilek, J.E., Deveci, O. and Yanikoglu, A. (2010) Effects of two temperature storage regimes on the efficacy of 3 commercial gel baits against the German cockroach, *Blattella germanica* L. (Dictyoptera: Blattellidae). *Iran Journal of Public Health* 39, 102–108.

Parmesan, C. (2007) Ecological and evolutionary responses to recent climate change. *Annual Review of Ecology, Evolution, and Systematics* 37, 637–669.

Patz, J.A., Githeko, A.K., McCarty, J.P., Hussein, S., Confalonieri, U. and de Wet, N. (2003) Climate change and infectious diseases. In: McMichael, A.J., Campbell-Lendrum, D.H., Corvalán, C.F., Ebi, K.L., Githeko, A., Scheraga, J.D. and Woodward, A. (eds) *Climate Change and Human Health – Risks and Responses.* World Health Organization, Geneva, Switzerland, pp. 103–132. Available at: http://www.who.int/globalchange/publications/climatechangechap6.pdf (accessed May 2016).

Pence, R.J. (1956) The tolerance of the drywood termite *Kalotermes minor* Hagen to desiccation. *Journal of Economic Entomology* 49, 55–554.

Petzoldt, C. and Seaman, A. (2010) Climate change effects on insects and pathogens. *Climate Change and Agriculture: Promoting Practical and Profitable Responses* III, 6–16. Available at: http://www.panna.org/sites/default/files/CC%20insects&pests.pdf (accessed May 2016).

Polson, K.A., Brogdon, W.G., Rawlins, S.C. and Chadee, D.D. (2011) Characterization of insecticide resistance in Trinidadian strains of *Aedes aegypti* mosquitoes. *Acta Tropica* 117(1), 31–38.

Potter, M.F. (1997) Termites. In: Mallis, A. (ed.) *Handbook of Pest Control*, 8th edn. Franzak and Foster Co., Cleveland, pp. 233–332.

Quarles, W. (2007) Global warming means more pests. *The IPM Practitioner* XXIX(9/10, September/October). Available at: http://www.birc.org/SepOct2007.pdf (accessed September 2015).

Régnière, J., Powell, J., Bentz, B. and Neali, V. (2012) Effects of temperature on development, survival and reproduction of insects: Experimental design, data analysis and modeling. *Journal of Insect Physiology* 58, 634–647.

Reiter, P. (2008) Global warming and malaria: knowing the horse before hitching the cart. *Malaria Journal* 7 Suppl. 1, S3. Available at: http://www.malariajournal.com/content/7/S1/S3 (accessed September 2015).

Rochlin, I., Ninivaggi, D.V., Hutchinson, M.L. and Farajolla, A. (2013) Climate change and range expansion of the Asian tiger mosquito (*Aedes albopictus*) in northeastern USA: implications for public health practitioners. *PLoS ONE* 8(4), e60874. DOI: 10.1371/journal.pone.0060874.

Roy, H., Beckmann, B., Comont, R., Halls, R., Harrington, R., Medlock, J., Purse, B. and Shortall, C. (2009) Nuisance insects and climate change. Available at: http://nora.nerc.ac.uk/8332 (accessed 15 September 2015).

Rust, M.K. (1995) Factors affecting control with insecticides. In: Rust, M.K., Owens, J.M. and Reierson, D.A. *Understanding and Controlling the German Cockroach.* Oxford University Press, Riverside, California, USA, pp. 149–170.

Scott, J.G. (1987) Effect of temperature on the toxicity of S-bioallethrin and cypermethrin to susceptible and *kdr*-resistant strains of *Blattella germanica* (L.) (Dictyoptera: Blattellidae). *Bulletin of Entomological Research* 77, 431–435.

Scott, J.G. (1995) Effects of temperature on insecticides toxicity. In: Roe, R.M. and Kuhr, R.J. (eds) *Reviews in Pesticide Toxicology*, Vol. 3. Toxicology Communications Inc., Raleigh, North Carolina, USA, pp. 111–135.

Sims, S.R. and Appel, A.G. (2013) What does climate change mean for pests and PMPs? Pest Control

Industry website. Available at: http://www.pcton-line.com/article/pct0413-mosquitoes-increase-climate-change (accessed 15 September 2015).

Smith, J.L. and Rust, M.K. (1994) Temperature preferences of the western subterranean termite, *Reticulitermes hesperus*. *Journal of Arid Environment* 28, 313–323.

Spomer N.A., Kamble, S.T., Warriner, R.A. and Davis, R.W. (2008) Influence of temperature on rate of uptake and subsequent horizontal transfer of [^{14}C]fipronil by eastern subterranean termites (Isoptera: Rhinotermitidae). *Journal of Economic Entomology* 10, 902–908.

Srinivasan, R., Kasinathan, G., Jambulingam, P. and Balaraman, K. (2006) Muscoid fly populations in tsunami-devastated villages of southern India. *Journal of Medical Entomology* 43, 631–633.

Sutherst, R.W. and Maywald, G.A. (2005) Climate model of the red imported fire ant, *Solenopsis invicta* Buren (Hymenoptera: Formicidae): implications for invasion of new regions, particularly Oceania. *Environmental Entomology* 34(2), 317–335.

Tun-Lin, W., Burkot, T.R. and Kay, B.H. (2000) Effects of temperature and larval diet on development rates and survival of the dengue vector *Aedes aegypti* in north Queensland, Australia. *Medical and Veterinary Entomology* 14, 31–37.

Uddin, M.A. and Ara, N. (2006) Temperature effect on the toxicity of six insecticides against red flour beetle, *Tribolium castaneum* (Herbst). *Journal of Life Earth Science* 1, 49–52.

USEPA (2010) *Public Health Consequences and Cost of Climate Change Impacts on Indoor Environments*. Available at: http://sustainablecities.org.nz/wp-content/uploads/US-EPA-Public-Health-Report.pdf (accessed September 2015).

Valles, S.M., Sánchez-Arroyo, H., Brenner, R.J. and Koehler, P.G. (1998) Temperature effects on λ-cyhalothrin toxicity in insecticide-susceptible and resistant German cockroaches (Dictyoptera: Blattellidae). *Florida Entomologist* 81, 193.

VCRC (2005) *Annual Report 2005*. Vector Control Research Institute, Pondicherry, India.

Wadleigh, R.W., Koehler, P.G., Preisler, H.K., Patterson, R.S. and Robertson, J.L. (1991) Effect of temperature on the toxicities of ten pyrethroids to German cockroach (Dictyoptera: Blattellidae). *Journal of Economic Entomology* 84, 1433–1436.

Wang, X.-Y. and Shen, Z.-R. (2007) Potency of some novel insecticides at various environmental temperatures on *Myzus persicae*. *Phytoparasitica* 35, 414–422.

WHO (1997) *Health and Environment in Sustainable Development*. World Health Organization, Geneva, Switzerland.

Wiltz, B.A. (2010) Laboratory evaluation of effects of soil properties on termiticide performance against Formosan subterranean termites (Isoptera: Rhinotermitidae). *Sociobiology* 56, 755–773.

Wood, T.G. and Johnson, R.A. (1986) The biology, physiology and ecology of termites. In: Vinson, S.B. (ed.) *Economic Impact and Control of Social Insects*. Praeger, New York, USA, pp. 1–68.

Yamamura, K. and Kiritani, K. (1998) A simple method to estimate the potential increase in the number of generations under global warming in temperate zones. *Applied Entomology and Zoology* 33, 289–298.

3 Climate Change and the New Dynamics of Urban Pest Management in North America

Steven R. Sims[1]* and Arthur G. Appel[2]

[1]Blue Imago LLC, Maryland Heights, Missouri, USA; [2]Department of Entomology and Plant Pathology, Auburn University, Auburn, Alabama, USA

3.1 Introduction

It's tough to make predictions, especially about the future.

(Yogi Berra)

There is nearly unanimous agreement in the scientific community that the temperature of the Earth is increasing. This is confirmed by ocean warming, sea level rises, glacial melting, sea ice receding in the Arctic, land ice loss in the Antarctic, and diminished snow cover in the Northern Hemisphere. Since 1970, temperatures across North America have warmed by approximately 0.25°C per decade (IPCC, 2007). The average rate of warming over inhabited continents in the 21st century is likely to be at least twice as much as that experienced during the 20th century. Advanced climate prediction models suggest that by the year 2100, annual surface temperatures will be 3–5°C warmer than they were in the 1960s. This seemingly minor warming translates to an increased growing season of 4 to 6 weeks, with a corresponding increase in the number of high temperature days and decrease in the number of low temperature days. Much of the observed increase in average global temperatures since 1950 is very likely to be due to an increase in anthropogenic (human-caused) greenhouse gas concentrations (IPCC, 2007).

Regional climate changes, especially temperature increases, have already had negative effects on physical and biological systems in many parts of the world. In addition, changes have been documented in the amounts, intensity, frequency and types of precipitation, and an increased number of extreme weather events, such as heatwaves, droughts, floods and hurricanes. Increases in the frequency and severity of extreme weather events are characteristics of climate change (Mirza, 2003; Harmeling and Eckstein, 2012). Extreme events can relate to temperatures, precipitation and storms. There have also been significant increases in the number of hot summer days. One example is the 2003 European heatwave, where prolonged temperatures >35°C resulted in at least 70,000 deaths across Europe (Beniston, 2004; Robine et al., 2008). Extreme storms leading to flooding and crippling droughts are also becoming more common (Mirza, 2003; Strzepek et al., 2010).

Rising sea level is another manifestation of climate change, a result of thermal expansion, melting glaciers and ice caps, and reduced polar ice sheets. Polar sea-ice alone has declined by more than 50% in the last 35 years (Vihma, 2014). The rate of global

*E-mail: steve.sims@blueimago.com

average sea-level rise has increased from 1.8 mm/year to 3.1 mm/year from 1961 to 1993. The projected sea-level rise at the end of the 21st century will be 18–59 cm. Rising sea levels are associated with freshwater degradation and loss of human structures and habitats. Since an estimated 39–52% of the US population lives in coastal shoreline locations, sea-level rise represents a substantial future threat to human health and property (Crossett *et al.*, 2004, Ache *et al.*, 2013).

These aspects of climate change will impact the pest management industry (PMI) and pest management professionals (PMPs) in many ways. This chapter evaluates these effects by first examining direct, or primary, impacts on the businesses and on the PMPs that are providing hands-on services. The chapter then looks at the potential impacts of climate change on individual and population responses of key pest species.

3.2 Human Health – Direct Impacts of Climate Change

The PMP industry has always been significantly influenced by weather and climate. In subtropical regions of the United States (e.g. southern Florida, Texas and Hawaii), residential pest management activity occurs year-round and pest pressure is generally high. At the most northerly locations of North America, most pest problems are condensed into a 5- to 6-month period, and during the other period pest pressure is usually low to moderate. Irrespective of the region, most PMP work is conducted outdoors, and PMPs will therefore be directly exposed to the weather components of a warming climate.

In general, weather and climate play a significant role in people's health. A warmer climate will increase the risk of heat-related illnesses. Changes in climate will affect both average, and extreme, weather conditions that PMPs experience. Warmer average temperatures are likely to lead to hotter days and more frequent and extended heatwaves.

3.2.1 Impacts from heatwaves

Heatwaves can lead to dehydration and heat stroke, and these are the most common causes of weather-related deaths. Excessive heat is more likely to impact PMPs and others in northern latitudes where people are less prepared to cope with higher temperatures. PMPs with unhealthy habits or physical conditions (smoking, obesity, diabetes, etc.) will be especially vulnerable to heat stress, as will older workers. Heatwaves are often accompanied by periods of unhealthy stagnant air, leading to increases in air pollution and the associated health effects. Heatwaves increase the demand for electricity in the summer needed to run air conditioning, which, in turn, would increase air pollution and greenhouse gas emissions from power plants.

More than 80% of the US population and the majority of PMPs are located in urban areas and these are especially sensitive to climate change impacts, especially extreme weather events. Climate change will lead to warmer temperatures in cities, which absorb more heat during the day and radiate less at night than rural areas. The impacts of future heatwaves could be severe in large metropolitan areas. For example, in Los Angeles, annual heat-related deaths are projected to increase two- to sevenfold by the end of the 21st century, depending on the future growth of greenhouse gas emissions (Luber and McGeehin, 2008). Indirect effects of heat are also important considerations. For example, the incidence of skin cancer is likely to increase from the combination of higher temperatures and UV radiation coupled with behavioural changes that result in greater exposure of unprotected skin.

3.2.2 Impacts from ozone

Despite significant improvements in US air quality since the 1970s, millions of Americans live in areas that do not meet national air-quality standards (Kinney, 2008). Warmer temperatures from climate change will increase the frequency of days with unhealthy levels of ground-level ozone, a harmful air pollutant, and a component of smog. Ground-level ozone is formed when air pollutants, such as carbon monoxide, oxides of nitrogen, and volatile organic compounds, are exposed to each other in sunlight. Warm, stagnant air tends to increase the formation of ozone, so climate change is likely to increase levels of ground-level ozone in already-polluted areas of the United States, thus increasing the number of days with poor air quality (Bell *et al.*, 2007). If emissions of air pollutants remain fixed at today's levels until 2050, warming from climate change alone could increase the number of Red Ozone Alert Days (when the air is unhealthy for everyone) by 68% in the 50 largest eastern US cities (Bell *et al.*, 2007). Ground-level ozone damages lung tissue, reducing lung function and inflaming airways. This can increase respiratory symptoms and aggravate asthma or other lung diseases. This is relevant to PMPs because ozone is more toxic to older adults, outdoor workers, and those with asthma and other chronic lung diseases.

3.2.3 Impacts from fine particulate matter

Particulate matter includes extremely small particles less than 2.5 μm and liquid droplets suspended in the atmosphere. These particles may be emitted directly or may be formed in the atmosphere from chemical reactions of gases such as sulfur dioxide, nitrogen dioxide and volatile organic compounds. Sources of fine-particle pollution include power plants, gasoline and diesel engines, wood combustion, high-temperature industrial processes such as smelters and steel mills, forest fires, and natural emissions such as volcanic eruptions. Inhaling fine particles can lead to a broad range of adverse health effects, including aggravation of cardiovascular and respiratory disease, development of chronic lung disease, exacerbation of asthma, and premature mortality (D'Amato *et al.*, 2010).

3.2.4 Impacts from pathogenic diseases

Changes in temperature, precipitation patterns and extreme events could enhance the spread of some non-arthropod-borne diseases, whose pathogenic agents are transmitted through food and water. Fungal-related diseases could become more common (Garcia-Solache and Casadevall, 2010). Higher air temperatures can increase cases of *Salmonella* and other bacteria-related food poisoning, because bacteria grow more rapidly in warm environments. Flooding and heavy rainfall can cause overflows from sewage treatment plants into fresh water sources. Overflows could contaminate food crops with pathogen-containing faeces. Heavy rainfall or flooding can increase water-borne parasites such as *Cryptosporidium* and *Giardia* that are sometimes found in drinking water.

3.2.5 Impacts from allergies

Climate change may affect allergies and respiratory health. The spring pollen season is already occurring earlier in the United States and the length of this season may have increased. In addition, climate change may facilitate the spread of ragweed (*Ambrosia artemisiifolia*), an invasive plant with very allergenic pollen. Increasing carbon dioxide concentrations and temperatures may increase the amount and affect the timing of ragweed pollen production (Blando *et al.*, 2012).

3.2.6 Location of PMP businesses

Location of businesses will influence their vulnerability to climate change and there will be significant variation in regional effects. Over the past 40 years, the population has grown rapidly in coastal areas in the southern and western regions of the United States. Counties directly on the shoreline constitute less than 10% of the total land area (not including Alaska), but account for >39% of the total population. From 1970 to

2010, the coastal populations of these counties increased by almost 40% and are projected to increase by an additional 8% by 2020 (National Ocean Service, 2013). These areas are relatively more sensitive to coastal storms, flooding, drought, air pollution and heatwaves. Increasing population and climate changes in coastal zones places more demands on transportation, water and energy infrastructure. Consequently, in the future it will cost relatively more to do business in these areas.

In contrast, the mountainous regions will be likely to face more frequent water shortages and increased wildfires in the future (Strzepek et al., 2010). Water shortages can lead to rationing, reduced water use outdoors, and reduced need for yard maintenance services, including pest control. Droughts may also reduce populations of certain pest species that require minimum amounts of water for survival and growth.

3.3 Urban PMP Business Issues

Climate change will make it more difficult and expensive for PMPs to insure their businesses, or other valuable assets, particularly in risk-prone areas. Insurance is the primary means used to protect a business against weather-related disasters. Climate change will increase the frequency and intensity of extreme weather events and these changes are likely to increase property losses and cause costly disruptions to operations. Escalating losses in many areas have already affected the availability and affordability of insurance.

3.4 Effects of Climate Change on Pests

Perhaps the greatest concerns among those in the PMI are the potential effects of climate change upon the pests that are currently being controlled and the emergence of new pests and/or novel pest situations.

To put the information about responses of specific pest groups into perspective, it is important to recognize that climate change is not simply global warming. Climate change models predict that the magnitude of recent and future warming will vary with latitude, continent, season and the diel cycle (IPCC, 2007). The increasing frequency of extreme temperature and precipitation events will vary both seasonally and regionally (Battisti and Naylor, 2009) and will be a greater challenge to the survival and fecundity of pests than are changes in mean conditions (Hoffmann et al., 2003; Stillman, 2003; Moreno and Møller, 2011). Temporal and spatial heterogeneity of climate change effects will essentially produce new climates in many areas. These will result in seasonal patterns of temperature and precipitation that previously did not exist (Williams et al., 2007).

The relatively rapid appearance of new climates provides a challenge for anyone trying to predict the ecological and evolutionary responses of organisms. For example, predicting the geographic ranges of species often involves environmental niche models, which depend on the correlation between different seasonal and diurnal components of temperature and precipitation (Meynard and Quinn, 2007). The accuracy of these correlations is likely to change over time and limit the predictive value of the models (Williams et al., 2007). Newer mechanistic models often use phenotypic and ecological information to predict ecological responses to climate changes and to relate environmental conditions to the physiological impact that they have on organisms (Kearney and Porter, 2009; Kingsolver et al., 2011).

Climate change is expected to have the following general effects on pests.

1. Existing pests – Common pests continue to remain problems for longer periods of time due to longer growth seasons. Populations may be larger or smaller and, if smaller, the pest status might be diminished. Examples include yellowjackets, ants and rodents.

2. Non-pest species from the existing area attain pest status – species previously occupying the area at low densities and not considered pests may become conspicuous and common and reach pest status. Examples include centipedes, millipedes and Collembola.

3. New pests (alien or invasive species) that invade from other areas – pests that have expanded their range, or invaded a new area from a distant area. The pests now occur in areas that were formerly unoccupied. Examples include the bronze birch borer (*Agrilus anxius*), emerald ash borer (*Agrilus planipennis*), kudzu bug (*Megacopta cribraria*) (Fig. 3.1), marmorated stinkbug (*Halyomorpha halys*) and Asian cockroach (*Blattella asahinai*).

A number of pest characteristics are likely to be affected by climate change, which is expected to result in potentially new problems for the pest control industry. These are as follows.

3.4.1 Outdoor habitat

Clearly, insect pests that complete their life cycles outdoors are more likely to be impacted by climate change. Pests with a dispersive phase outdoors (e.g. earwigs, millipedes and dermestid beetles) are more likely to be impacted by climate changes than those that live exclusively inside houses and other structures (e.g. bed bugs and German cockroaches). Pests with aquatic or semi-aquatic breeding sites are likely to be negatively impacted by drought and positively impacted by average or excess precipitation and temperature.

3.4.2 Overwintering survival

Species that overwinter outside and are adapted to relatively low minimum temperatures may be more likely to increase in abundance as the climate warms. Some pest species such as the Asian multicolored ladybeetle (*Harmonia axyridis* Pallas) survive cold temperatures by a process called supercooling. Adults seek out dark concealed microclimate areas that protect them from temperatures below their fatal super-cooling point (Koch *et al.*, 2004; Labrie *et al.*, 2008).

Warmer winters will increase the overwintering survival of this species in the northern USA and southern Canada. Species lacking an obligate diapause (such as house flies) or those with facultative dormancy can be active, temperatures permitting, year-round. One example is the periodic occurrence of 'giant' yellowjacket (*Vespula squamosa*) nests in Europe and North America (Fig 3.2). These nests result from both the survival and activity of multiple queens and the availability of prey and honeydew during warm winters (Darrouzet *et al.*, 2007; Monceau *et al.*, 2014). Populations of other species with obligate diapause may be negatively impacted.

3.4.3 Species with high reproductive potential

Short generation times and high reproductive potential can allow some species to take advantage of increasingly warm conditions by producing large numbers of offspring quickly. For example, *Drosophila melanogaster* Meigen can develop large populations in late summer and autumn and also evolve quickly in response to environmental stress (Rodriguez-Trelles *et al.*, 2013).

3.4.4 Resource specialization

Some pest insect species that are highly specialized in resource utilization (e.g. diet, hosts or habitat use) may be less likely to increase due to climate change than generalists because their distribution and abundance is closely tied to that of their hosts. This group includes species such as boxelder bugs (*Boisea trivittata*) (limited by distribution of their *Acer* spp. host trees) and cluster flies (limited by the presence of earthworm hosts).

3.4.5 Dispersal

Pest species that are introduced or dispersed through human activity or on hosts may

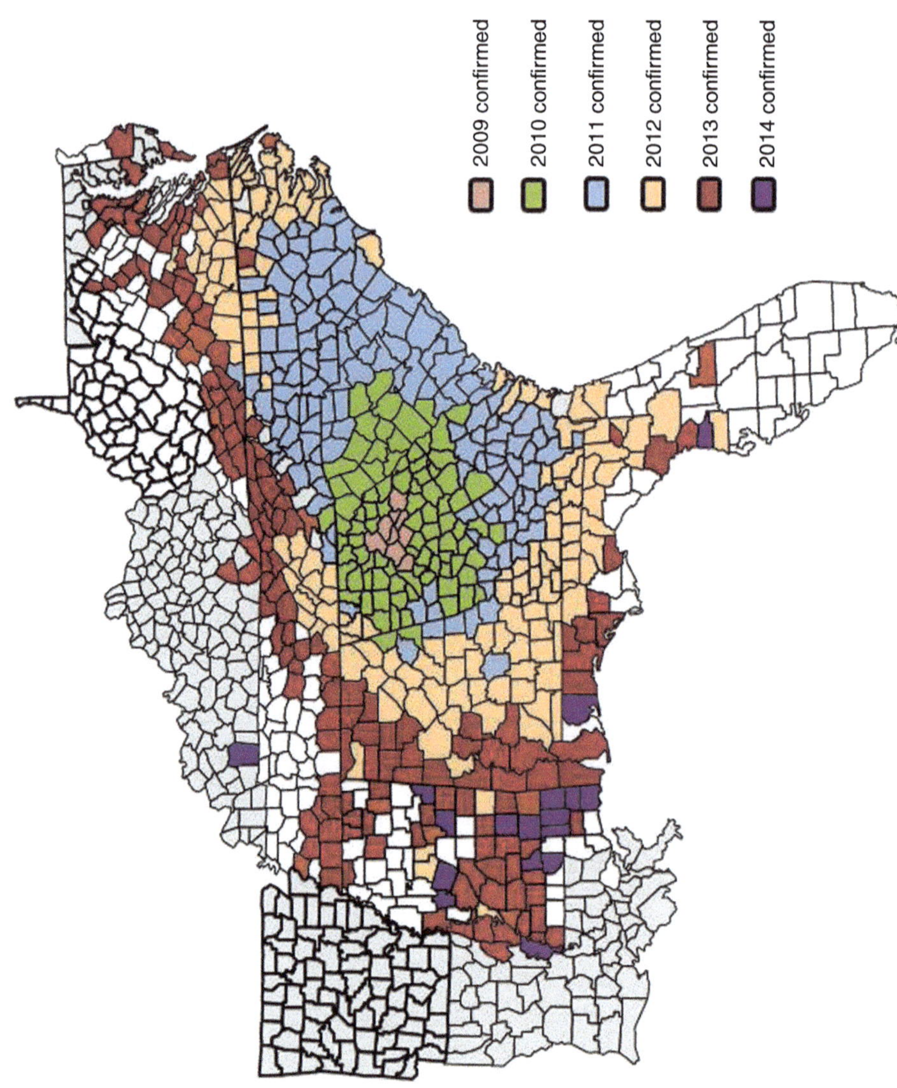

Fig. 3.1. Expanding distribution of kudzu bug (*Megacopta cribraria*) in the south-eastern United States. Compiled by Wayne A. Gardner, University of Georgia, Athens, Georgia, USA.

Fig. 3.2. Giant nest of *Vespula squamosa* in Chilton Co., Alabama, USA. Photo by A.G. Appel, Auburn University, Auburn, Alabama, USA.

respond to climate change-induced increases in habitat availability. This group includes both native and alien (e.g. red imported fire ant (*Solenopsis invicta* Buren), kudzu bug and Asian cockroach (*Blattella asahinai* Mitzukubo)) species. As climate changes increase the extent of climatically moderate areas, the regions suitable for initial invasion increase and the probability of successful invasions increases. This situation should be true for 'alien susceptible' states such as California, Florida, Hawaii and perhaps the gulf regions of the south-eastern United States.

3.4.6 Temperature and moisture conditions associated with pest development

All pest species have optimal conditions of temperature and humidity under which they can successfully complete development. Climate change could provide new habitats with these favourable conditions and increase the range currently occupied by the pest. This point is illustrated by millipedes, occasional invaders whose northern ranges in Europe are generally increasing with warmer temperatures (David, 2009) and with soil-inhabiting Collembola (springtails) whose population levels and species richness are positively related to optimum soil moisture and temperature (Kardol *et al.*, 2011).

3.5 Impact of Pests on Humans Under Climate Change

The impact of climate change on vectors of disease organisms (mosquitoes, ticks, blackflies, fleas, etc.) and other medically important arthropods (stinging insects, spiders and bed bugs) is a topic of great concern. Warming climates increase the risk or prevalence of mosquito and tick-borne disease in North America, as has been projected in many scientific publications and governmental reports. However, predictions must adequately consider the current and historical distribution of the vectors and diseases, their local ecology and epidemiology, and the impact of societal features and the capacity for public health interventions.

Evaluation of potential effects of climate change will require detailed examination of site-specific vector, host and other factors likely to influence human health outcomes. Within specific geographical areas there may be many new problems to deal with and existing problems have the potential to either increase or decrease in severity.

Many pest groups, in addition to vectors, will respond to climate change and, for the purpose of this review, the following section is presented from an economic (market share) perspective.

3.5.1 Wood-destroying insects

Wood-destroying insects, ants and cockroaches are broad pest categories that, combined, dominate the professional pest control market. Wood-destroying pests include termites, beetles, carpenter ants and carpenter bees. Termites occupy, by far, the largest share of the North American pest-control market and the total value of control services, control materials, and damage exceeds US$ 1.5 billion/year (Su and Scheffrahn, 1998). As such, any changes in their distribution, abundance and swarming due to climate change could have a major impact on the pest-control industry. Termites generally do not disperse over long distances without the help of humans, such as moving infested materials (Bordereau and Pasteels, 2010) and, on their own, slower geographic expansion is more typical. In colder areas of North America, such as Wisconsin and southern Canada, populations of *Reticulitermes flavipes* Kollar appear to be gradually expanding their range and even swarming under natural outdoor conditions (Scaduto *et al.*, 2012; Arango *et al.*, 2014). However, colony growth and slow dispersal without the need for swarming has been associated with these northern colonies on the fringe of their range (Arango *et al.*, 2014). Swarming events can occur in all months of the year. With increasing temperatures, both the total yearly number of swarms and frequency of swarms in cooler months can be expected to increase.

Warmer or longer reproductive seasons are likely to produce extended seasonal feeding activity and result in greater structural damage. In contrast, lack of swarming or changes in traditional swarming times due to drought (Nagro, 2012) or other weather conditions have already been noted in many regions of the country. Modelling studies done on the Formosan subterranean termite (*Coptotermes formosanus* Shiraki) suggest that climate change can influence the distribution of termite territory sizes, although the direction of this distribution (i.e. fewer but larger colonies vs more mid-size and small colonies) remains unclear (Lee and Chon, 2011). Extreme weather events can be detrimental to termite populations. For example, prolonged flooding may drown subterranean termite colonies located in flood-prone areas, in part because the termites do not move to escape the flooded soil zones (Forschler and Henderson, 1995).

3.5.2 Ants

Ants have the second-largest share of the professional pest-control market. This broad category includes wood-destroying ants (carpenter), medically important species (fire ants, Pharaoh ants) and many nuisance species (Argentine, crazy, odorous house, pavement, etc.). Ants are very sensitive to changes in temperature and humidity, because these changes affect colony survival (Walters and Mackay, 2004), foraging activity and networks (Heller and Gordon, 2006; Brightwell *et al.*, 2010) and dominance hierarchies (Cerda *et al.*, 1997). Climate is one of the most important factors influencing the distribution of ants (Jenkins *et al.*, 2011) and climatic suitability may be the most important factor responsible for the current global distribution of the invasive Argentine ant, *Linepithema humile* (Mayr) (Roura-Pascual *et al.*, 2011). Temperature affects the respiration pattern of ants and other insects. Warmer temperatures may reduce the frequency of discontinuous gas exchange and increase respiratory water loss. Climate can therefore serve as an important proxy to estimate the potential distribution of invasive ants worldwide. It is generally recognized that climate change will be a major determinant of species physiology, phenology and range shifts during the 21st century (Bertelsmeier *et al.*, 2013c). With continuing climate change, many invasive ant species will progressively colonize higher latitudes and altitudes, where the present cold climatic conditions are expected to moderate and become more suitable (Walther *et al.*, 2009; Rabitsch, 2011).

The range of the red imported fire ant (*Solenopsis invicta*) in the eastern United States is expected to increase by 5% to 21% between now and 2100 (Morrison *et al.*, 2005). If this occurs, *S. invicta* could occupy locations in Delaware and Maryland and might even cause problems on the White House lawn! The potential extent of the problem increases when one considers the black imported fire ant, *Solenopsis richteri* Forel, which occupies higher latitudes than *S. invicta* and could have a range expansion equal to, or greater than, *S. invicta*.

The Argentine ant (*L. humile*) is a widespread invader native to South America. It is now established in more than 15 countries on six continents and on many oceanic islands (Suarez *et al.*, 2001). It is associated with human-modified habitats throughout its non-native range, but it can also colonize natural areas. Worldwide, *L. humile* causes urban problems such as food contamination, damage to infrastructure, and nuisance issues. Ecological impacts include disturbance of native ant faunas leading to changes in arthropod communities, ant–vertebrate interactions, and ant–plant relationships. The Argentine ant will also be likely to expand its range in North America, especially in areas of the West and Northeast as temperatures increase, but, curiously, its distribution in the tropics may shrink (Roura-Pascual *et al.*, 2004).

The rasberry crazy ant, now positively identified as *Nylanderia fulva* (Mayr) (Gotzek *et al.*, 2012), was introduced into the southeastern United States and its distribution continues to rapidly expand. It should eventually occupy the entire Gulf Coast but the northern limits of its range are unclear and

will be influenced by weather conditions associated with climate change.

Other subtropical species such as leaf-cutter ants in the genus *Atta* may extend their range northward through winter warming and adaptations involving increased cold-tolerance of their associated fungus (*Attamyces* spp.) and seasonal changes in the depth at which fungus gardens are located by the ants (Mueller *et al.*, 2011).

The Asian needle ant, *Pachycondyla chinensis* Emery, was introduced from Asia to eastern North America in the early 20th century, but the invasive potential of this species has only recently been realized (Bertelsmeier *et al.*, 2013b). It forms comparatively small colonies within both disturbed urban and undisturbed natural habitats and, like *L. humile*, *P. chinensis* has impermanent nests (Guenard and Dunn, 2010). This species has a venomous sting which can cause human anaphylaxis, and poses a health threat in human-occupied habitats (Nelder *et al.*, 2006). Of great interest is the finding that the broader climatic tolerances of *P. chinensis* enables it to establish earlier in the year than *L. humile* and, although *L. humile* is usually numerically superior and behaviourally dominant, *P. chinensis* is currently displacing *L. humile* across the invaded landscape where these species co-occur (Rice and Silverman, 2013).

Not all pest ant species may be favourably impacted by global climate change. One example is the big-headed ant, *Pheidole megacephala* (Fab.). Currently, almost one-fifth (18.5%) of Earth's land mass has suitable climatic conditions for *P. megacephala*. This species has invaded tropical and subtropical, warm-temperate and Mediterranean regions. Surprisingly, climate modelling research indicates that the invasion of big-headed ants is not only unlikely to benefit from climate change, but may even suffer from it as the extent of its preferred habitats decline. Projections show a global decrease in the invasive potential of big-headed ants during the 21st century (Bertelsmeier *et al.*, 2013a).

One might think that climate change is not likely to affect carpenter ants, but species in the genus *Camponotus* undergo winter dormancy, during which they live off food stored as body fat, which is slowly metabolized during dormancy. At low (near-freezing) temperatures, metabolism is slow and food energy is conserved, allowing for enhanced probability of survival until spring. At warmer temperatures, energy is depleted more rapidly and winter survival may decline. A situation similar to this has been documented in the European wood ant, *Formica aquilonia* Yarrow (Sorvari *et al.*, 2011). If a similar situation is true for carpenter ants, then warmer winters may cause colony decline and ultimately lead to fewer pest problems. Adverse effects may occur on other pest species that undergo diapause or dormancy during the winter. These effects may result from loss of winter snow cover, resulting in exposure to more severe air temperatures and increased frequency of freeze–thaw cycles. The dominant diapause-inducing cue (photoperiod) will not be affected by climate change, but higher temperatures may modify normal development rates and lead to asynchrony between diapause-sensitive life-cycle stages and critical photoperiods for diapause induction (Bale and Hayward, 2010).

3.5.3 Cockroaches

Cockroach control has the third-largest market share and activity is focused mainly on domestic species that reproduce indoors, where they are buffered from most outdoor weather. With earlier springs and later autumns, one might expect to see more invasions of houses from peridomestic species such as *Parcoblatta* spp. (Blattaria: Blattellidae), but there is no documentation of this happening yet. Peridomestic cockroaches include species living in natural or semi-natural habitats within urban settings. They normally do not reproduce within structures but often enter houses in search of food and are a concern to human occupants. Warmer winters may allow increased survival and the northern movement of these native species.

Domestic cockroaches, living in human-made structures, may not be greatly affected by climate change, but exotic species, introduced into North America and Europe from more tropical areas, might expand their range. One example is the Asian cockroach (*Blattella asahinai*), which was introduced into Florida in the 1980s and has since expanded its range northward into many southern states, including Alabama and Georgia (Snoddy and Appel, 2008). Further expansion might only be limited by the severity of winter weather.

Some introduced species of cockroaches represent alien invaders, but the relevance of climate change to these is yet to be determined. One example is *Periplaneta japonica* Karny, which is native to, and a common pest in, central and northern Japan. It has spread to China, Korea and south-eastern Russia. The first record of *P. japonica* in the United States was from a suburb of New York City in 2013 (Evangelista *et al.*, 2013). The ability of *P. japonica* nymphs to survive subfreezing winter temperatures preadapts it to survival in temperate zone climates (Tanaka and Tanaka, 1997).

In general, climate change may increase the potential for tropical cockroach species, if accidentally introduced, to become established in North America. The areas where this is most likely to happen include California, Hawaii, Florida, southern Texas and the Gulf Coast states. Florida may be the 'poster child' for the topic of introduced cockroaches, because 13 of 38 cockroach species in the state were originally introduced from tropical and subtropical areas (Atkinson *et al.*, 1990). One of these species (*Epilampra maya* Rehn) is currently reported to be expanding its range into northern Florida.

3.5.4 Mice and rats

Norway and black rats (*Rattus norvegicus* and *Rattus rattus*) and the house mouse (*Mus musculus*) are commensal rodents with worldwide distributions. They are highly adapted to coexisting with humans and are ubiquitous in urban environments. Infestations are problematic in urban areas because rats are the source of a number of zoonotic pathogens (pathogens transmissible from animals to people) responsible for significant human morbidity and mortality in cities around the world. These pathogens include *Leptospira interrogans*, *Rickettsia typhi*, *Yersinia pestis*, *Streptobacillus moniliformis*, and Seoul hantavirus, among others (Begon, 2003). Urban rat infestations and rat-associated public health risks increase in association with the growth of cities and increased urban poverty (Hotez, 2008). Given current unprecedented rates of global urbanization (over half the global population now resides in urban centres), rat-related disease issues are likely to increase in the future. Fewer diseases are transmitted to humans by mice, but one major disease that is harboured and transmitted by mice is leptospirosis.

The commensal nature of rats makes projected effects of climate change on their populations unclear. Habitats of pest rodents in urban areas often include human-made structures that can serve to buffer the effects of temperature and precipitation. However, in many areas, rodents live and breed in outdoor habitats and these populations are more subject to climate-related stressors (van Aarde and Jackson, 2007). For example, if mice are not temperature-stressed and do not experience temperature-induced die-offs during winter, then moderate conditions could facilitate higher survival rates in winter, leading to higher densities at the start of the breeding season and potentially higher average or peak densities. Climate change has the potential to increase mouse densities by enhancing terrestrial productivity, through longer growing seasons for plants or longer breeding seasons for mice and invertebrates. For example, substantial rainfall can produce abundant plant growth and food supplies, supporting population growth (Fitzgerald *et al.*, 1996). In contrast, lengthy droughts reduce food and water, thereby increasing mortality and reducing reproduction and population increases.

Rat and mouse population sizes can vary between and within years but establishing links to climatic variables is difficult. Climate clearly influences the population dynamics of mammals such as rodents, but there is difficulty in determining what specific effects short- and long-term changes in climate have on the demography of species. Climate can affect rodent populations directly, e.g. by excessive temperatures or rainfall, but this is uncommon and the effects are usually indirect. These could be either bottom-up effects on food plant productivity or top-down effects on predator efficiency.

Mice and rats have high intrinsic rates of increase (Batzli, 1999) so in the absence of competition or natural enemies, populations should respond rapidly to favourable environmental conditions, such as those with an abundance of food. Warmer winters could increase overwintering success and lead to an earlier start of reproduction. A combination of abundant food and favourable temperatures sets the stage for increased population sizes and greater pest problems. It will be difficult to identify the combination of environmental factors that lead to these larger rodent populations and so the problems they cause will need to be addressed in a symptomatic manner.

3.5.5 Nuisance pests

A number of minor pests can also be influenced by change in climate. Spiders are among the most important nuisance pests and there is evidence that they can respond to climate change. For example, the spider *Argiope bruennichi* (Scopoli) is rapidly expanding its range in central and northern Europe, at least partially due to climate change by the creation of favourable habitat (Kumschick *et al.*, 2011). Similar examples of native spiders expanding their previous range may be occurring in North America, but these are not yet documented. Europe has also been invaded by at least 87 alien spider species, mainly resulting from commerce, and these represent species that could expand their ranges and increase their pest status with a warming climate (Kobelt and Nentwig, 2008). Most spider bites are due to species living synanthropically, and this is promoted by climate and habitat changes. The annual frequency of spider bites in Switzerland is estimated at 10–100 bites per million inhabitants, but this is predicted to increase due to the continuous arrival of new alien species, many of which have a high potential to establish in urban areas (Nentwig *et al.*, 2013).

Loxosceles reclusa Gertsch and Mulaik (brown recluse spider, Sicariidae), causes the majority of necrotic wounds induced by the Araneae. At present, the spider's range is in the south and central USA, 'from south-eastern Nebraska to south-western Ohio and south through most of Texas and into north-western Georgia' (Vetter, 2015, p. 74). This range includes at least parts of 15 states. In addition to *L. reclusa*, there are at least five other species of *Loxosceles* in the USA and about 100 worldwide (Vetter, 2015). Ecological niche modelling was used to investigate the present and future distributional potential of this species, and showed that under future climate-change scenarios, the distribution of the brown recluse spider may expand northward in the next 30–50 years, invading previously unaffected regions of the USA (Saupe *et al.*, 2011). New areas of infestation may include parts of Nebraska, Minnesota, Wisconsin, Michigan, South Dakota, Ohio and Pennsylvania. In addition to long-term range expansion, conditions such as mild winters followed by hot and dry summer weather can drive spiders such as the brown recluse, and other species, indoors, where they are more likely to interact with humans.

3.5.6 Stinging insects

The spread of Africanized honeybees (AHB) throughout the Americas represents one of the best-documented examples of invasion by an alien biotype (Schneider *et al.*, 2004). The AHB can now be found in many states including California, Arizona, New Mexico, Texas and parts of Florida. The initial rate of spread into North America, and especially into the Southeast United States, has declined for various reasons, but in the long run, global warming will result in a greater northern extension of the AHB. By 2100, AHB distributions could extend up to several hundred additional kilometres northward in the interior, and by a greater distance along both coasts of the United States (Harrison *et al.*, 2006).

Stinging insects, such as the red imported fire ant (Morrison *et al.*, 2005), the little fire ant (Wetterer and Porter, 2003) and needle ant (Bertelsmeier *et al.*, 2013b), could expand their distributions in North America as the climate continues to warm and more overwintering habitat becomes suitable for survival.

3.5.7 Flies

Climate change is also predicted to have major effects on flies of public health significance in domestic premises, and climate change models predict that populations of the housefly, *Musca domestica* L., and blowflies, *Calliphora* spp., could increase by up to 244% by 2080 (Goulson *et al.*, 2005). With this predicted increase in numbers, the role of houseflies as vectors of pathogenic organisms is likely to take on greater importance. Laboratory tests show that houseflies are able to mechanically transfer *Clostridium difficile*, one of the so-called 'hospital superbugs', as well as a great many other disease organisms (Goulson *et al.*, 2005; Davies and Anderson, 2012).

3.5.8 Kissing bugs

Kissing bugs of the subfamily Triatominae (family Reduviidae) are mostly found in the New World. They are nuisance invaders of human habitations and medically important because they are vectors of *Trypanosoma cruzi*, a protozoan parasite that causes Chagas disease (CD). *T. cruzi* has been found in many North American mammal hosts of kissing bugs but disease transmission is rare in North America because North American triatomine species do not share the South American species' habit of rapid defecation during feeding, leaving protozoans on the skin which can invade the body via abrasions caused by scratching (Klotz *et al.*, 2014). However, with an increase in temperatures, the North American triatomines could expand their ranges and represent a greater allergic reaction and CD transmission problem to both humans and dogs. One simulation study predicts that, by 2060, in addition to retaining its current distribution in Central America, CD could find a foothold in northern South America, mid-Africa, Southeast Asia, and Malaysia (Horner, 2013). Garza *et al.* (2014) show that the ranges of two triatomine vectors of CD could greatly expand to the north of Texas by 2050.

3.5.9 Fleas

In years with favourable temperatures, fleas may become a greater nuisance simply because populations of wild hosts may increase and alter the number of wild fleas that pet animals can pick up. Fleas in urban areas normally represent only a nuisance pest, but fleas can transmit plague (*Yersinia pestis*), typhus and other human disease organisms (Bitam *et al.*, 2010). These diseases represent part of a three-way interaction between the flea, the pathogen disease agent and mammalian hosts. Climate impacts all components of the disease cycle in various ways and over a wide range of scales, from individual flea life cycles to the extent of the human disease area (Ben-Ari *et al.*, 2011). Flea-borne pathogens are widely distributed worldwide in endemic disease foci, where they persist as components of enzootic cycles. However, flea-borne diseases could re-emerge in epidemic form, from changes in vector–host ecology caused by environmental and human behaviour modifications. Global climate change may influence parameters of flea development, distribution and disease transmission on a large scale. For many fleas, temperature and humidity are important for development and survival. Warmer temperatures

resulting from climate change could lead to an increased expansion of flea vectors into the Northern Hemisphere (Gage *et al.*, 2008). In addition, climate change and human encroachment into natural areas may allow transmission of poorly known pathogens to humans, as well as flea-bite allergy dermatitis to humans and companion animals.

3.5.10 Ticks

Ticks are a medically important arthropod group that commonly occurs in and around human habitats. Their health significance involves transmission of several zoonotic diseases, including Lyme borreliosis (LB) and tick-borne encephalitis (TBE), which are major health problems in North America and Europe. There is substantial evidence that, within the last 40 years, the incidence of these diseases has increased and that the tick vectors (*Ixodes* spp.) have generally become more abundant and expanded their ranges (Ogden *et al.*, 2006; Gray *et al.*, 2009). The epidemiology and life histories of ticks are complex and ticks are dependent on favourable biotic and abiotic conditions for their survival and reproduction. Ticks are usually associated with their hosts only during the blood meal. During the free-living phase of their life cycle, ticks are very vulnerable to environmental conditions, often requiring very specific conditions of humidity and temperature; climatic conditions may well limit tick distributions more than the presence and abundance of specific host types (Estrada-Peña *et al.*, 2004). The controversy surrounding changes in tick vector and disease distributions revolves around the relative importance of climate change – specifically, warming temperatures. Most studies conclude that temperature is only one of several environmental influences on tick distributions (Legér *et al.*, 2013). Other factors include rainfall, humidity, host availability and human-created habitat modifications. Despite the lack of a clear connection between climate change and tick populations, we can be reasonably certain that ticks, and the diseases that they transmit, will continue to be problems that will increase in the future.

3.5.11 Mosquitoes

Mosquito control is presently a niche market in North American professional pest control. Most control efforts are expended by state and local mosquito-control agencies that use pesticides, biological control, surveillance and breeding source reduction or prevention to manage populations. Mosquito-transmitted pathogens are among the most important sources of human disease worldwide. They include the viruses responsible for outbreaks of yellow fever, Rift Valley fever, eastern equine encephalitis, Japanese encephalitis, dengue, and other serious illnesses such as malaria.

West Nile virus (WNV) was first detected in the United States in 1999. Since that time, it has become the most prevalent cause of arthropod-borne viral (arboviral) disease in the United States. However, several other arboviruses continue to cause sporadic and seasonal outbreaks of encephalitis and other neurological diseases. In the United States, dengue and malaria are frequently brought back from tropical and subtropical countries by travellers. Despite previous elimination from the United States, dengue transmission has recently recurred in southern Florida (Radke *et al.*, 2012). The outbreak and its continuation into 2010 demonstrates the potential for re-emergence of dengue in subtropical areas of the United States where *Aedes aegypti* mosquitoes are present.

Changes in mosquito population levels and disease incidence(s) are difficult to predict. Some reports claim that, as global temperatures rise, more areas will be affected by diseases that are spread by mosquitoes, such as malaria, dengue and yellow fever (Murray-Smith and Weinstein, 1993; Martens *et al.*, 1999). As temperatures rise, the logic goes, mosquito habitat will increase and more people will be exposed to them, making them potential disease victims.

However, the actual situation is much more complicated. Mosquito population growth requires the availability of water, and warmer temperatures may also mean reduced precipitation, allowing fewer mosquitoes to survive. A climate-based mosquito population model used to simulate the abundance of *Culex* vectors of West Nile virus across the southern United States indicated that, under a future predicted climate, many locations will have a longer mosquito season but the populations will be smaller, because warm and dry conditions will be unfavourable for larval survival (Morin and Comrie, 2013). And so, water, and the lack of it, can influence mosquito populations as well as temperatures, which can exceed the tolerances of mosquito species and so cause increased larval and adult mortality.

Going forward, mosquito control activity by PMPs and agencies will probably be most effective by focusing on climate conditions within specific geographical areas. For example, the south-western USA may have hotter and drier summers that would delay onset of the mosquito season, but late summer and autumn rains could extend the season. Sporadic rainfall may increase the need to rapidly locate and treat ephemeral bodies of water for mosquitoes. Other areas of the USA, such as south and central states, may see lower mosquito populations during summer and autumn due to less rain. Warmer temperatures during spring and autumn shoulder seasons could make for a longer mosquito season across much of the USA, except in the Southwest during spring, where severe drought would inhibit population growth.

Not all mosquitoes and disease will be equally affected by climate change. On a global scale, the incidence of malaria, transmitted by *Anopheles* mosquitoes, has declined in most previously affected areas during the last 100 years, and there is little evidence to support an increase in the disease associated with a warming climate (Gething *et al.*, 2010). These findings are consistent with other work, demonstrating that warmer temperatures can reduce the effective transmission of *Plasmodium* parasites (Paaijmans *et al.*, 2012).

The question about which locations are most likely to experience epidemics in the future remains unanswered despite the insight provided by modelling because these diseases involve a complex interlinking between mosquito vectors, virus, intermediate hosts (birds in the case of West Nile virus) and elements of climate.

3.6 Conclusion

Unabated, the current projection for global climate change is an increase in average temperatures, as well as increasing variability of local weather events. Peridomestic pests will be more directly affected by temperatures, rainfall and more violent weather than domestic (synathropic) species that live in protected locations such as indoors. Undoubtedly, more species will invade and become established as North American temperatures increase. New tropical species and increasing numbers of generations of well-adapted existing pests will make future urban pest management increasingly challenging.

References

Ache, B.W., Crossett, K.M., Pacheco, P.A., Adkins, J.E. and Wiley, P.C. (2013) 'The coast' is complicated: a model to consistently describe the nation's coastal population. *Estuaries and Coasts* 36(June), 1–5. DOI: 10.1007/s12237-013-9629-9.

Arango, R.A., Green III, F., Esenther, G.R., Marschalek, D.A., Berres, M.E. and Raffa, K.F. (2014) Mechanisms of termite spread in Wisconsin and potential consequences as a result of changing climate trends. In: Wilson, C. *Proceedings of the 109th American Wood Protection Association, April 28–May 1, 2013, Honolulu, Hawaii.* American Wood Protection Association, Birmingham, Alabama, USA, pp. 111–114.

Atkinson, T.H., Koehler, P.G. and Patterson, R.S. (1990) Research reports: annotated checklist of the cockroaches of Florida (Dictyoptera: Blattaria: Blattidae, Polyphagidae, Blattellidae, Blaberidae). *Florida Entomologist* 73, 303–327.

Bale, J.S. and Hayward, S.A.L. (2010) Insect overwintering in a changing climate. *The Journal of Experimental Biology* 213, 980–994.

Battisti, D.S. and Naylor, R.L. (2009) Historical warnings of future food security with unprecedented seasonal heat. *Science* 323, 240–244.

Batzli, G.O. (1999) Can seasonal changes in density dependence drive population cycles? *Trends in Ecology and Evolution* 14, 129–131.

Begon, M. (2003) Disease: health effects on humans, population effects on rodents. *Australian Centre for International Agricultural Research Monograph Series* 96, 13–19.

Bell, M.L., Goldberg, R., Hogrefe, C., Kinney, P.L., Knowlton, K., Lynn, B., Rosenthal, J., Rosenzweig, C. and Patz, J.A. (2007) Climate change, ambient ozone, and health in 50 US cities. *Climatic Change* 82, 61–76.

Ben-Ari, T., Neerinckx, S., Gage, K.L., Kreppel, K., Laudisoit, A., Herwig, L. and Stenseth, N.C. (2011) Plague and climate: scales matter. *PLoS Pathogens* 7, e1002160. DOI: 10.1371/journal.ppat.1002160.

Beniston, M. (2004) The 2003 heat wave in Europe: A shape of things to come? An analysis based on Swiss climatological data and model simulations, *Geophysical Research Letters* 31, L02202. DOI: 10.1029/2003GL018857. 4 pp.

Bertelsmeier, C., Luque, G.M. and Courchamp, F. (2013a) Global warming may freeze the invasion of big-headed ants. *Biological Invasions* 15, 1561–1572.

Bertelsmeier, C., Guenard, B. and Courchamp F. (2013b) Climate change may boost the invasion of the Asian needle ant. *PLoS ONE* 8(10), e75438. DOI: 10.1371/journal.pone.0075438.

Bertelsmeier, C., Luque, G.M. and Courchamp, F. (2013c) Increase in quantity and quality of suitable areas for invasive species as climate changes. *Conservation Biology* 27, 1458–1467.

Bitam, I., Dittmar, K., Parola, P., Whiting, M.F. and Raoult, D. (2010) Fleas and flea-borne diseases. *International Journal of Infectious Diseases* 14, E667–E676.

Blando, J., Bielory, L., Nguyen, V., Diaz, R. and Jeng, H.A. (2012) Anthropogenic climate change and allergic diseases. *Atmosphere* 2012, 3, 200–212.

Bordereau, C. and Pasteels, J.M. (2010) Pheromones and chemical ecology of dispersal and foraging in termites. In: Bignell, D.E., Roisin, Y. and Lo, N. (eds) *Biology of Termites: A Modern Synthesis.* Springer, Dordrecht, The Netherlands, pp. 279–320.

Brightwell, R., Labadie, P. and Silverman, J. (2010) Northward expansion of the invasive *Linepithema humile* (Hymenoptera: Formicidae) in the eastern United States is constrained by winter soil temperatures. *Environmental Entomology* 39, 1659–1665.

Cerda, X., Retana, J. and Cros, S. (1997) Thermal disruption of transitive hierarchies in Mediterranean ant communities. *Journal of Animal Ecology* 66, 363–374.

Crossett, K.M., Culliton, T.J., Wiley, P.C. and Goodspeed, T.R. (2004) *Population Trends Along the Coastal United States: 1980–2008*. Coastal Trends Report. NOAA, National Ocean Service, Silver Spring, Maryland, USA. 47 pp.

David, J.F. (2009) Ecology of millipedes (Diplopoda) in the context of global change. *Soil Organisms* 81, 719–733.

Davies, M.P. and Anderson, M. (2012) Insecticide use in houses. *Outlooks on Pest Management* 23, 199–203.

D'Amato, G., Cecchi, L., D'Amato, M. and Liccardi, G. (2010) Urban air pollution and climate change as environmental risk factors of respiratory allergy: an update. *Journal of Investigational Allergology and Clinical Immunology* 20, 95–102.

Darrouzet, E., Villemant, C. and Ray, C. (2007) Des nids de guêpes géants aux *États-Unis*. *Insectes* 145, 37–39.

Estrada-Peña, A., Martinez, J.M., Acedo, C.S., Quilez, J. and Del Cacho, E. (2004) Phenology of the tick, *Ixodes ricinus*, in its southern distribution range (central Spain). *Medical and Veterinary Entomology* 18, 387–397.

Evangelista, D., Buss, L. and Ware, J.L. (2013) Using DNA barcodes to confirm the presence of a new invasive cockroach pest in New York City. *Journal of Economic Entomology* 106, 2275–2279.

Fitzgerald, B.M., Daniel, M.J., Fitzgerald, A.E., Karl, B.J., Meads, M.J. and Notman, P.R. (1996) Factors affecting the numbers of house mice (*Mus musculus*) in hard beech (*Nothofagus truncata*) forest. *Journal of the Royal Society of New Zealand* 26, 237–249.

Forschler, B. and Henderson, G. (1995) Subterranean termite behavioral reaction to water and survival of inundation: implications for field populations. *Environmental Entomology* 24, 1592–1597.

Gage, K.L., Burkot, T.R., Eisen, R.J. and Hayes, E.B. (2008) Climate and vectorborne diseases. *American Journal of Preventive Medicine* 35, 436–450.

Garcia-Solache, M.A. and Casadevall, A. (2010) Global warming will bring new fungal diseases for mammals. *mBio* 1, e00061-10. DOI: 10.1128/mBio.00061-10.

Garza, M., Feria-Arroyo, T.P., Casillas, E.A., Sanchez-Cordero, V., Rivaldi, C.-L. and Sarkar, S. (2014) Projected future distributions of vectors of *Trypanosoma cruzi* in North America under climate change scenarios. *PLoS Neglected Tropical Diseases* 8, e2818. DOI: 10.1371/journal.pntd.0002818

Gething, P.W., Smith, D.L., Patil, A.P., Tatem, A.J., Snow, R.W. and Hay, S.I. (2010) Climate change and the global malaria recession. *Nature* 465(7296), 342–345.

Goulson, D., Derwent, L.C., Hanley, M.E., Dunn, D.W. and Abolins, S.R. (2005) Predicting calyptrate fly populations from the weather, and probable consequences of climate change. *Journal of Applied Ecology* 42, 795–804.

Gotzek, D., Brady, S.G., Kallal, R.J. and LaPolla, J.S. (2012) The importance of using multiple approaches for identifying emerging invasive species: The case of the rasberry crazy ant in the United States. *PLoS ONE* 7(9), e45314. DOI: 10.1371/journal.pone.0045314.

Gray, J.S., Dautel, H., Estrada-Peña, A., Kahl, O. and Lindgren, E. (2009) Effects of climate change on ticks and tick-borne diseases in Europe. *Interdisciplinary Perspectives on Infectious Diseases* 2009, Article ID 593232. 12 pp.

Guenard, B. and Dunn, R.R. (2010) A new (old), invasive ant in the hardwood forests of eastern North America and its potentially widespread impacts. *PLoS ONE* 5(7), e11614. DOI: 10.1371/journal.pone.0011614.

Harmeling, S. and Eckstein, D. (2012) *Global Climate Risk Index 2013*. Available at: http://germanwatch.org/de/5696.

Harrison, J.F., Fewell, J.H., Anderson, K.E. and Loper, G.M. (2006) Environmental physiology of the invasion of the Americas by Africanized honeybees. *Integrative and Comparative Biology* 46, 1110–1122.

Heller, N.E. and Gordon, D.M. (2006) Seasonal spatial dynamics and causes of nest movement in colonies of the invasive Argentine ant (*Linepithema humile*). *Ecological Entomology* 31, 499–510.

Hoffmann, A.A., Sorensen, J.G. and Loeschcke, V. (2003) Adaptation of *Drosophila* to temperature extremes: bringing together quantitative and molecular approaches. *Journal of Thermal Biology* 28, 175–216.

Horner, J.K. (2013) A maximum entropy/environmental niche modeling prediction of the potential distribution of Chagas disease under climate change. *Proceedings of the International Conference on Bioinformatics & Computational Biology (BIOCOMP)*. Steering Committee of The World Congress in Computer Science, Computer Engineering and Applied Computing (WorldComp), Athens, Greece, pp. 1–6.

Hotez, P.J. (2008) Neglected infections of poverty in the United States of America. *PLoS Neglected Tropical Diseases* 2, e256. DOI: 10.1371/journal.pntd.0000256.

IPCC (2007) Summary for policymakers. In: Solomon, S., Qin, D., Manning, M., Chen, Z. and Marquis, M. (eds) *Climate Change 2007: The*

Physical Science Basis. Contribution of Working Group I to the Fourth Assessment Report of the Intergovernmental Panel on Climate Change. Cambridge University Press, Cambridge, UK, pp. 1–18.

Jenkins, C.N., Sanders, N.J., Andersen, A.N., Arnan, X., Brühl, C.A., Cerda, X., Ellison, A.M., Fisher, B.L., Fitzpatrick, M.C., Gotelli, N.J., Gove, A.D., Guénard, B., Lattke, J.E., Lessard, J.-P., McGlynn, T.P., Menke, S.B., Parr, C.L., Philpott, S.M., Vasconcelos, H.L., Weiser, M.D. and Dunn, R.R. (2011) Global diversity in light of climate change: the case of ants. *Diversity and Distributions* 17, 652–662.

Kardol, P., Reynolds, W.N., Norby, R.J. and Classen, A.T. (2011) Climate change effects on soil microarthropod abundance and community structure. *Applied Soil Ecology* 47, 37–44.

Kearney, M. and Porter, W.P. (2009) Mechanistic niche modelling: combining physiological and spatial data to predict species' ranges. *Ecology Letters* 12, 334–350.

Kinney, P.L. (2008) Climate change, air quality, and human health. *American Journal of Preventive Medicine* 35, 459–467.

Kingsolver, J.G., Woods, H.A., Buckley, L.B., Potter, K.A., MacLean, H.J. and Higgins, J.K. (2011) Complex life cycles and the responses of insects to climate change. *Integrative and Comparative Biology* 51, 719–732.

Klotz, S.A., Dorn, P.L., Mosbacher, M. and Schmidt, J.O. (2014) Kissing bugs in the United States: risk for vector-borne disease in humans. *Environmental Health Insights* 8(S2), 49–59.

Kobelt, M. and Nentwig, W. (2008) Alien spider introductions to Europe supported by global trade. *Diversity and Distributions* 14, 273–280.

Koch, R.L., Carrillo, M.A., Venette, R.C., Cannon, C.A. and Hutchison, W.D. (2004) Cold hardiness of the multicolored Asian lady beetle (Coleoptera: Coccinellidae). *Environmental Entomology* 33, 815–822.

Kumschick, S., Fronzek, S., Entling, M.H. and Nentwig, W. (2011) Rapid spread of the wasp spider *Argiope bruennichi* across Europe: a consequence of climate change? *Climatic Change* 109, 319–329.

Labrie, G., Coderre, D. and Lucas, E. (2008) Overwintering strategy of multicolored Asian lady beetle (Coleoptera: Coccinellidae): cold-free space as a factor of invasive success. *Annals of the Entomological Society of America* 101, 860–866.

Lee, S-H. and Chon, T-S. (2011) Effects of climate change on subterranean termite territory size: A simulation study. *Journal of Insect Science* 11, 80.

Léger, E., Vourc'h, G., Vial, L., Chevillon, C. and McCoy, K.D. (2013) Changing distributions of ticks: causes and consequences. *Experimental and Applied Acarology* 59, 219–244.

Luber, G. and McGeehin, M. (2008) Climate change and extreme heat events. *American Journal of Preventive Medicine* 35, 429–435.

Martens, P., Kovats, R., Nijhof, S.S., de Vries, P., Livermore, M.T.J., Bradley, D.J., Cox, J. and McMichael, A.J. (1999) Climate change and future populations at risk of malaria. *Global Environmental Change* 9, Suppl.1, S89–S107.

Meynard, C.N. and Quinn, J.F. (2007) Predicting species distributions: a critical comparison of the most common statistical models using artificial species. *Journal of Biogeography* 34, 1455–1469.

Mirza, M.M.Q. (2003) Climate change and extreme weather events: can developing countries adapt? *Climate Policy* 3, 233–248.

Monceau, K., Bonnar, O. and Thiéry, D. (2014) *Vespa velutina*: a new invasive predator of honeybees in Europe. *Journal of Pesticide Science* 87, 1–16.

Moreno, J. and Møller, A.P. (2011) Extreme climatic events in relation to global change and their impact on life histories. *Current Zoology* 57, 375–389.

Morin, C.W. and Comrie, A.C. (2013) Regional and seasonal response of a West Nile virus vector to climate change. *Proceedings of the National Academy of Sciences* 110, 15620–15625.

Morrison, L.W., Korzukhin, M.D. and Porter, S.D. (2005) Predicted range expansion of the invasive fire ant, *Solenopsis invicta*, in the eastern United States based on the VEMAP global warming scenario. *Diversity and Distributions* 11, 199–204.

Mueller, U.G., Mikheyev, A.S., Hong, E., Sen, R., Warren, D.L., Solomon, S.E., Ishak, H.D., Cooper, M., Miller, J.L., Shaffer, K.A. and Juenger, T.E. (2011) Evolution of cold-tolerant fungal symbionts permits winter fungiculture by leafcutter ants at the northern frontier of a tropical ant–fungus symbiosis. *Proceedings of the National Academy of Sciences* 108, 4053–4056.

Murray-Smith, S. and Weinstein, P. (1993) A time bomb in north Queensland: a case report of introduced malaria south of the nineteenth parallel. *Communicable Diseases Intelligence* 17, 211–213.

Nagro, A. (2012) A new termite reality. *Pest Control Technology*, February, 34–39.

National Ocean Service (2013) *National Coastal Population Report: Population Trends from 1970 to 2020.* NOAA's State of the Coast Report Series. National Oceanic and Atmospheric Administration, Silver Spring, Maryland, US.

Available at: http://oceanservice.noaa.gov/facts/coastal-population-report.pdf.

Nelder, M.P., Paysen, E.S., Zungoli, P.A. and Benson, E.P. (2006) Emergence of the introduced ant *Pachycondyla chinensis* (Formicidae: Ponerinae) as a public health threat in the southeastern United States. *Journal of Medical Entomology* 43, 1094–1098.

Nentwig, W., Gnädinger, M., Fuchs, J. and Ceschi, A. (2013) A two year study of verified spider bites in Switzerland and a review of the European spider bite literature. *Toxicon* 73, 104–110.

Ogden, N.H., Maarouf, A., Barker, I.K., Bigras-Poulin, M., Lindsay, L.R., Morshed, M.G., O'Callaghan, C.J., Ramay, F., Waltner-Toews, D. and Charron, D.F. (2006) Climate change and the potential for range expansion of the Lyme disease vector *Ixodes scapularis* in Canada. *International Journal for Parasitology* 36, 63–70.

Paaijmans, K.P., Blanford, S., Chan, B.H.K. and Thomas, M.B. (2012) Warmer temperatures reduce the vectorial capacity of malaria mosquitoes. *Biology Letters* 8, 465–468.

Rabitsch, W. (2011) The hitchhiker's guide to alien ant invasions. *BioControl* 56, 551–572.

Radke, E.G., Gregory, C.J., Kintziger, K.W., Sauber-Schatz, E.K., Hunsperger, E.A., Gallagher, G.R., Barber, J.M., Biggerstaff, B.J., Stanek, D.R., Tomashek, K.M. and Blackmore, C.G.M. (2012) Dengue outbreak in Key West, Florida, USA, 2009. *Emerging Infectious Diseases* 18, 135–137.

Rice, E.S. and Silverman, J. (2013) Propagule pressure and climate contribute to the displacement of *Linepithema humile* by *Pachycondyla chinensis*. *PLoS ONE* 8, e56281. DOI: 10.1371/journal. pone.0056281.

Robine, J.M., Cheung, S.L.K., Le Roy, S., Van Oyen, H., Griffiths, C., Michel, J.P. and Herrmann, F.R. (2008) Death toll exceeded 70,000 in Europe during the summer of 2003. *Comptes rendus biologies* 331, 171–178.

Rodriguez-Trelles, F., Tarrio, R. and Santos, M. (2013) Genome-wide evolutionary response to a heat wave in *Drosophila*. *Biology Letters* 9, article ID 20130228. DOI: 10.1098/rsbl.2013.0228

Roura-Pascual, N., Suarez, A.V., Gómez, C., Pons, P., Touyama, Y., Wild, A.L. and Peterson, A.T. (2004) Geographical potential of Argentine ants (*Linepithema humile* Mayr) in the face of global climate change. *Proceedings of the Royal Society of London. Series B: Biological Sciences* 271(1557), 2527–2535.

Roura-Pascual, N., Hui, C., Ikeda, T., Leday, G., Richardson, D.M., Carpintero, S., Espadaler, X., Gómez, C., Guénard, B., Hartley, S., Krushelnycky, P., Lester, P.J., McGeoch, M.A., Menke, S.B., Pedersen, J.S., Pitt, J.P.W., Reyes, J., Sanders, N.J., Suarez, A.V., Touyama, Y., Ward, D., Ward, P.S. and Worner, S.P. (2011) Relative roles of climatic suitability and anthropogenic influence in determining the pattern of spread in a global invader. *Proceedings of the National Academy of Sciences* 108, 220–225.

Saupe, E.E., Papes, M., Selden, P.A. and Vetter, R.S. (2011) Tracking a medically important spider: climate change, ecological niche modeling, and the brown recluse (*Loxosceles reclusa*). *PLoS ONE* 6(3), e17731. DOI: 10.1371/journal. pone.0017731.

Scaduto, D.A., Garner, S.R., Leach. E.L. and Thompson, G.J. (2012) Genetic evidence for multiple invasions of the eastern subterranean termite into Canada. *Environmental Entomology* 41, 1680–1686.

Schneider, S.S., Hoffman, G.D. and Smith, D.R. (2004) The African honey bee: Factors contributing to a successful biological invasion. *Annual Review of Entomology* 49, 351–376.

Snoddy, E.T. and Appel, A.G. (2008) Distribution of *Blattella asahinai* (Dictyoptera: Blattellidae) in southern Alabama and Georgia. *Annals of the Entomological Society of America* 101, 397–401.

Sorvari, J., Haatanen, M.K. and Vesterlund, M.R. (2011) Combined effects of overwintering temperature and habitat degradation on survival of the boreal wood ant. *Journal of Insect Conservation* 15, 727–731.

Stillman, J.H. (2003) Acclimation capacity underlies susceptibility to climate change. *Science* 301, 65–66.

Strzepek, K., Yohe, G., Neumann, J. and Boehlert, B. (2010) Characterizing changes in drought risk for the United States from climate change. *Environmental Research Letters* 5, 044012. 9 pp.

Su, N.-Y. and Scheffrahn, R.H. (1998) A review of subterranean termite control practices and prospects for integrated pest management programmes. *Integrated Pest Management Reviews* 3, 1–13.

Suarez, A.V., Holway, D.A. and Case, T.J. (2001) Patterns of spread in biological invasions dominated by long-distance jump dispersal: insights from Argentine ants. *Proceedings of the National Academy of Sciences USA* 98, 1095–1100.

Tanaka, K. and Tanaka, S. (1997) Winter survival and freeze tolerance in a northern cockroach, *Periplaneta japonica* (Blattidae: Dictyoptera). *Zoological Science (The Zoological Society of Japan)* 14, 849–853.

van Aarde, R.J. and Jackson, T.P. (2007) Food, reproduction and survival in mice on sub-Antarctic Marion Island. *Polar Biology* 30, 503–511.

Vetter, R.S. (2015) *The Brown Recluse Spider*. Cornell University Press, Ithaca, New York, USA.

Vihma, T. (2014) Effects of Arctic Sea ice decline on weather and climate: a review. *Surveys in Geophysics* 35, 1175–1214.

Walther, G-R., Roques, A., Hulme, P.E., Sykes, M.T., Pyšek, P., Kühn, I., Zobel, M., Bacher, S., Botta-Dukát, Z., Bugmann, H., Czúcz, B., Dauber, J., Hickler, T., Jarosík, V., Kenis, M., Klotz, S., Minchin, D., Moora, M., Nentwig, W., Ott, J., Panov, V.E., Reineking, B., Robinet, C., Semenchenko, V., Solarz, W., Thuiller, W., Vilà, M., Vohland, K. and Settele, J. (2009) Alien species in a warmer world: risks and opportunities. *Trends in Ecology and Evolution* 24, 686–693.

Walters, A.C. and Mackay, D.A. (2004) Comparisons of upper thermal tolerances between the invasive Argentine ant (Hymenoptera: Formicidae) and two native Australian ant species. *Annals of the Entomological Society of America* 97, 971–975.

Wetterer, J.K. and Porter, S.D. (2003) The little fire ant, *Wasmannia auropunctata*: distribution, impact, and control. *Sociobiology* 42, 1–42.

Williams, J.W., Jackson, S.T. and Kutzbach, J.E. (2007) Projected distributions of novel and disappearing climates by 2100 AD. *Proceedings of the National Academy of Sciences USA* 104, 5738–5742.

4 Natural Disasters, Extreme Events and Vector-borne Diseases: Impact on Urban Systems

Chester G. Moore*

Department of Microbiology, Immunology & Pathology, Arthropod-borne & Infectious Diseases Laboratory (AIDL), Colorado State University, Fort Collins, Colorado, USA

4.1 Introduction

It is generally accepted by the scientific community (IPCC, 2014) that climate change is, and has been, occurring for some time. The impacts of change are likely to continue and even to increase over time. Because the arthropod disease vectors (insects, ticks, and mites) are cold-blooded organisms, they are greatly impacted by changing climatic conditions such as temperature and precipitation. Therefore, it is not surprising to expect that vector-borne diseases will be impacted in various ways by changing local, regional or global climate. But what is the rationale for selecting natural disasters and extreme events as a focus area in the discussion of climate change? One of the major predictions of climate-change models is that extreme environmental events will become more frequent (i.e. that the variance for any given class of events will increase according to model predictions (IPCC, 2013; Knutson *et al.*, 2010; Elsner *et al.*, 2015). Within the past several decades, there have been multiple natural disasters that provide examples for study and examination in terms of potential involvement of vector-borne disease. This chapter makes an effort to answer the following questions.

- What evidence is there that a particular class of extreme event or natural disaster will increase or decrease the likelihood of outbreaks of vector-borne disease, either in the near term or long term?
- How are vector-borne disease systems impacted by weather or climate events, and are some disease systems more or less likely to be impacted by these events than other disease systems (e.g. malaria vs dengue; filariasis vs leishmaniasis)?
- What are some of the issues that arise in dealing/responding to vector-borne events that emerge in the wake of extreme events/natural disasters?
- What sorts of data should be collected, and at what scale, to provide definitive answers to remaining issues related to these topics?

In order to have a meaningful discussion of the impact of natural disasters on vector-borne diseases, we must first define what is meant by the terms 'natural disaster' and 'extreme event'. An extreme event can be defined in many ways. Generally speaking, it is a highly atypical event. The US National Oceanic and Atmospheric Administration

*E-mail: chester.moore@colostate.edu

(NOAA) notes: 'extreme events are defined as lying in the outermost ("most unusual") 10 percent of a place's history' (http://www.ncdc.noaa.gov/climate-information/extreme-events). A natural disaster is a major environmental event resulting from natural processes such as floods, tsunamis or hurricanes, tornadoes, earthquakes, volcanic eruptions, drought, heatwaves, or other physical processes. Natural disasters can cause loss of life, property damage, destruction of physical or social infrastructure, industry and agriculture, and other changes to a region that can limit the local population's ability to survive or recover. Given these definitions, what are some examples from the world of vector-borne disease that can instruct us about the potential impact of changing climate? What evidence is there that a particular class of extreme event/natural disaster will increase or decrease the likelihood of outbreaks of vector-borne disease, either in the near term or long term?

Various host–vector–pathogen systems will be impacted in different ways by particular aspects of climate. The likelihood of a particular class of climatic event causing or leading to an increase (or decrease) in vector-borne disease outbreaks will depend heavily on the ecology of the vectors and vertebrate hosts, as well as the behaviour and other characteristics of the local human population. For example, diseases that have mosquitoes as their vectors will be strongly dependent on water because the larval stages of the vector are aquatic. For flea-transmitted diseases, this is less likely to be the case. Different vector species are more or less closely tied to the human domicile (e.g. *Culex* vs *Anopheles* vs *Aedes*; exophilic vs endophilic feeding behaviour, etc.). In addition, local socioeconomic and cultural considerations have an impact on disease dynamics. The following section examines several major environmental causes of disasters, and reviews evidence for impact, or lack of impact, on vector-borne disease.

4.2 Wind and Water (Hurricane/Tornado/Tsunami/Storm/Flood)

Nasci and Moore (1998) reviewed data for a series of 12 floods, tropical storms and hurricanes occurring in the United States between 1975 and 1997. They found only one major epidemic event (Red River flood of 1975) that involved large numbers of human and veterinary cases of arboviral disease (St Louis encephalitis and western equine encephalomyelitis). While surveillance activities indicated arbovirus activity in vectors and sentinel flocks, and occasional veterinary infections in the other extreme events, no significant disease outbreaks were documented. Caillouët *et al.* (2008), on the other hand, found a twofold or more increase in neuroinvasive cases of West Nile virus infection in Louisiana and Mississippi counties impacted by Hurricane Katrina in 2005. They found a pattern of increased West Nile neuroinvasive disease (WNND) cases both spatially (within the hurricane damage path area) and temporally (in comparison to the same periods in previous years: 2002–2004, and again in 2006). A broadly based survey of the global literature (Ahern *et al.*, 2005) revealed a mixture of impact levels of flooding on vector-borne disease. These differing observations indicate that severe storms vary and their impact on vectors and vector-borne disease may depend on many factors other than the storm itself. It is also possible, at least in the United States, where active local surveillance and control programmes are common in some localities, that early implementation of vector-control activities prevents some outbreaks that would otherwise arise following storm activity.

Water, in the form of tsunamis, can have another, perhaps less obvious, impact on vector-borne disease. Ocean-wave incursion can alter the salinity and other features of near-shore water bodies, enhancing or reducing suitability for vector species. Waves from a tsunami that struck the Andaman and Nicobar Islands in 2004 created increased brackish water larval habitat for *Anopheles sundaicus*, the major local vector of *Plasmodium vivax* and *P. falciparum* (Krishnamoorthy *et al.*, 2005), with a resulting increase in reported malaria incidence as indicated by the slide positivity rate.

Disease is not the only issue following major hurricanes or similar events. 'Nuisance' species can become an issue in post-disaster recovery when they prevent first responders and other recovery personnel from functioning in the recovery operation (Breidenbaugh *et al.*, 2008). An example of this problem occurred following Hurricane Andrew in Louisiana in 1992, when nuisance mosquitoes (mainly *Aedes*, *Psorophora* and *Coquillettidia* spp.) became so numerous that electrical workers were unable to replace downed power lines in the impacted area (author's personal observation). This led to the Federal Emergency Management Agency (FEMA) authorizing emergency nuisance mosquito control to permit the reconstruction work to continue.

Wind associated with storm fronts and other meteorological events may be a significant factor in the spread of vectors and pathogens (Sellers, 1980; Sellers and Maarouf, 1988; Hendrickx *et al.*, 2008). Vectors can be carried long distances on storm fronts or on major wind patterns of the Intertropical Convergence Zone or other major circulatory systems (Sellers, 1980). Vectors as small and fragile as *Culicoides* can be moved long distances. Although wind-borne dispersal of several hundred kilometres has been documented over water, smaller distances seem to be typical. For example, in Europe wind-borne spread over land was estimated at 35–85 km (Hendrickx *et al.*, 2008).

4.3 Drought

The ecological impact of drought lasting over one to several seasons is likely to be quite different from the impact of a multi-decadal drought. For example, the current (2015) drought in the western United States is drying up streams and irrigation canals, resulting in impounded rather than flowing water, which provides increased larval habitat for *Culex* vectors of West Nile virus. With long-term drought, such habitats would possibly not exist.

One prediction of climate models (Fig. 4.1) is for decreased precipitation throughout a large band of geography that includes the Middle East, southern Europe, Mexico and the south-western United States. How drought will affect vector-borne diseases in

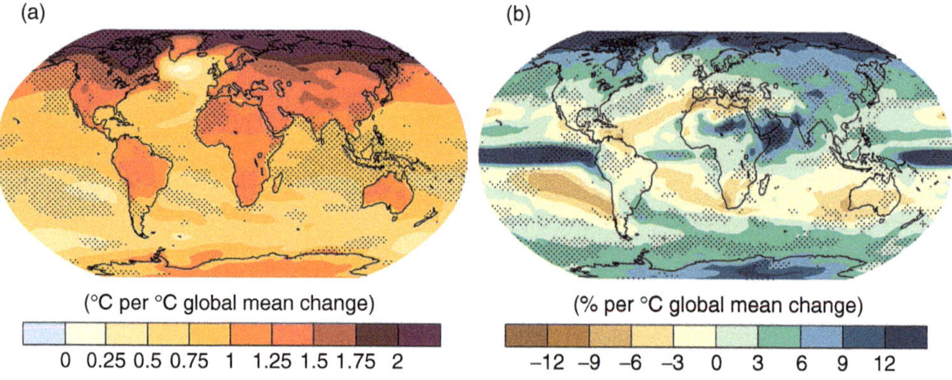

Fig. 4.1. Predictions of the Coupled Model Inter-comparison Project (CMIP5) show the broad predicted patterns of change in global temperature and precipitation in Middle East, North Africa, southern Europe, Mexico and southern United States, 2081–2100. (a) Temperatures are predicted to increase poleward (red shades) in the Northern Hemisphere, perhaps less so in the Southern Hemisphere. (b) Severe drought (brown shades). Source: IPCC Technical Summary Box TS 6.1, used with permission.

these areas is unknown. It seems likely that rodent-borne agents such as plague and leishmaniasis will increase, and mosquito-transmitted agents will become less common or more localized (more focal). One question that arises is whether the development of extensive dry zones will interrupt the northward movement of tropical species toward the north in Europe and North America.

However, before drying possibly interrupts the northward dispersal of tropical vectors and pathogens into more northern habitats, there are many opportunities for northward expansion. For example, Chagas disease, caused by the protozoan parasite *Trypanosoma cruzi* has been slowly extending its range northward in North America for many years (Sarkar *et al.*, 2010). This zoonotic parasite normally cycles between woodrats (*Neotoma* sp.) and triatome bugs of the order Hemiptera. Recent modelling studies using multiple general circulation models (GCMs) to develop a consensus model suggest that known *T. cruzi* vectors may move far north within the USA, given the predicted warming and drying of the southwestern and central portions of the USA.

There is evidence from historical research that historical periods of excessive warming (CE 800–1900) led to demographic catastrophes in the dry zones of the Northern Hemisphere, with 'Malthusian checks (famines, wars and epidemics' (Zhang *et al.*, 2011). See section 4.9 below, for additional discussion on this topic.

4.4 Temperature – Extreme Cold or Heat

In general, assuming predicted change scenarios, vector life-cycle parameters will be speeded up (Levi *et al.*, 2015), and doubling times of pathogens will be reduced, thus increasing the likelihood that a particular disease system can be maintained in a particular area under predicted warming scenarios. Dog heartworm, *Dirofilaria repens*, may provide an example and an early warning of changes to come. It is becoming an increasing problem in Europe (Svobodová *et al.*, 2006), both in canines and in humans (Genchi *et al.*, 2011). A growing degree-day model was developed that forecasts the distribution of *D. repens* in north-eastern European countries. The increase in *Dirofilaria* infections in Europe is no doubt due to climatic change, plus increased movement of dogs across Europe (Genchi *et al.*, 2011).

Bluetongue virus (BTV) is a devastating viral infection of a variety of ruminants, causing massive economic losses due to economic embargoes and other issues. BTV circulated in sub-Saharan Africa and the Middle East, but only rarely passed into southern Europe, and then quickly disappeared. Beginning in the autumn of 1998, however (Purse *et al.*, 2005), BTV-9 (one of 24 different serotypes) appeared on four Greek islands, spreading into Turkey, Bulgaria, Kosovo, Albania, Bosnia and Herzegovina, westward into mainland Greece, Italy, Sicily, Sardinia and Corsica. Subsequently, three more serotypes (BTV-1, BTV-4, and BTV-16) also entered through Greece. A BTV-2 introduction from Tunisia or Algeria into Italy and Sardinia occurred in 2000, and in 2004 BTV-4 appeared in Spain and Portugal. In short, within a short period of just 7 years, at least 12 countries recorded bluetongue outbreaks up to 800 km farther north (up to 44° 30′N), of five different serotypes, with overwintering transmission. In addition, the range of the main vector, *Culicoides imicola*, was found to have extended substantially northward into France, Switzerland and eastern Spain. Purse *et al.* (2005) presented evidence linking these changes to changes in regional temperature and precipitation. As Maclachlan (2010) points out, the more recent extension of BTV into northern Europe around 2006 has involved different BTV serotypes (BTV-6, 8 and 11), although the systems appear to be merging over time. Of particular concern is the ability of a number of Palaearctic *Culicoides* species to serve as efficient BTV vectors: *C. obsoletus*, *C. dewulffi* and *C. chiropterus* (Maclachlan, 2010). On the other hand, general warming may have a substantial impact on the course of dengue according to recent modelling studies. Kuzera, working in

Thailand, found a link between recent climatic change and change in dengue hemorrhagic fever (DHF) rates (Kuzera, 2011). In particular, there has been a rise in winter DHF cases. Kuzera speculates that as minimum temperatures increase, dengue transmission rates will also increase.

4.5 Earthquake/Volcanic Eruption/ Landslide

The most likely impacts of earthquake, volcanic eruption and subsequent landslide would seem to be in the alteration of waterways and floodplains creating impounded water habitats. One example is a malaria outbreak following the 1991 earthquake in the Limón region of Costa Rica (Sáenz *et al.*, 1995). Major physical changes in topography due to the earthquake, followed approximately 4 months later by heavy rains, destroyed or badly damaged homes, washed out bridges, and otherwise destroyed the region. This left people without proper housing and exposed them to mosquitoes. With increased larval habitat for *Anopheles* mosquitoes, and reduced access to the area by the national malaria control programme, a malaria outbreak was inevitable. Comparing malaria rates for the 22 months prior to the earthquake to the 13 months after the earthquake, Sáenz *et al.* (1995) calculated increases as high as 1600% and 4700% above the average monthly rate for the pre-earthquake rate ($P<0.01$). They postulated human exposure (sleeping outside without proper housing) and interruption of malaria control, increased mosquito habitats and flooding as the major contributors to the epidemic.

In Iran, cutaneous leishmaniasis infection rates were found to differ before and after an earthquake (Sharifi *et al.*, 2011). Following a devastating earthquake in Bam, Iran, in December, 2003, there was a sharp increase in cutaneous leishmaniasis cases, reaching 7.6 per 1000 population (compared to 1.9 per 1000 for the 5 years prior to the earthquake). Also, land uplift and subsidence in the Andaman Islands may be leading to increased environmental disturbance and

malaria vector proliferation (*An. sundaicus* and *An. subpictus*) (Krishnamoorthy *et al.*, 2005).

4.6 Unknown Causes

The trombiculid mite vectors of scrub typhus, *Orientia* (= *Rickettsia*) *tsutsugamushi*, are typically found in open, grassy, 'scrub' vegetation, frequently along the sandy margins of rivers and streams. They can rapidly colonize disturbed environments, whether disturbed by human or natural events (Traub and Wisseman, 1974). Already, reports of expanded distribution of scrub typhus are appearing (Currie, 2001; Zhang *et al.*, 2010, 2013). To what extent these extensions are a reflection of changing climate/weather as opposed to physical damage to ecosystems is unknown.

4.7 Climatic Regulation of Vectors

If climate change is to have any effect at all on vector-borne diseases, it must do so through the intermediaries of the parameters and variables of the R_0 equation.
(Rogers *et al.*, 2006)

R_0 is defined as:

$$R_0 = \frac{mbca^2 e^{-\mu T}}{\mu r}$$

where:

m = the ratio of vector numbers (V) to host numbers (N), e.g. bites per night,
b = the transmission coefficient from vertebrate to vector
c = the transmission coefficient from vector to vertebrate
a = the biting rate of the vector on the host of interest
μ = the (daily) mortality rate of the vector
T = the extrinsic incubation period of the pathogen in the vector
r = the recovery rate of the host from infection.

Thus, R_0 defines the number of new cases of disease that will arise at some time in the future from one case of the disease at the present time when introduced into a population of totally susceptible hosts. How then might extreme events related to climate change have an impact on these components of the vector-borne disease equation?

4.7.1 Change in *m*, vector density

Change due to *m* could be either absolute or relative. Absolute change would be, for example, an explosion of mosquitoes, muscoid flies, or other species following a hurricane, as occurred after Hurricane Katrina (Breidenbaugh and Haagsma, 2008). Relative change can occur when there is a change in the rate (or opportunity) of contact between host and vector. It is not uncommon in severe storm events for housing to be severely compromised, such that residents are no longer protected by door or window screens. Also, they are likely to be outdoors, involved in clean-up and reconstruction activities. In many cases, people may be crowded into evacuation centres with little protection from bites. Thus, their relative exposure to mosquitoes or other vectors will be increased. The most frequent impact is likely to be due to flooding and its impact on mosquito larval habitat. However, this can be either positive or negative. Widespread flooding can dramatically increase the area available for ovipositing females to lay their eggs, and these new areas will not be rapidly colonized by predators. Conversely, intense rainfall and flooding can have a negative impact by flushing larval habitats, washing away eggs of *Aedes* species, and removing nutrient resources within the habitat. Flooding may have a major impact on *Simulium* vectors of onchocerciasis as well, if stream flow becomes strong enough to dislodge rocks that form larval attachment sites. Habitat destruction, such as deforestation, can lead to increased exposure to vectors or vertebrate hosts of a variety of pathogens. Drought affects *m* by reducing available habitat for aquatic groups, and may decrease survival rates due to the effects of lower environmental humidity.

Drought also may serve to concentrate vectors and hosts at limited aquatic sites, increasing probability of contact.

4.7.2 Change in *b* and *c*, transmission efficiency

Not every contact between an infected vector and a susceptible host or an infected host and a susceptible vector results in transmission. The values *b* and *c* reflect those probabilities. The probabilities involve genetic characteristics as well as physiological processes in the vertebrate host (*b*) and vector (*c*) that govern – among other things – the efficiency of the transmission process, and may be particularly important in the case of the vector (*c*). In addition, temperature, humidity and wind might indirectly affect *b* or *c*.

4.7.3 Change in *a*, biting rate

The biting rate plays a major role in the equation (because it is squared) – does this vector feed only once in each gonotrophic cycle or does it take multiple meals (Arredondo-Jimenez *et al.*, 1998; Conway *et al.*, 1974)? Is that behaviour dependent on or altered by temperature? This can make a massive difference in the value of R_0. There is a need for more data on this subject.

4.7.4 Change in μ, daily mortality (survival)

The daily mortality rate (μ) is the most important variable in the equation, but also the hardest to measure (Dye, 1986). Because daily mortality (or survival) is difficult to measure directly, a wide range of methods have been developed to estimate it. The most common methods utilize changes in the female reproductive system (Detinova, 1968). Other techniques utilize changes in hydrocarbon coatings of the cuticle, pteridines (Wall *et al.*, 1990), changes in internal apodemes (Moore *et al.*, 1986) or other physical structures (Hayes and Wall, 1999). Examples of field application include sand fly vectors of *Leishmania* in France (Dye *et al.*, 1987) and *Anopheles* vectors of malaria in Mali (Diuk-Wasser *et al.*, 2005).

4.7.5 Change in *T*, extrinsic incubation

The extrinsic incubation period, the time needed for the pathogen to complete development within the vector and become infective, is strongly temperature-dependent. Therefore, we expect warmer climates to lead to shorter values of T in general. It is important to note that there is not a straight-line relationship between temperature and T. At least for some viruses, modulation occurs above a certain temperature, and infection falls off (Kramer *et al.*, 1983, 1998). Some arboviruses might even disappear at higher ambient temperatures (Reeves *et al.*, 1994).

4.7.6 Change in *r*, host recovery rate

The recovery rate of the human or other vertebrate host, r (or average duration of infectiousness, $1/r$), seems somewhat less likely to be impacted by changing climate. This is the period when the pathogen remains at an infectious level in the host. The vertebrate host, whether human or other mammal or bird, will typically only be infectious for a certain length of time (except for diseases such as filariasis where microfilariae are shed over a long period). After that, the host may become immune to further infection, as with many arboviruses, or may be reinfected, as with malaria.

4.8 The Impact of Extreme Events on Vector-borne Diseases

There seem to be no all-encompassing principles/statements that govern the effect of weather/climate on vector-borne disease activity – not when comparing two diseases, when comparing two vectors, or perhaps even the same vector on two different continents. In general, with respect to water (flood, precipitation, etc.), disease systems such as malaria, Rift Valley fever, and Japanese encephalitis, whose vectors utilize open, flooded habitats (many *Anopheles*, and *Culex* species, floodwater *Aedes*, etc.) will be most likely to be impacted by an increase in rainfall and flooding. Disease systems such as dengue and Chikungunya, whose vectors (e.g. *Aedes aegypti*, *Ae. albopictus*) primarily utilize domestic containers and related indoor habitats as the larval habitat, are less likely to be impacted by changes in precipitation. Those same species, when found in outdoor habitats, may be negatively impacted if containers are flushed by excess rainfall. The relationship with these latter 'domestic' vectors can be complicated, for example when water storage in homes is increased during the dry season, or when local piped water delivery systems have been interrupted by failure of local infrastructure (author's personal observation). The impact of weather on the *Culex pipiens* complex of species will depend on local conditions, with regard to livestock, sewage disposal, etc., as well as the details of weather.

As an example of the difficulty of predicting the outcome of climate change on a single disease, the following example can be considered. Leishmaniasis is a widespread zoonotic disease common in both the New World and Old World tropical regions. A recent study modelled the spatial distribution of the two major vectors of visceral leishmaniasis (*Leishmania infantum*) in Spain, *Phlebotomus perniciosus* and *Phlebotomus ariasi*, to predict their distribution under several climate-change scenarios (Galvez *et al.*, 2011). Using two different models, the authors found substantial and increasing expansion of the vectors in each third of the 21st century, under both the IPCC Atmosphere-Ocean General Circulation Model A2 and B2 emission scenarios (the two scenarios differ in the assumed rate of rise in greenhouse gas emissions). However, leishmaniasis in the New World and the Old World may have very different futures under changing climate. It appears that *Leishmania infantum* cases have been declining in recent years in Colombia. Maxent modelling, using both Hadley B2 and CSIRO A2 climate scenarios, showed an overall reduction in the spatial distribution of the two vector species with recent warming (González *et al.*, 2014), due to a shift in altitudinal distribution of the vectors. It is important to note that there were significant methodological differences in the two studies described here, and that could impact the comparison, but it does appear that there may be genuine ecological differences in the ecology of this parasite.

4.9 Issues in Vector-borne Events Following Extreme Events/Natural Disasters

The issues that we can expect to deal with following extreme events or natural disasters typically fall into three classes: lack of staffing and resources; lack of concern or lack of will; and lack of information or data. In addition, as a result of climate change, there may be additional impacts brought on by major societal changes (Burke *et al.*, 2009; Zhang *et al.*, 2011) that will further complicate any response.

In the case of large systems such as vector–host–pathogen systems, there is seldom sufficient involvement of multiple groups (stakeholders) in the development of response plans. Crucial players are often left out of the planning stages, only to be found missing once the action is underway, and when they are no longer able to do their own pre-planning and staging. Thus rigorous application of a process called 'disaster incubation analysis' and more advanced forms of environmental analysis have considerable potential to improve the pattern of planning and decision making (Mulvihill and Ali, 2007).

There is often a reluctance to assign staff and budget to known new or arising problems, whether at the local, national or international level. For example, an evaluation of progress made after 5 years on objectives of the health-sector component of the US 2000 National Assessment on climate change pointed out that: 'relatively little progress has been noted in the literature on implementing adaptive strategies and measures' (Ebi *et al.*, 2006). There is the famous 'predator–prey' model of administrative response: preparedness declines following an epidemic, to the point where a response is no longer possible when the next epidemic strikes (Hadler *et al.*, 2015). When the next outbreak hits, the 'powers that be' throw resources at the problem – but too late to have an impact on the current outbreak. They now rebuild the programme to its original (or nearly so) level, for the cycle to be repeated again and again. Disruption of social services after a serious weather event also may lead to accumulation of garbage, faecal material and other items in crowded conditions, allowing muscoid flies to multiply to very high densities, following disasters such as tsunamis (Srinivasan *et al.*, 2006) and other similar events (Breidenbaugh and Haagsma, 2008; Srinivasan *et al.*, 2009), with the potential for transmission of bacterial pathogens (Kobayashi *et al.*, 1999). The 5 Ps (Prior Planning Prevents Poor Performance) is a good reminder in regional planning exercises. Thousands of confined animal-feeding operations (CAFOs) have been constructed in eastern North Carolina in the United States. The faecal waste pit and spray field waste management systems used by these operations are susceptible to flooding in this low-lying region (Wing *et al.*, 2002). This was evident following Hurricane Floyd in 1999, resulting in creation of extensive habitat for polluted-water mosquitoes such as *Culex quinquefasciatus*, a major vector of West Nile and St Louis encephalitis viruses, and flies.

Zhang *et al.* (2011) explored the effects of long-term climate change in pre-industrial countries (800–1900 CE) in the Northern Hemisphere. They found clear evidence of impacts due to climate change. Deteriorating climate led to reduced carrying capacity of ecosystems, leading to population collapse via 'Malthusian checks (famines, wars, and epidemics)'. Perhaps most interesting in the current context is their finding that all of the population collapses in periods of warm climate occurred in dry and tropical humid zones (Zhang *et al.*, 2011). As of 2000, nearly all the world's episodes of political violence were occurring in drylands (Sachs, 2008).

Response to vector-borne disease outbreaks under these conditions would be severely limited if not impossible (e.g. in wartime, mass evacuee/refugee situations) (Burke *et al.*, 2009). One modelling study of future climatic conditions for 2020–2050 and 2070–2099 in the Gulf countries predicted: 'increased frequency of infectious vector borne diseases in the region between 2070 and 2099' (Husain and Chaudhary, 2008). In fact, the current combined impacts of drought and political instability in the Middle East is leading to massive human migrations, with significant potential for disease movement (Taylor, 2015). In the age of modern transportation, a major problem during and following disasters is the dispersal of vertebrates, vectors and, in many cases, affected human populations. Not only do animals disperse on their own, but they may be trapped and transported to international destinations with potential disease consequences. The international market in exotic animals is poorly monitored at the best of times (Ehnert and Galland, 2009). It is likely that, following a major disaster, regulatory processes are at their lowest level, creating opportunities for dispersal of exotic vectors, pathogens and hosts across major geographic regions. There is a large literature on the international dispersal of vectors by human activities (Pratt *et al.*, 1946; Evans *et al.*, 1963; Ross, 1964; Belton, 1980; Gagne, 1981; Le Maitre and Chadee, 1983; Goh *et al.*, 1985; Sprenger and Wuithiranyagool, 1986). In the past, some countries have instituted disinsection programmes to try to prevent such introductions (Laird *et al.*, 1994; Gratz *et al.*, 2000). It seems fairly likely that, under the extreme conditions of a major environmental catastrophe, most, if not all, preventive measures would be abandoned and borders would be open to incursion for exotic vectors, hosts and pathogens alike. Another mechanism for moving pathogens is via the disaster responders. Responders may bring in 'exotic' diseases. For example, a journalist visiting the Japan earthquake site in 2011 was infected with the D genotype of measles virus, not found in Japan (Takahashi *et al.*, 2013). An additional 11 measles cases were identified, some of which were of the D genotype. But relief workers can avoid being part of the problem if they take proper precautions (O'Leary *et al.*, 2002). Relief workers (and others) travelling to areas recovering from disaster situations should receive specific pre-travel counselling to cover the likely conditions they will face (Esposito *et al.*, 2012). In addition to the known vector-borne infections, there are likely to be additional pathogens that will become significant as ecosystems change. Recent discoveries of new viruses in the USA, such as Heartland virus (Bosco-Lauth *et al.*, 2015) and Bourbon virus (Kosoy *et al.*, 2015), and Nipah (Chua *et al.*, 2000), Hendra (Halpin *et al.*, 2000), and related bat-borne viruses in Asia, suggest there may be many more zoonotic pathogens that can pass from wild animal reservoirs to the human population, given the right conditions.

A recent study of transmission pathways (Loh *et al.*, 2015), using a database of 183 zoonotic pathogens, found that over 50% of agents associated with land-use change were transmitted by vectors. This may have major significance in terms of climate change, given the expected massive changes in landscape that are predicted by the end of the 21st century (see Fig. 4.1 above). There is also the example of the expanding distribution of the *Culicoides* vector into new areas, which remained undetected until an outbreak of BTV-8 (Weaver and Reisen, 2010). It is unknown how many more vectors and pathogens are waiting to be discovered in new locations.

4.10 Additional Data Needed to Provide Definitive Answers to Climate Change Issues

The most crucial need is to collect data in areas where little or no data are currently available. This will help to better understand the global impact of climate change impacts. Active surveillance for animal diseases is poor or absent in areas such as Latin America (Pinto *et al.*, 2008). Given that many human pathogens are zoonotic in origin, this is a major shortcoming. In general, data should be collected at the finest practical scale, since data can always be aggregated upward. The precise scale will depend greatly on the particular host–vector–pathogen system in question; birds, for example, typically have much larger foraging ranges than rodents, and the same for mosquitoes vs fleas. Ideally, disease monitoring programmes would be designed to gather data along, across or within different predicted zones of the changing climate, so as to better document change as it occurs. Data on the most significant disease/vector issues in potentially dry, arid regions (Middle East, south-western United States, etc.) are particularly important. This will be a major concern for future climate change (Sachs, 2008; Zhang *et al.*, 2011) and should be a major focus of data collection and study. Some of this monitoring can be done by using remote-sensing satellite platforms, but there are issues of pixel size, return time, bands sampled, and related issues to consider (Guptill and Moore, 2013; Moore and Freier, 2013). There are opportunities for drone-based trapping systems that can fly into study sites, land, collect samples and fly back to a field laboratory. This would be valuable in many areas. Citizen science projects may be one way to economically gather this type of data at the scale needed to have a meaningfully detailed analysis of patterns in time and space (Ryder *et al.*, 2010). This is fairly easy for environmental data (weather, soil, plants) and vector monitoring, but less so for pathogen and vertebrate host (bird, rodent, etc.), and needs to be given some thought. Schools are the ideal focus for such programmes (Doesken, 2007). In the United States, youth clubs such as 4-H have taken on summer projects to monitor the spread of *Aedes albopictus* by placing ovitraps each week and sending the collected egg papers to a laboratory for counting.

Large-scale, intensively collected data sets are needed to establish the current distribution and geographic limits of the major and minor vectors. For example, a recent international collaborative project has assembled a database of the global distribution of *Aedes aegypti* (19,930 records) and *Ae. albopictus* (22,137 records) (Kraemer *et al.*, 2015a). Those data were used in combination with environmental and other data to construct maps of the most likely current distribution of these two important vectors (Kraemer *et al.*, 2015b). This is crucial going forward as we map changes over the next decades. Intensive studies of specific diseases may reveal geographic differences in ecology and behaviour that impact, for example, malaria control strategies (Conn *et al.*, 2015). Work should continue to explore and develop a database on 'natural disaster cycles', such as those related to El Niño (ENSO) (Bouma *et al.*, 1997). Collaborative surveillance and reporting programmes like ProMED-mail (Program for Monitoring Emerging Diseases) also need to be expanded. Sponsored by the international Society for Infectious Diseases, ProMED-mail (http://www.promedmail.org) provides rapid global dissemination of information on infectious disease outbreaks, toxin exposures, and similar issues affecting humans, animals and food and forage plants. A group of highly experienced moderators provides valuable background and context for each episode, along with questions remaining to be answered, such as national or international response, extent of the outbreak, types of confirmatory testing, etc. Because ProMED-mail is not connected to any governmental agency, reports tend to be posted in near 'real time', subject to review by the moderators. This can be extremely valuable in the case of highly contagious pathogens. Unfortunately, funding for programmes such as ProMED is generally tenuous and their long-term survival is questionable, in spite of their obvious value.

Extreme events, strictly speaking, are weather events, not climate events. However, extreme events are predicted to become more frequent as a result of a generally warming climate. Since extreme events are clearly identifiable and measurable, they provide an opportunity to search for any possible effects of future climatic factors on disease vectors, their hosts or the pathogens they transmit.

During an extreme event, such as a hurricane or tsunami, there is little time for planning. Planning should have been done before the event, and, hopefully, resources are sufficient to respond following the event. This should be the first priority for every local and state health department and local vector-control programme. Simulation exercises, coordinated through the state emergency response office or other resource can be a valuable tool in identifying missing components in the response plan. Finally, there is a need at the global level for the World Health Organization (WHO) and other international agencies, as well as the individual nations of the world, to ensure both political and financial support (Campbell-Lendrum *et al.*, 2015). This will help continue the prevention and control of vector-borne diseases globally, in spite of whatever new developments come about as a result of changing climate.

References

Ahern, M., Kovats, R.S., Wilkinson, P., Few, R. and Matthies, F. (2005) Global health impacts of floods: epidemiologic evidence. *Epidemiologic Reviews* 27, 36–46.

Arredondo-Jimenez, J.I., Rodriguez, M.H. and Washino, R.K. (1998) Gonotrophic cycle and survivorship of *Anopheles vestitipennis* (Diptera: Culicidae) in two different ecological areas of southern Mexico. *Journal of Medical Entomology* 35, 937–942.

Belton, P. (1980) The first record of *Aedes togoi* (Theo.) in the United States – aboriginal or ferry passenger? *Mosquito News* 40, 624–626.

Bosco-Lauth, A.M., Panella, N.A., Root, J.J., Gidlewski, T., Lash, R.R. Harmon, J.R., Burkhalter, K.L., Godsey, M.S., Savage, H.M., Nicholson, W.L., Komar, N. and Brault, A.C. (2015) Serological Investigation of Heartland virus (Bunyaviridae: *Phlebovirus*) exposure in wild and domestic animals adjacent to human case sites in Missouri 2012-2013. *The American Journal of Tropical Medicine and Hygiene* 92, 1163–1167.

Bouma, M.J., Kovats, R.S., Goubet, S.A., Cox, J.S.H. and Haines, A. (1997) Global assessment of El Niño's disaster burden. *The Lancet* 350, 1435–1438.

Breidenbaugh, M. and Haagsma, K. (2008) The US Air Force Aerial Spray Unit: a history of large area disease vector control operations, WWII through Katrina. *US Army Medical Department Journal* (April–June), 54–61.

Breidenbaugh, M.S., Haagsma, K.A., Walker, W.W. and Sanders, D.M. (2008) Post-Hurricane Rita mosquito surveillance and the efficacy of Air Force aerial applications for mosquito control in east Texas. *Journal of the American Mosquito Control Association* 24, 327–330.

Burke, M.B., Miguel, E., Satyanath, S., Dykema, J.A. and Lobell, D.B. (2009) Warming increases the risk of civil war in Africa. *Proceedings of The National Academy of Sciences* 106, 20670–20674.

Caillouët, K.A., Michaels, S.R., Xiong, X., Foppa, I. and Wesson, D.M. (2008) Increase in West Nile Neuroinvasive Disease after Hurricane Katrina. *Emerging Infectious Disease Journal* 14, 804–807.

Campbell-Lendrum, D., Manga, L., Bagayoko, M. & Sommerfeld, J. (2015) Climate change and vector-borne diseases: what are the implications for public health research and policy? *Philosophical Transactions of the Royal Society B: Biological Sciences* 370, article ID 20130552.

Chua, K.B., Bellini, W.J., Rota, P.A., Harcourt, B.H., Tamin, A., Lam, S.K., Ksiazek, T.G., Rollin, P.E., Zaki, S.R., Shieh, W., Goldsmith, C.S., Gubler, D.J., Roehrig, J.T., Eaton, B., Gould, A.R., Olson, J., Field, H., Daniels, P., Ling, A.E., Peters, C.J., Anderson, L.J. and Mahy, B.W. (2000) Nipah virus: a recently emergent deadly paramyxovirus. *Science* 288, 1432–1435.

Conn, J.E., Norris, D.E., Donnelly, M.J., Beebe, N.W., Burkot, T.R., Coulibaly, M.B., Chery, L., Eapen, A., Keven, J.B. and Kilama, M. (2015) Entomological monitoring and evaluation: diverse transmission settings of ICEMR projects will require local and regional malaria elimination strategies. *American Journal of Tropical Medicine and Hygiene* 93, s3, 28–41. DOI: 10.4269/ajtmh.15-0009.

Conway, G.R., Trpis, M. and McClelland, G.A.H. (1974) Population parameters of the mosquito *Aedes aegypti* (L.) estimated by mark-release-recapture in a suburban habitat in Tanzania. *Journal of Animal Ecology* 43, 289–304.

Currie, B.J. (2001) Environmental change, global warming and infectious diseases in northern Australia. *Environmental Health* 1, 35–44.

Detinova, T. (1968) Age structure of insect populations of medical importance. *Annual Review of Entomology* 13, 427–450.

Diuk-Wasser, M.A., Toure, M.B., Dolo, G., Bagayoko, M., Sogoba, N., Traore, S.F., Manoukis, N. and Taylor, C.E. (2005) Vector abundance and malaria transmission in rice-growing villages in Mali. *American Journal of Tropical Medicine and Hygiene* 72, 725–731.

Doesken, N. (2007) Let it rain: how one grassroots effort brought weather into America's backyards. *Weatherwise* 60, 50–55.

Dye, C. (1986) Vectorial capacity: must we measure all its components? *Parasitology Today* 2, 203–209.

Dye, C., Guy, M.W., Elkins, D.B., Wilkes, T. J. and Killick-Kendrick, R. (1987) The life expectancy of phlebotomine sandflies: first field estimates from southern France. *Medical and Veterinary Entomology* 1, 417–425.

Ebi, K.L., Mills, D.M., Smith, J.B. and Grambsch, A. (2006) Climate change and human health impacts in the United States: an update on the results of the U.S. national assessment. *Environmental Health Perspectives* 114(9), 1318–1324.

Ehnert, K. and Galland, G.G. (2009) Border health: who's guarding the gate? *Veterinary Clinics of North America: Small Animal Practice* 39, 359–372.

Elsner, J., Elsner, S. and Jagger, T. (2015) The increasing efficiency of tornado days in the United States. *Climate Dynamics* 45, 651–659.

Esposito, D.H., Han, P.V., Kozarsky, P.E., Walker, P.F., Gkrania-Klotsas, E., Barnett, E.D., Libman, M., McCarthy, A.E., Field, V., Connor, B.A., Schwartz, E., MacDonald, S., Sotir, M.J. and GeoSentinel Surveillance Network (2012) Characteristics and spectrum of disease among ill returned travelers from pre- and post-earthquake Haiti: the GeoSentinel experience. *American Journal of Tropical Medicine and Hygiene* 86, 23–28.

Evans, B.R., Joyce, C.R. and Porter, J.E. (1963) Mosquitoes and other arthropods found in baggage compartments of international aircraft. *Mosquito News* 23, 9–12.

Gagne, R.J. (1981) *Chrysomya* spp., Old World blow flies recently established in the Americas. *Bulletin of the Entomological Society of America* 27, 21–22.

Galvez, R., Descalzo, M.A., Guerrero, I., Miro, G. and Molina, R. (2011) Mapping the current distribution and predicted spread of the leishmaniosis sand fly vector in the Madrid region (Spain) based on environmental variables and expected climate change. *Vector-borne and Zoonotic Diseases* 11, 799–806.

Genchi, C., Mortarino, M., Rinaldi, L., Cringoli, G., Traldi, G. and Genchi, M. (2011) Changing climate and changing vector-borne disease distribution: the example of *Dirofilaria* in Europe. *Veterinary Parasitology* 176, 295–299.

Goh, K.T., Ng, S.K. and Kumarapathy, S. (1985) Disease-bearing insects brought in by international aircraft into Singapore. *Southeast Asian Journal of Tropical Medicine and Public Health* 16, 49–53.

González, C., Paz, A. and Ferro, C. (2014) Predicted altitudinal shifts and reduced spatial distribution of *Leishmania infantum* vector species under climate change scenarios in Colombia. *Acta Tropica* 129, 83–90.

Gratz, N.G., Steffen, R. and Cocksedge, W. (2000) Why aircraft disinsection? *Bulletin of the World Health Organization* 78, 995–1004.

Guptill, S.C. and Moore, C.G. (2013) Investigating vector-borne and zoonotic diseases with remote sensing and GIS. In: Selinus, O. (ed.) *Essentials of Medical Geology,* 2nd edn. Springer, Dordrecht, Netherlands.

Hadler, J., Patel, D., Nasci, R.S., Petersen, L.R., Hughes, J.M., Bradley, K., Etkind, P., Kan, L. and Engel, J. (2015) Assessment of arbovirus surveillance 13 years after introduction of West Nile Virus, United States. *Emerging Infectious Disease Journal* 21, 1159–1166.

Halpin, K., Young, P.L., Field, H.E. and Mackenzie, J.S. (2000) Isolation of Hendra virus from pteropid bats: a natural reservoir of Hendra virus. *Journal of General Virology* 81, 1927–1932.

Hayes, E.J. and Wall, R. (1999) Age-grading adult insects: a review of techniques. *Physiological Entomology* 24, 1–10.

Hendrickx, G., Gilbert, M., Staubach, C., Elbers, A., Mintiens, K., Gerbier, G. and Ducheyne, E. (2008) A wind density model to quantify the airborne spread of *Culicoides* species during north-western Europe bluetongue epidemic, 2006. *Preventive Veterinary Medicine* 87, 162–81.

Husain, T. and Chaudhary, J.R. (2008) Human health risk assessment due to global warming – a case study of the Gulf countries. *International Journal of Environmental Research and Public Health* 5, 204–212.

IPCC (2013) Climate change 2013: the physical science basis. Contribution of Working Group I to the Fifth Assessment Report of the Intergovernmental Panel on Climate Change. In: Stocker, T.F., Qin, D., Plattner, G.-K., Tignor, M., Allen, S.K., Boschung, J., Nauels, A., Xia, Y., Bex, V., Midgley, P.M. (eds) *Climate Change 2013.* Cambridge University Press, Cambridge, UK.

IPCC (2014) Climate change 2014: synthesis report. Contribution of Working Groups I, II and III to the Fifth Assessment Report of the Intergovernmental Panel on Climate Change. Core Writing Team, Pachauri, R.K. and Meyer, L.A. (eds) IPCC , Geneva, Switzerland.

Knutson, T.R., McBride, J.L., Chan, J., Emanuel, K., Holland, G., Landsea, C., Held, I., Kossin, J.P., Srivastava, A.K. and Sugi, M. (2010) Tropical cyclones and climate change. *Nature Geoscience* 3, 157–163.

Kobayashi, M., Sasaki, T., Saito, N., Tamura, K., Suzuki, K., Watanabe, H. and Agui, N. (1999) Houseflies: not simple mechanical vectors of enterohemorrhagic *Escherichia coli* O157:H7. *American Journal of Tropical Medicine and Hygiene* 61, 625–629.

Kosoy, O.I., Lambert, A.J., Hawkinson, D.J., Pastula, D.M., Goldsmith, C.S., Hunt, D.C. and Staples, J.E. (2015) Novel *Thogotovirus* species associated with febrile illness and death, United States, 2014. *Emerging Infectious Diseases* 21(5). DOI: 10.3201/eid2105.150150.

Kraemer, M.U., Sinka, M.E., Duda, K.A., Mylne, A., Shearer, F.M., Brady, O.J., Messina, J.P., Barker, C.M., Moore, C.G., Carvalho, R.G., Coelho, G.E., Van Bortel, W., Hendrickx, G., Schaffner, F., Wint, G.R.W., Elyazar, I.R.F., Teng, H.-J. and Hay, S.I. (2015a) The global compendium of *Aedes aegypti* and *Ae. albopictus* occurrence. *Scientific Data* 2, article ID 150035.

Kraemer, M.U., Sinka, M.E., Duda, K.A., Mylne, A.Q., Shearer, F.M., Barker, C.M., Moore, C.G., Carvalho, R.G., Coelho, G.E., Van Bortel, W., Hendrickx, G., Schaffner, F., Elyazar, I.R., Teng, H.-J., Brady, O.J., Messina, J.P., Pigott, D.M., Scott, T.W., Smith, D.L., Wint, G.W., Golding, N. and Hay, S.I. (2015b) The global distribution of the arbovirus vectors *Aedes aegypti* and *Ae. albopictus*. *eLife* 4, article ID e08347.

Kramer, L.D., Hardy, J.L. and Presser, S.B. (1983) Effect of temperature of extrinsic incubation on the vector competence of *Culex tarsalis* for western equine encephalomyelitis virus. *The American Journal of Tropical Medicine and Hygiene* 32, 1130–9.

Kramer, L.D., Hardy, J.L. and Presser, S.B. (1998) Characterization of modulation of western equine encephalomyelitis virus by *Culex tarsalis* (Diptera: Culicidae) maintained at 32 degrees C following parenteral infection. *Journal of Medical Entomology* 35, 289–95.

Krishnamoorthy, K., Jambulingam, P., Natarajan, R., Shriram, A.N., Das, P.K. and Sehgal, S.C. (2005) Altered environment and risk of malaria outbreak in South Andaman, Andaman & Nicobar Islands, India affected by tsunami disaster. *Malaria Journal* 4, 32.

Kuzera, K. (2011) Climate and climate change and infectious disease risk in Thailand: A spatial study of dengue hemorrhagic fever using GIS and remotely-sensed imagery. PhD thesis, San Diego State University and University of California Santa Barbara.

Laird, M., Calder, L., Thornton, R.C., Syme, R., Holder, P.W. and Mogi, M. (1994) Japanese *Aedes albopictus* among four mosquito species reaching New Zealand in used tires. *Journal of the American Mosquito Control Association* 10, 14–23.

Le Maitre, A. and Chadee, D.D. (1983) Arthropods collected from aircraft at Piarco International Airport, Trinidad, West Indies. *Mosquito News* 43, 21–23.

Levi, T., Keesing, F., Oggenfuss, K. and Ostfeld, R.S. (2015) Accelerated phenology of black-legged ticks under climate warming. *Philosophical Transactions of the Royal Society B: Biological Sciences* 370, article ID 20130556.

Loh, E.H., Zambrana-Torrelio, C., Olival, K.J., Bogich, T.L., Johnson, C.K., Mazet, J.A.K., Karesh, W. and Daszak, P. (2015) Targeting transmission pathways for emerging zoonotic disease surveillance and control. *Vector-borne and Zoonotic Diseases* 15, 432–437.

Maclachlan, N.J. (2010) Global implications of the recent emergence of bluetongue virus in Europe. *Veterinary Clinics of North America: Food Animal Practice* 26, 163–171.

Moore, C.G. and Freier, J.E. (2013) Geospatial technologies and spatial data analysis. In: Mikanatha, N., Lynfield, R., Van Beneden, C. and De Valk, H. (eds) *Infectious Disease Surveillance*, 2nd edn. Wiley-Blackwell, Oxford, UK.

Moore, C.G., Reiter, P. and Xu, J.J. (1986) Determination of chronological age in *Culex pipiens* s.l. *Journal of the American Mosquito Control Association* 2, 204–208.

Mulvihill, P.R. and Ali, S.H. (2007) Disaster incubation, cumulative impacts and the urban/ex-urban/rural dynamic. *Environmental Impact Assessment Review* 27, 343–358.

Nasci, R.S. and Moore, C.G. (1998) Vector-borne disease surveillance and natural disasters. *Emerging Infectious Disease Journal* 4, 333–334.

O'Leary, D.R., Rigau-Perez, J.G., Hayes, E.B., Vorndam, A.V., Clark, G.G. and Gubler, D.J. (2002) Assessment of dengue risk in relief workers in Puerto Rico after Hurricane Georges, 1998. *American Journal of Tropical Medicine and Hygiene* 66, 35–39.

Pinto, J., Bonacic, C., Hamilton-West, C., Romero, J. and Lubroth, J. (2008) Climate change and animal diseases in South America. *Revue scientifique et technique* 27, 599–613.

Pratt, J.J. Jr, Heterick, R.H., Harrison, J.B. and Haber, L. (1946) Tires as a factor in the transportation of mosquitoes by ships. *Military Surgeon* 99, 785–788.

Purse, B.V., Mellor, P.S., Rogers, D.J., Samuel, A.R., Mertens, P.P. and Baylis, M. (2005) Climate change and the recent emergence of bluetongue in Europe. *Nature Reviews Microbiology* 3, 171–181.

Reeves, W.C., Hardy, J.L., Reisen, W.K. and Milby, M.M. (1994) Potential effect of global warming on mosquito-borne arboviruses. *Journal of Medical Entomology* 31, 323–332.

Rogers, D.J. and Randolph, S.E. (2006) Climate change and vector-borne diseases. *Advances in Parasitology* 62, 345–381.

Ross, H.H. (1964) The colonization of temperate North America by mosquitoes and man. *Mosquito News* 24, 103–118.

Ryder, T.B., Reitsma, R., Evans, B. and Marra, P.P. (2010) Quantifying avian nest survival along an urbanization gradient using citizen- and scientist-generated data. *Ecological Applications* 20, 419–426.

Sachs, J.D. (2008) *Common Wealth: Economics for a Crowded Planet.* Penguin Books, New York, USA.

Sáenz, R., Bissell, R.A. and Paniagua, F. (1995) Post-disaster malaria in Costa Rica. *Prehospital and Disaster Medicine* 10, 154–160.

Sarkar, S., Strutz, S.E., Frank, D.M., Rivaldi, C.L., Sissel, B. and Sánchez-Cordero, V. (2010) Chagas disease risk in Texas. *PLoS Neglected Tropical Diseases* 4, article ID e836.

Sellers, R.F. (1980) Weather, host and vector – their interplay in the spread of insect-borne animal virus diseases. *The Journal of Hygiene* 85, 65–102.

Sellers, R.F. and Maarouf, A.R. (1988) Impact of climate on western equine encephalitis in Manitoba, Minnesota and North Dakota, 1980–1983. *Epidemiology and Infection* 101, 511–535.

Sharifi, I., Nakhaei, N., Aflatoonian, M., Parizi, M.H., Fekri, A., Safizadeh, H., Shirzadi, M., Gooya, M., Khamesipour, A. and Nadim, A. (2011) Cutaneous leishmaniasis in Bam: a comparative evaluation of pre- and post-earthquake years (1999–2008). *Iranian Journal of Public Health* 40, 49–56.

Sprenger, D. and Wuithiranyagool, T. (1986) The discovery and distribution of *Aedes albopictus* in Harris County, Texas. *Journal of the American Mosquito Control Association* 2, 217–219.

Srinivasan, R., Gunasekaran, K., Jambulingam, P. and Balaraman, K. (2006) Muscoid fly populations in tsunami-devastated villages of southern India. *Journal of Medical Entomology* 43, 631–633.

Srinivasan, R., Jambulingam, P., Gunasekaran, K. and Basker, P. (2009) Abundance & distribution of muscoid flies in tsunami-hit coastal villages of southern India during post-disaster management period. *Indian Journal of Medical Research,* 129, 658–664.

Svobodová, Z., Svobodová, V., Genchi, C. and Forejtek, P. (2006) The first report of authochthonous dirofilariosis in dogs in the Czech Republic. *Helminthologia* 43, 242–245.

Takahashi, K., Kodama, M. and Kanda, H. (2013) Call for action for setting up an infectious disease control action plan for disaster area activities: learning from the experience of checking suffering volunteers in the field after the Great East Japan Earthquake. *Bioscience Trends* 7, 294–295.

Taylor, A. (2015) 17 ways the unprecedented migrant crisis is reshaping our world. *Washington Post*, 20 June, Washington DC, USA. Available at: http://www.washingtonpost.com/blogs/worldviews/wp/2015/06/20/17-ways-the-unprecedented-migrant-crisis-is-reshaping-our-world/ (accessed 20 June 2015).

Traub, R. and Wisseman Jr, C.L. (1974) The ecology of chigger-borne rickettsiosis (scrub typhus). *Journal of Medical Entomology* 11, 237–303.

Wall, R., Langley, P.A., Stevens, J. and Clarke, G.M. (1990) Age-determination in the old-world screw-worm fly *Chrysomya bezziana* by pteridine fluorescence. *Journal of Insect Physiology* 36, 213–218.

Weaver, S.C. and Reisen, W.K. (2010) Present and future arboviral threats. *Antiviral Research* 85, 328–345.

Wing, S., Freedman, S. and Band, L. (2002) The potential impact of flooding on confined animal feeding operations in eastern North Carolina. *Environmental Health Perspectives* 110, 387–391.

Zhang, S., Song, H., Liu, Y., Li, Q., Wang, Y., Wu, J., Wan, J., Li, G., Yu, C. and Li, X. (2010) Scrub typhus in previously unrecognized areas of endemicity in China. *Journal of Clinical Microbiology* 48, 1241–1244.

Zhang, D.D., Lee, H.F., Wang, C., Li, B., Zhang, J., Pei, Q. and Chen, J. (2011) Climate change and large-scale human population collapses in the pre-industrial era. *Global Ecology and Biogeography* 20, 520–531.

Zhang, W.-Y., Wang, L.-Y., Ding, F., Hu, W.-B., Soares Magalhaes, R.J., Sun, H.-L., Liu, Y.-X., Liu, Q.-Y., Huang, L.-Y., Clements, A.C.A., Li, S.-L. and Li, C.-Y. (2013) Scrub typhus in mainland China, 2006–2012: the need for targeted public health interventions. *PLoS Neglected Tropical Diseases* 7, article ID e2493.

5 Survival of Formosan Subterranean Termite Colonies during Periods of Flooding

Carrie Cottone[1]* and Nan-Yao Su[2]

[1]*City of New Orleans Mosquito, Termite & Rodent Control Board, New Orleans, Louisiana, USA;* [2]*Fort Lauderdale Research and Education Center, University of Florida, Fort Lauderdale, Florida, USA*

5.1 Introduction

The Formosan subterranean termite, *Coptotermes formosanus* Shiraki (Isoptera: Rhinotermitidae), is an invasive pest of the United States and has become successfully established in areas of the south-eastern United States and Hawaii. In New Orleans, Louisiana, *C. formosanus* populations are ubiquitous and well established. This invasive species is now the primary structural pest in the greater New Orleans area (Su, 2003). Termite pressure within the city is so high that neighbouring termite colonies readily invade areas where existing termite colonies have been eliminated (Husseneder *et al.*, 2007).

After Hurricane Katrina made landfall along the Gulf Coast on 29 August 2005, approximately 80% of New Orleans was inundated due to levee breaches (FEMA, 2005). Immediately after this flooding, the question of subterranean termite colony survivorship was raised. *Coptotermes formosanus* colonies are known to shift their foraging areas in soil disturbed by landscaping, construction or treatment (Aluko and Husseneder, 2007). The prolonged inundation would also be a form of soil disturbance for termite colonies, so it may be hypothesized that colonies in flooded areas would also shift their foraging territories. However, a survey of operational research sites of City Park, New Orleans, prior to and following Hurricane Katrina revealed that the overall distribution of *C. formosanus* within the park was unchanged (Cornelius *et al.*, 2007).

The determination of colony survival mechanisms is important for those that live in flood-prone areas in which *C. formosanus* has become established. Understanding the survival and behaviour of *C. formosanus* colonies during flooding will help pest management professionals convey the need of retreating flooded structures for termites and educate the public of the importance of vigilance in termite control practices after natural disasters. To fully understand the success of this invasive species, it is important to understand its biology and distribution, as well as potential survival mechanisms.

5.2 Economic Impact of the Formosan Subterranean Termite

What most homeowners and pest management professionals don't realize is that termites are actually important to numerous ecosystems, from tropical and temperate regions. They are beneficial because of their

*E-mail: cbowens@nola.gov

ability to consume cellulose, which allows stored nutrients in wood to be released back into the ecosystem. When subterranean termites excavate their gallery system below the soil surface, they facilitate soil turnover and aerate the soil (Lobry de Bruyn and Conacher, 1990). It is when termites cause damage to structures, historic live trees, ornamental landscaping, and anything else considered valuable to man that they are considered pests.

Coptotermes formosanus is a pest of major economic importance in areas where it has become successfully established. An estimate by Osbrink *et al.* (1999) claimed that *C. formosanus* is responsible for $66 million in structural repairs and $37.5 million in insecticide treatments annually in New Orleans alone. Lax and Osbrink (2003) estimated that $300 million is spent each year in the city of New Orleans on preventive and remedial termite treatments as well as structural repairs. Given that the last estimate of the cost of termite control in New Orleans was conducted over 10 years ago, it could be speculated that this estimated cost has increased since then.

In New Orleans, *C. formosanus* has become the primary structural pest species and easily outcompetes native subterranean termites, *Reticulitermes* spp. (Isoptera: Rhinotermitidae). It has been suggested that the city of New Orleans has one of the highest termite pressures in North America (Lax and Osbrink, 2003). The damage caused by *C. formosanus* infestations in New Orleans has been so significant that, in 1998, the United States legislature passed a termite population management initiative which remained effective until 2011. This area-wide termite treatment programme, called 'Operation Full Stop', employed monitoring and baiting technology throughout New Orleans' historic French Quarter to reduce subterranean termite populations, thus reducing the overall termite pressure in the area (Lax and Osbrink, 2003).

The extensive damage caused by *C. formosanus* is not attributed to individual workers within the colony consuming more cellulose than other species of subterranean termites. Rather, it is because *C. formosanus*

colonies are more aggressive and can consist of 1–4 million individuals (Su *et al.*, 1984), with foraging territories that can be 100 m in length and encompass multiple feeding sites (King and Spink, 1969).

5.3 Distribution of Formosan Subterranean Termites

The distribution of *C. formosanus* is limited by its temperature and humidity requirements. *Coptotermes formosanus* thrives in warm, humid areas, making it successful in temperate and subtropical regions (Su, 2003). In general, populations of *C. formosanus* are observed around the world, in areas between 35°C north and south of the Equator. *Coptotermes formosanus* populations are currently established in China, Taiwan, Japan, South Africa, and multiple areas within the continental United States and Hawaii (Su and Tamashiro, 1987).

The distribution of this species throughout so many countries can be attributed to unintentional anthropogenic movement of infested materials (La Fage, 1987; Scheffrahn *et al.*, 2001; Messenger *et al.*, 2002; Su, 2003). For example, *C. formosanus* was most likely introduced into Japan before the 1600s by trading ships from southern China to Nagasaki (Mori, 1987). Sandalwood trade ships from China probably introduced *C. formosanus* into Hawaii during the 1800s (Su, 2003). Following the Second World War, *C. formosanus* was introduced into United States port cities, such as New Orleans, via infested wooded cargo crates and pallets shipped from eastern Asia (La Fage, 1987). Within the continental United States, reports of *C. formosanus* have come from Texas, Florida, California, Mississippi, Alabama, Tennessee, North Carolina, South Carolina, Georgia, and of course, Louisiana (Su and Tamashiro, 1987; Scheffrahn *et al.*, 2001; Woodson *et al.*, 2001).

There has never been a documented case of *C. formosanus* being eradicated from an area once it has become established (Su, 2003). Therefore, it is extremely important to take measures to prevent further spread of this invasive species. The Louisiana Department of Agriculture put a quarantine into effect in 1967 in the hope of preventing the spread of infested materials into areas where *C. formosanus* had not yet been introduced (La Fage, 1987). Quarantined materials included wood products, poles, sawdust, contaminated soil, and pilings. A more recent additional quarantine was made effective in 2005, in response to Hurricane Katrina, which resulted in millions of tons of wood debris from felled trees and hurricane-damaged structures within Louisiana (LDAF, 2005).

5.4 Biology and Behaviour of Formosan Subterranean Termites

As with all termite species, *C. formosanus* is a eusocial insect with parental care, overlapping generations within the colony and a distinct division of labour, or caste system, among individuals within the colony. Each caste serves a unique function necessary for the survival of the colony. These castes include primary reproductives, replacement and supplementary reproductives, workers, soldiers, nymphs and larvae (Imms, 1931; Edwards and Mill, 1986; Lainé and Wright, 2003). Each caste is comprised of both male and female individuals (Imms, 1931).

The function of the reproductive castes is to produce new members of the colony (Edwards and Mill, 1986; Imms, 1931). The primary reproductives are the winged termites that are produced by mature colonies. The reproductive dispersal flights, commonly known as 'swarms', of *C. formosanus* typically take place in the evenings from May until July in Louisiana. These termites are attracted to light sources, where they pair up and drop to the ground. Males and females form tandem pairs as the female seeks a new nest site. Once a suitable environment is located, both will shed their wings and excavate a chamber in which they will mate and start a new colony (Edwards and Mill, 1986). The king and the queen of the colony are monogamous, mating throughout their lifetimes (Imms, 1931). A

colony may be headed by a single reproducing pair or multiple reproducing pairs. Those headed by the founding queen and king are called simple family colonies, while those headed by multiple breeding pairs are considered extended family colonies (Thorne *et al.*, 1999). Additional reproductives, produced when one or both of the founding reproductive die or are isolated from the colony, are called replacement reproductives. Those produced when the fecundity of the primary reproductives decreases are called supplemental reproductives (Thorne *et al.*, 1999). During the life cycle of subterranean termite colonies, the breeding system shifts from a simple family to an extended family.

The functions of the worker caste are to forage, expand the gallery and nest system, feed members of other castes, and tend to eggs, newly hatched larvae and reproductives (Imms, 1931; Edwards and Mill, 1986). The worker caste comprises the highest proportion of individuals within the colony, *c.*80–90%. Approximately 10% of individuals within a *C. formosanus* colony are soldiers (Lax and Osbrink, 2003). The function of the soldier caste is to defend the colony against invaders and predators. They do this via sharp, slender, pinching mandibles and a glue-like defensive secretion that is expelled from their fontanelle, located on the anterior frons. In a mature colony, some individuals will moult into nymphs, which can again moult to form winged reproductives.

Direct observations of subterranean termite colonies are not feasible because of their cryptic nature. However, it has been determined that *C. formosanus* colonies create an extensive gallery system, connecting multiple nests and multiple feeding sites (King and Spink, 1969). In addition to excavating extensive gallery systems, *C. formosanus* also creates hard carton nest material, which provides protection and structural support. They create this nest material primarily from soil, saliva and excrement (Edwards and Mill, 1986). These nests can be located both below the soil surface (King and Spink, 1969) and above ground level within live trees (Osbrink *et al.*, 1999) or within structures.

5.5 Formosan Subterranean Termites and Flooding

5.5.1 Climate change and precipitation

When considering how the global climate will impact the distribution of *C. formosanus*, it is important to realize that both temperature and precipitation are both key environmental factors. *Coptotermes formosanus* thrives in warm, humid climates. According to Hu and Appel (2004), *C. formosanus* can survive a wide range of temperatures, from 7°C to 46°C. The distribution of *C. formosanus* is often limited by cooler temperatures in the northern regions of the United States. There could be cases of *C. formosanus* being introduced to an area via infested goods; however, if the climate is too cold, *C. formosanus* could not become successfully established outside a climate-controlled or insulated human-made structure. However, the average global temperature has increased by 0.6°C since the early 1900s (Nicholls *et al.*, 1996), and if our climate continues to warm, it could be speculated that the distribution of *C. formosanus* could potentially spread.

By the end of this century, it is predicted that not only will the global climate be warmer, but precipitation will increase in areas along the Equator and poles while decreasing in subtropical areas (Held and Vecchi, 2007) (Fig. 5.1). A change in precipitation within the United States can already be observed, as there has been an overall increase in precipitation by 10% over the past century (Karl and Knight, 1998). This increase in precipitation is caused by both an increased frequency of days in which precipitation occurs and an increased intensity of these precipitation events (Karl and Knight, 1998; Easterling *et al.*, 2000).

Figure 5.1 shows that many areas of the south-eastern United States in which *C. formosanus* has become established will experience less precipitation in the future. Areas of south-east China, where *C. formosanus* is native, and Japan, where *C. formosanus* was successfully established, will experience more precipitation. For the purposes of this chapter, it should be noted that flood conditions will be considered the same as heavy precipitation. This is because heavy rains could produce saturated soil conditions similar to what would be experienced by termites during times of flooding. Subterranean termites would be in the same microenvironment of saturated soil, whether it is

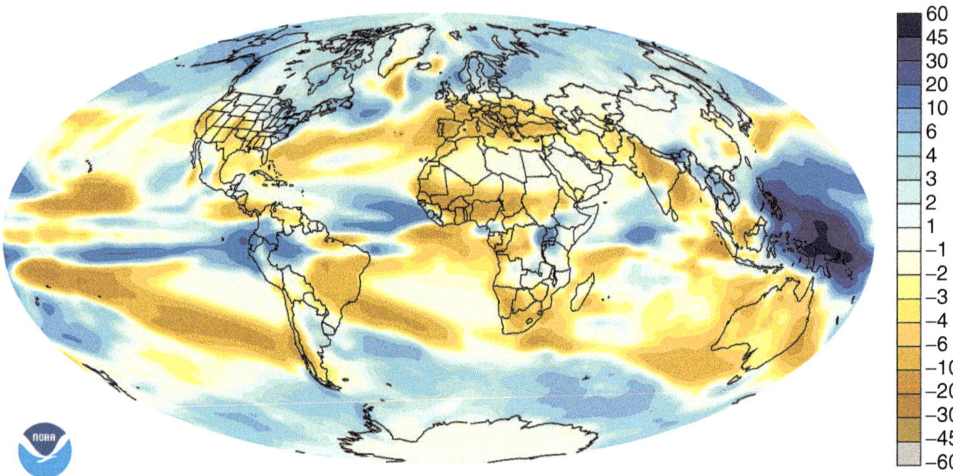

Fig. 5.1. The projected change in global precipitation (in inches of liquid water per year) by the end of the 21st century. Areas in blue are expected to experience more precipitation, while the areas in brown are expected to experience less precipitation. Reproduced with permission of the National Oceanic and Atmospheric Administration Geophysical Fluid Dynamics Laboratory (NOAA GFDL).

caused by heavy rainfall or flood waters rising several feet above the soil surface for a period of time. Heavy rainfall can cause a significant reduction in subterranean termite foraging populations, though entire termite colonies are not eliminated (Forschler and Henderson, 1995). Therefore, consistent saturated soil conditions caused by heavy precipitation or sustained flooding would temporarily impede foraging activity of *C. formosanus* colonies, but only until the soil conditions dried. The projected decrease of precipitation in the subtropical areas in which *C. formosanus* has been established should not hinder *C. formosanus* activity, as long as the environment remains relatively humid.

5.5.2 Termite colonies in flooded areas

Some previously established termite research sites within the city of New Orleans were flooded after Hurricane Katrina for up to 2 weeks. In-ground termite monitoring stations located at these sites contained live foraging Formosan subterranean termites just 2 months after the flood (Cornelius *et al.*, 2007). This initial observation indicated that at least some *C. formosanus* colonies survived. Both Cornelius *et al.* (2007) and Osbrink *et al.* (2008) have shown that the overall distribution of *C. formosanus* colonies was not altered from the flooding. Owens *et al.* (2012) showed that *C. formosanus* colonies present at research sites before Hurricane Katrina were the same colonies present at these sites following the flood, thereby determining that *C. formosanus* colonies did survive the weeks of prolonged inundation following Hurricane Katrina.

Coptotermes formosanus originated from areas of southern China, which experiences regular droughts and floods related to monsoons (Kripalani and Kulkarni, 1997; Lau and Li, 1984). Despite these flood conditions, *C. formosanus* populations continue to thrive in their native areas. It is therefore understandable that populations can survive in temporary and seasonally flooded environments in other areas of the world.

A variety of survival mechanisms during times of flooding have been documented for ants, another highly successful group of social insects. Colonies of the acrobat ant, *Crematogaster cerasi* (Fitch) (Hymenoptera: Formicidae), survive inundation by evacuating to trees when waters rise (Ellis *et al.*, 2001). Colonies of the carpenter ant, *Camponotus anderseni* McArthur and Shattuck (Hymenoptera: Formicidae), prevent their nests from filling with water by blocking the entrance holes (Nielsen *et al.*, 2006). The mangrove mud-nesting ant, *Polyrhachis sokolova* Forel (Hymenoptera: Formicidae), remains within the nest when it floods, and survives by locating and exploiting pockets of air trapped within the nest chambers until conditions become more favourable (Nielsen, 1997). Rafting is a behaviour exhibited by the red imported fire ant, *Solenopsis invicta* Buren (Hymenoptera: Formicidae), in which individual ants link their bodies together to form a raft that floats on the surface of the water until they contact a solid surface on which they can climb to escape the flood waters until the waters recede (Anderson *et al.*, 2002).

Some termite species other than *C. formosanus* are also established in seasonally flooded areas. A survey conducted in Darwin, northern Australia, showed that *Nasutitermes graveolus* (Hill) (Isoptera: Termitidae) can inhabit mangrove swamps partially due to their arboreal nesting behaviour (Dawes-Gromadzki, 2005). Constantino (1992) showed that several termite species can survive within periodically inundated swamp forests in Brazilian Amazonia, though termite species diversity was not as great as in forested areas that did not experience inundation. The following termite species were documented as inhabiting these periodically saturated areas: *Rhinotermes marginalis* (L.) (Isoptera: Rhinotermitidae), *Anoplotermes* spp., *Cavitermes tuberosus* Emerson, *Termes hispaniolae* (Banks), *Termes medioculatus* (Emerson), *Coatitermes clevelandi* (Snyder), *Ereymatermes rotundiceps* Constantino, *Nasutitermes* spp., and *Rotunditermes bragantinus* (Roonwal and Rathore) (Isoptera: Termitidae). The mechanisms by which these termite species survive inundation are unknown.

For *C. formosanus*, the presence of carton material may be a key factor in colony survival during inundation. This nest material with hydrophobic properties may provide protection against water-saturated soil (Cornelius *et al.*, 2007). The study by Cornelius *et al.* (2007) tested the effect of carton material on termite survival during inundation, and revealed that termites given longer periods of time to construct carton nests within containers in a laboratory setting had a higher survival rate following inundation. During the study, termites that survived had remained above the water line, or were within the carton material. No survivors were found in the inundated foraging galleries themselves.

Coptotermes formosanus colonies have even been observed infesting naturally inundated trees. A study by Delaplane *et al.* (1991) of foraging behaviour of *C. formosanus* isolated in cypress trees in the Calcasieu River near Lake Charles, Louisiana, revealed that foraging activity varied seasonally. The termite colonies isolated within these trees were also smaller than nearby colonies within the soil (Su and La Fage, 1999). It was hypothesized that this was due to restricted populations having limited food resources and space. Because *C. formosanus* colonies are known to survive in isolated trees that are constantly surrounded by water, colonies were shown to survive in New Orleans after the prolonged inundation following Hurricane Katrina, and because colonies experience seasonal inundation and saturated soil conditions in their native lands, it follows that there must be distinct survival mechanisms used by *C. formosanus* to survive flooding, rather than surviving by fortuitous means.

5.5.3 Colony survival mechanisms

For eusocial insects, such as termites that live in colonies, the colony itself needs to be considered as a unit or organism, rather than considering only individual termites within the colony. This is especially true when discussing termite colony survival after some type of treatment or disturbance, including flooding or heavy precipitation. Only a portion of the colony would need to survive to produce more termite foragers and reproductives, thus allowing the colony to eventually recover (Su and Scheffrahn, 1990).

There are logically four methods by which *C. formosanus* colonies could survive flooding. These hypotheses are:

1. Termite colonies survive flooding simply by tolerating being submerged.
2. Termite colonies survive flooding by exploiting air pockets formed and trapped within their nesting system.
3. Termite colonies move away from rising flood waters, or evacuate, to escape inundation.
4. Termite colonies create a sealed environment within their nest system and remain there until flood waters recede.

5.5.3.1 Hypothesis 1: termite colonies survive by tolerating inundation

A study conducted by Forschler and Henderson (1995) examined termite survivorship after being submerged under water. Termites were maintained under water for increasing periods of time to determine the lethal time (LT) values for different termite species. Forschler and Henderson (1995) showed that the lethal time for 50% of the test population, LT_{50}, for *C. formosanus* maintained under water is 11.1 hours. This same study showed the LT_{90} for *C. formosanus* is 15.8 hours. The authors determined that the LT values for *C. formosanus* were much lower than those yielded from native *Reticulitermes* spp. They speculated that this is because *C. formosanus* has the ability to nest above ground as well as under the soil surface. For this reason, *C. formosanus* may not have the need to tolerate inundation as much as native subterranean termite species that are dependent solely on nesting within the soil. The lethal times exhibited by *C. formosanus* are much shorter than what termite colonies experienced in New Orleans following

Hurricane Katrina. Also, these lethal times could not account for this species' success in seasonally inundated areas. For these reasons, we must reject the hypothesis that *C. formosanus* survives by simply tolerating inundation. *Coptotermes formosanus* colonies must survive flooding by other means.

5.5.3.2 Hypothesis 2: termite colonies survive by exploiting air pockets

Cottone *et al.* (2015) tested the hypothesis that *C. formosanus* colonies survive inundation by exploiting trapped oxygen within air pockets by studying the mortality of groups of termites inside identical airtight containers held at different temperatures. Groups containing different numbers of termites were used in this bioassay and compared to each other because termite survivorship in airtight environments (i.e. within air pockets) is dependent upon the amount of air available for respiration. The more termites exploiting the air within the container, the less air available for the termites within that container, and the higher the mortality rate. Survivorship is also dependent upon temperature, because termites increase their respiration rates at warmer temperatures (Cotton, 1932). The higher the respiration rate, the faster the available air will be depleted and the higher the mortality.

In the study by Cottone *et al.* (2015), groups of termites were maintained within the airtight containers at either 10°C, 21°C or 32°C. At 10°C, groups of 20, 40 and 60 termites reached 100% mortality at 89.5, 89.5 and 57.5 days, respectively. At 21°C, groups of 20, 40 and 60 termites reached 100% mortality at 52.0, 51.0 and 22.5 days, respectively. At 32°C, groups of 20, 40 and 60 termites reached 100% mortality at 3.5, 3.5 and 1.0 day(s), respectively. The LT_{50} values for groups of 20, 40 and 60 termites at 10°C were 23.00, 20.31 and 16.80 days, respectively. Groups of 20, 40 and 60 termites maintained at 21°C yielded LT_{50} values of 6.99, 4.41 and 1.92 days, respectively. Groups of 20, 40 and 60 termites at 32°C yielded LT_{50} values of 1.14, 0.43 and 0.27 days, respectively.

These results showed that both the amount of available air and temperature are major factors in termite survivorship within airtight environments. The greater the amount of available air, combined with relatively few termites confined to a given space, the greater the survivorship. Termites existing at relatively warmer temperatures not only increase their respiration (Cotton, 1932), but they also increase their foraging activity and wood consumption (Fei and Henderson, 2004; Messenger and Su, 2005; Cornelius and Osbrink, 2011; Delaplane *et al.*, 1991). Therefore, it makes sense that this increase in activity and oxygen requirements at warmer temperatures would cause greater mortality for termites maintained within airtight containers at relatively warmer temperatures.

In the study by Cottone *et al.* (2015), groups of termites survived within their respective airtight containers for extended periods of time. Furthermore, *C. formosanus* experiences environments with relatively low levels of oxygen and relatively high levels of carbon dioxide within their subterranean nesting system (Wheeler *et al.*, 1996). Therefore, it can be concluded that *C. formosanus* colonies could survive saturated soil conditions by remaining within and exploiting pockets of air. However, this also depends upon whether pockets of air would actually form within their nest material.

5.5.3.3 Hypothesis 3: termite colonies survive by evacuating

A study of the behaviour of *C. formosanus* colonies during seasonal inundation was conducted by Cottone and Su (Owens, 2011) to test the hypothesis that termite colonies move away from rising flood waters to survive inundation. The objectives of this study were to determine whether termite colonies shift their foraging territories, or evacuate, during seasonal inundation, and to determine whether termite colonies move vertically within trees to escape rising flood waters.

For this study, two research sites were established along the river batture in New

Orleans, Louisiana. This area is located between the Mississippi River and the levee system. These sites were chosen because the trees located here were already infested with *C. formosanus* and the sites experience inundation annually between February and June. The flooding ranged from 0.3 m to 1.8 m during this study (NOAA, 2005).

Five infested black willow trees (*Salix* sp.) were located at one of the river batture sites (N 29.93942, W 90.05397), and one infested black willow tree was located at the second river batture site (N 29.91834, W 90.06686). Three trees that did not experience flooding were used as control sites for this study. All trees were examined for termite activity using a video borescope before and after each flood season.

In-ground monitoring stations were installed at the two seasonally flooded sites and the three control sites. These stations were round irrigation valve boxes buried to lie flush with the ground and filled with a wood resource. They were installed in a grid pattern on 3 m centres and extending at least 9 m from the infested trees. At the two river batture sites, rows of stations were installed from the river to 4.5 m from the base of the levee wall. All stations were checked monthly for 2 years for evidence of wood consumption and the presence of live foraging termites. Once the stations located along the river battue were flooded, they could not be checked.

During the course of this study, two infested black willow trees located at one of the river batture sites had fallen due to strong storm winds coupled with termite damage within the trees. These trees were sectioned at 15 cm increments and evaluated for the presence of carton material at increasing heights to give evidence of possible means of vertical evacuation during flooding.

Cottone and Su (Owens, 2011) observed live foraging termites within in-ground monitoring stations extending from trees in areas that did not experience flooding, while there was no such foraging observed within in-ground monitoring stations outside the infested trees at the seasonally inundated river batture. However, live termites were observed inside these seasonally inundated trees between 0.3 and 1.0 m from ground level using a video borescope before and after inundation for both flood seasons. No live termites were observed in the seasonally inundated trees above 1.0 m from ground level. Examination of the tree sections for the presence of carton material revealed that foraging galleries were present within the trees above what would have been the water line (1.8 m from ground level), while carton material was only present in areas of the tree that would have been below the water line (i.e. in tree sections that would have been inundated).

Termite colonies located along the river batture may adjust to seasonal inundation by remaining within trees, while those located in areas that do not experience seasonal inundation actively forage away from their main food source, thus creating a larger foraging territory. This is similar to what was observed by Su and La Fage (1999), who determined that *C. formosanus* colonies restricted to cypress trees in the Calcasieu River, Louisiana, have smaller foraging populations and foraging range than those observed on dry land. Because *C. formosanus* colonies remained within seasonally inundated trees, the assumption that *C. formosanus* colonies shift their foraging areas to escape rising water is incorrect, and the hypothesis that *C. formosanus* colonies move away from rising water to survive inundation is rejected.

At least in areas that experience seasonal inundation, *C. formosanus* colonies do not shift their foraging areas to escape rising flood waters. Instead, termites remained confined to their food source throughout this study. This behaviour was not observed for *C. formosanus* colonies which experienced prolonged inundation following Hurricane Katrina, as foraging termites were observed within inundated in-ground monitoring stations before flooding and after the flood waters receded (Cornelius *et al.*, 2007; Osbrink *et al.*, 2008). It may be that *C. formosanus* colonies that experienced seasonal inundation have adjusted behaviourally to survive repeated flood events, while those infesting areas that had not been subjected to inundation prior to Hurricane Katrina do not exhibit such behavioural changes and readily forage away from protective structures. Whether termites moved from the soil into trees during the flooding following Hurricane Katrina is unknown, as carton nests within trees could not be observed during the flooding. This is unlikely, though, as previous studies conducted by Cornelius and Osbrink (2010) have shown that in laboratory tests, *C. formosanus* does not attempt to evacuate from its gallery system within soil to seek higher elevation during flooding.

Carton material was not present above the water line within trees that experienced seasonal inundation. This is not indicative of vertical movement of entire *C. formosanus* colonies, though foraging galleries located above the water line of seasonally inundated trees could provide the means of evacuation for individual termites. Cornelius and Osbrink (2010) have shown that *C. formosanus*, when inside hollowed wood, may attempt vertical evacuation to escape rising waters. However, a relatively small percentage of the colony surviving flooding by vertically evacuating within trees does not explain the survival mechanisms of entire *C. formosanus* colonies during the 4-month flood season along the river batture, unless reproductives vertically evacuated to survive and propagate the colony. This is because termites were observed within seasonally inundated trees immediately after the flood waters receded, and this is not indicative of significant termite mortality.

Because *C. formosanus* colonies nesting within seasonally inundated trees do not evacuate flooded trees, the nesting system within trees may be the mechanism by which *C. formosanus* survives inundation. Carton material, observed in the bases of seasonally inundated trees, may provide protection against saturated soil because of its hydrophobic properties (Cornelius *et al.*, 2007). If carton material provides a sealed, water-resistant environment for termites, it would explain how *C. formosanus* colonies could survive within inundated trees for extended periods.

5.5.3.4 Hypothesis 4: termite colonies survive by creating a sealed environment

Cottone and Su (Owens, 2011) tested the hypothesis that *C. formosanus* colonies survive prolonged inundation by creating a sealed environment filled with trapped air within their nesting and gallery system and remaining there until flood waters recede and soil conditions become drier and more favourable to tunnelling. This may be the most plausible hypothesis of the survival mechanism for *C. formosanus* colonies surrounded by saturated soil, as this species does not simply tolerate inundation for more than approximately 16 hours (Forschler and Henderson, 1995), nor does it evacuate to escape rising flood waters (Owens, 2011). However, *C. formosanus* is capable of remaining within airtight environments for extended periods, depending on temperature and the number of termites occupying the space (Cottone *et al.*, 2015).

To test whether *C. formosanus* colonies survive inundation by creating a watertight environment within their nest, Cottone and Su (Owens, 2011) conducted a bioassay in which five logs were manually infested with termite foragers and exposed to periodic inundation. Artificial nests were made by hollowing the centres of 50-cm sections of logs that ranged from 20 to 40 cm in diameter (Fig. 5.2a). For each log, the hollowed centre had a diameter equal to half that of the tree section. Three tree sections were from cypress trees (*Taxodium* sp.) and two were from live oak trees (*Quercus* sp.).

The bottom end of each tree section was capped with 1.3-cm thick polyethylene sheeting. Holes were drilled into the bottom cap to allow termites to forage outside the tree sections. Each log was placed within a galvanized steel container (69 cm height × 52 cm diameter) on top of play sand and gravel to help facilitate drainage. A spigot was installed for drainage. Moistened play sand was added to the containers and the hollowed centres of tree sections.

Twenty thousand termites were placed into the hollowed centre of each log. The top of each log section was capped with 1.3-cm

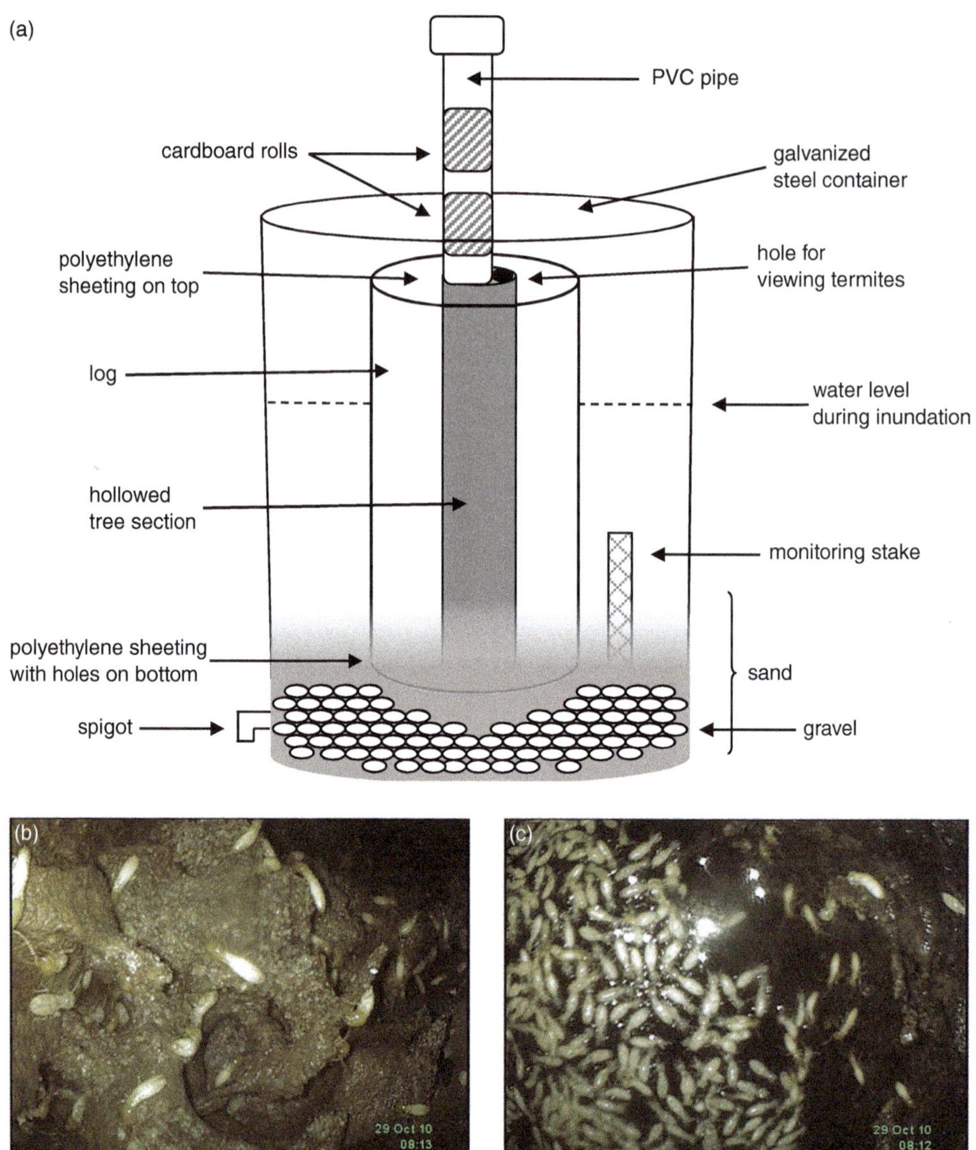

Fig. 5.2. (a) Schematic of bioassay in which *C. formosanus* foragers are nesting within the hollowed log and subjected to periodic inundation. (b) Images taken using a video borescope system to determine whether water entered the logs during periods of inundation. This is the inside of a log during inundation in which water did not enter into the log. (c) This log filled with water during inundation, and termites were observed floating on the water surface.

thick polyethylene sheeting and sealed. The top cap included a 50-cm section of 5-cm polyvinyl chloride (PVC) pipe protruding upwards. This pipe was filled with corrugated cardboard moistened with water. This was to allow termites to travel vertically, as they would in a natural environment. A viewing hole was drilled through the top polyethylene sheeting to allow a video borescope to be inserted for monitoring termites. When not in use, the hole was plugged with a rubber stopper.

Termites within the logs were allowed to forage and tunnel for 4 weeks. At the end of 4 weeks, infested logs were inundated every other week at increasing time intervals until inundation time reached 48 hours. One log was used as a control and was not subjected to inundation. Each week, the cypress stakes were checked for the presence of active foraging termites to determine whether they were leaving the protection of the tree trunk after the initial inundation. Also, at weekly intervals, termites were monitored inside the logs using a video borescope system. The interiors of inundated artificial nests were monitored for the presence of water during periods of inundation. Vertical movement of termites was measured by removing corrugated cardboard from the PVC pipe protruding from the tree sections and counting the number of termites located within the cardboard.

A second bioassay was performed to determine whether termites move away from or repair tunnels that are being inundated. For this study, Cottone and Su (Owens, 2011) constructed 12 two-dimensional foraging arenas, similar to those used by Su (2005) as shown in Fig. 5.3. Arenas consisted of two transparent Plexiglas sheets (24 × 24 × 0.6 cm thick). Between these sheets, four Plexiglas laminates (three that were 2 cm in width, one 7 cm in width) that were 0.2 cm thick were placed along the outer edges. The foraging area within each arena was 20 cm × 15 cm in size. To maintain structural integrity, an additional 2 cm × 2 cm piece of Plexiglas laminate was inserted in the centre of the arena. All pieces of Plexiglas and Plexiglas laminate were bolted into place. An access hole (0.5 cm in diameter) was included on the upper sheet, near one of the corners, to allow termites to enter the foraging area. A termite release chamber, consisting of a Plexiglas cup (4.5 cm diameter × 3 cm height) with a lid was attached onto the access hole.

Sand was added to each arena and moistened. The edges of each arena were made watertight by sealing them with silicone plumbing caulk. Pieces of wood were moistened and placed inside each termite release chamber to serve as a food source. One thousand termites (900 workers and 100 soldiers) were added to the termite release chamber of each foraging arena. Termites were allowed to forage within the arenas for 6 weeks. At the end of this period, water was added to each arena at a rate of 3 ml per minute. Water was added so that only the bottom half of each arena was inundated.

During the artificial nest inundation bioassay, termites were observed foraging outside only in one infested log. However, after the initial inundation of artificial nests, the foraging activity was no longer observed. Live termites were observed inside each log using the video borescope every week. Three out of four logs did not exhibit water entering into their hollowed interior during periods of inundation (see Fig. 5.2b). As water entered into the interior of the fourth log, termites were observed floating on the water surface (see Fig. 5.2c).

The lack of foraging outside any of the artificial trees following initial inundation was not due to termite mortality following inundation, as live termites were observed inside artificial trees each week during the bioassay. Rather, this behaviour is more representative of what was observed of field colonies along the Mississippi River batture, in which *C. formosanus* colonies infesting seasonally inundated trees did not exhibit any foraging away from their food source.

Because water was not observed within three out of four logs during periods of inundation, it can be concluded that foraging termites had created a protected and watertight environment within the hollowed logs. This method of protection may not necessarily be solely used for protection against flooding. Creating a sealed environment could also provide protection against predation and other insects that would otherwise exploit the termite nest and resources. However, the watertight seal was not observed for all colonies, even following the initial inundation period. Even without the sealed protection, there was no evidence of high termite mortality within any log. Live termites were viewed inside each artificial tree every week during the inundation bioassay. Termites inhabiting the log that was not sealed from rising water may have survived either by floating on the water surface until the water receded, as was observed, or by possibly remaining within hidden pockets of air at the base of the tree section. There is no evidence from our data to suggest that termites evacuate vertically within trees to escape rising flood waters. A significantly high number of termites was not observed moving vertically within the inundated logs, including the log that was not sufficiently protected against rising water.

In contrast to these observations, termites in the foraging arenas were observed moving away from water during inundation (Fig. 5.3a, b). Those that were overcome by rising water did not exhibit any movement once they were inundated. However, termites that became surrounded by water but were located within a pocket of air were observed antennating the water's edge (Fig. 5.3c). The foraging tunnels remained intact, though water passed through the galleries. Termites were not observed repairing or constructing any tunnels during inundation to further protect themselves from rising water.

Our findings differ greatly from those discussed by Cornelius and Osbrink (2010), who concluded that *C. formosanus* moves vertically within wood in response to rising water, but does not move from the galleries to escape rising water. This lack of movement from the gallery system in response to flooding was also observed for *Reticulitermes flavipes* (Kollar) (Forschler and Henderson, 1995). These differing conclusions are most likely due to varying methodologies employed for the different bioassays.

Although foraging tunnels within arenas appeared to remain intact during inundation, water readily passed through the galleries. There were no observations made of termites attempting to repair tunnels or prevent water from rising further. These observations would indicate that either the interphase between the Plexiglas and sand was not watertight, or foraging galleries may have at least some hydrophobic properties,

but would not provide as effective protection under the soil surface as that provided by carton material within the nest inside trees during periods of inundation.

The hypothesis that *C. formosanus* colonies survive prolonged inundation by

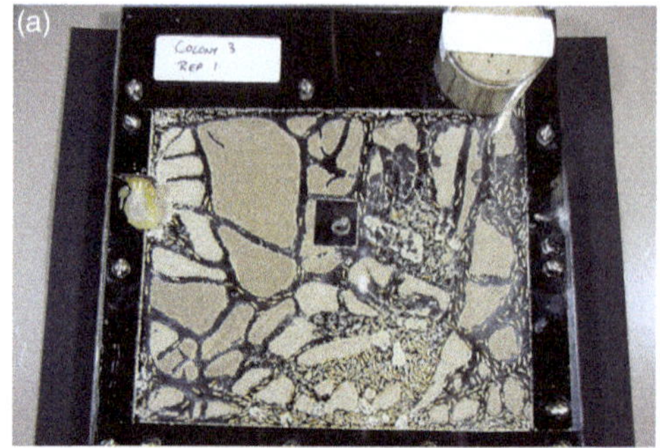

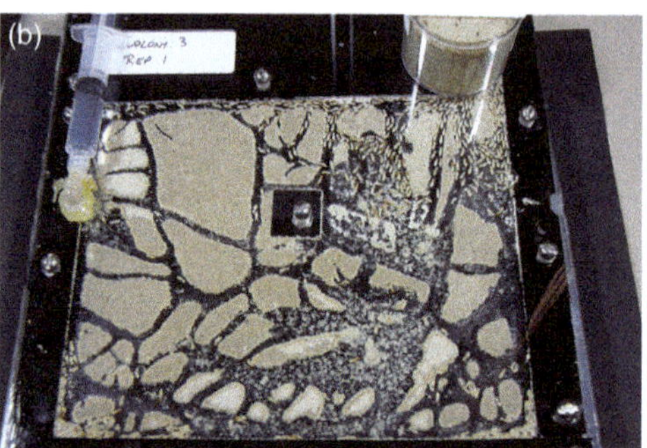

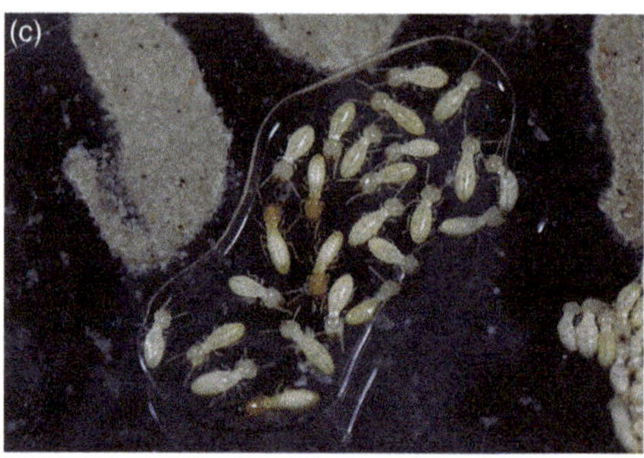

Fig. 5.3. Observations of termite movement within a foraging arena subjected to inundation. (a) Prior to flooding, termites are evenly distributed throughout the arena. (b) During inundation, water filled only the bottom half of the arena and termites were observed moving away from rising water. (c) At least some termites within each arena became restricted within a pocket of air surrounded by water.

creating a sealed environment filled with trapped air and remaining there until flood water recedes or saturated soil conditions become more favourable is accepted.

5.4 Conclusion

Coptotermes formosanus is known to readily infest trunks of live trees and create carton material that can extend metres above ground level (Osbrink *et al.*, 1999, 2008). It has been suggested that these voids can trap air to sustain termites below the water level during inundation (Osbrink *et al.*, 2008) and *C. formosanus* colonies can survive prolonged inundation by remaining protected within their hydrophobic carton nests and gallery systems (Cornelius *et al.*, 2007; Osbrink *et al.*, 2008). Galleries located above the water line may also serve to trap air that can be exploited by termites during periods of inundation. This is the most likely survival mechanism of *C. formosanus* during prolonged inundation for two reasons. First, Cottone *et al.* (2015) concluded that termites can survive within an airtight environment for an extended period of time. Second, it has been shown that termites create a watertight and protective environment within trees that traps air, which can be exploited by termites during periods of inundation.

Flooding of urban areas gains much notoriety because of the devastation to property and lives it causes, whether the flooding is caused by hurricanes, tropical storms, tsunamis or rising rivers due to ice melt. It is important to realize that not only will prolonged and devastating floods create saturated soil conditions for subterranean termites, but heavy prolonged rainfall will create these conditions as well. As precipitation increases in the coming years, it is important for those who reside in areas in which *C. formosanus* has become successfully established to understand that a natural disaster that causes floods or increased precipitation will not eradicate *C. formosanus* populations.

References

Aluko, G.A. and Husseneder, C. (2007) Colony dynamics of the Formosan subterranean termite in a frequently disturbed urban landscape. *Journal of Economic Entomology* 100, 1037–1046.

Anderson, C., Theraulaz, G. and Deneubourg, J.-L. (2002) Self-assemblages in insect societies. *Insectes Sociaux* 49, 99–110.

Constantino, R. (1992) Abundance and diversity of termites (Insecta: Isoptera) in two sites of primary rain forest in Brazilian Amazonia. *Biotropica* 24, 420–430.

Cornelius, M.L. and Osbrink, W.L.A. (2010) Effect of flooding on the survival of Formosan subterranean termites (Isoptera: Rhinotermitidae) in laboratory tests. *Sociobiology* 56, 699–711.

Cornelius, M.L. and Osbrink, W.L.A. (2011) Effect of seasonal changes in soil temperature and moisture on wood consumption and foraging activity of Formosan subterranean termite (Isoptera: Rhinotermitidae). *Journal of Economic Entomology* 104, 1024–1030.

Cornelius, M.L., Duplessis, L.M. and Osbrink, W.L.A. (2007) The impact of Hurricane Katrina on the distribution of subterranean termite colonies (Isoptera: Rhinotermitidae) in City Park, New Orleans, Louisiana. *Sociobiology* 50, 311–335.

Cotton, R.T. (1932) The relation of respiratory metabolism of insects to their susceptibility to fumigants. *Journal of Economic Entomology* 25, 1088–1103.

Cottone, C.B., Su, N.-Y., Scheffrahn, R. and Riegel, C. (2015) Survivorship of the Formosan subterranean termite (Isoptera: Rhinotermitidae) in a hypoxic environment. *Sociobiology* 62, 76–81.

Dawes-Gromadzki, T.Z. (2005) The termite (Isoptera) fauna of a monsoonal rainforest near Darwin, northern Australia. *Australian Journal of Entomology* 44, 152–157.

Delaplane, K.S., Saxton, A.M. and La Fage, J.P. (1991) Foraging phenology of the Formosan subterranean termite (Isoptera: Rhinotermitidae) in Louisiana. *American Midland Naturalist* 125, 222–230.

Easterling, D.R., Meehl, G.A., Parmesan, C., Changnon, S.A., Karl, T.A. and Mearns, L.O. (2000) Climate extremes: observations, modeling, and impacts. *Science* 289, 2068–2074.

Edwards, R. and Mill, A.E. (1986) *Termites in Buildings. Their Biology and Control.* The Rentokil Library, Rentokil Ltd, East Grinstead, UK.

Ellis, L.M., Crawford, C.S. and Molles Jr, M.C. (2001) Influence of annual flooding on terrestrial arthropod assemblages of a Rio Grande riparian forest. *Regulated Rivers: Research and Management* 17, 1–20.

Fei, H. and Henderson, G. (2004) Effects of temperature, directional aspects, light, conditions, and termite species on subterranean termite activity (Isoptera: Rhinotermitidae). *Environmental Entomology* 33, 242-248.

FEMA (2005) Federal Emergency Management Agency website. Available at: http://www.fema.gov (accessed 5 May 2016).

Forschler, B.T. and Henderson, G. (1995) Subterranean termite behavioral reaction to water and survival of inundation: implications for field populations. *Environmental Entomology* 24, 1592–1597.

Held, I. and Vecchi, G. (2007) Will the wet get wetter and the dry drier? *GFDL Climate Modeling Research Highlights* 1(5), 1.

Hu, X.P. and Appel, A.G. (2004) Seasonal variation of critical thermal limits and temperature tolerance in Formosan and eastern subterranean termites (Isoptera: Rhinotermitidae). *Environmental Entomology* 33, 197–205.

Husseneder, C., Simms, D.M. and Riegel, C. (2007) Evaluation of treatment success and patterns of reinfestation of the Formosan subterranean termite (Isoptera: Rhinotermitidae). *Journal of Economic Entomology* 100, 1370–1380.

Imms, A.D. (1931) *Social Behaviour in Insects.* Methuen & Co., London, UK.

Karl, T.R. and Knight, R.W. (1998) Secular trends of precipitation amount, frequency, and intensity in the United States. *Bulletin of the American Meteorological Society* 79, 231–241.

King, E.G. and Spink, W.T. (1969) Foraging galleries of the Formosan subterranean termite, *Coptotermes formosanus*, in Louisiana. *Annals of the Entomological Society of America* 62, 536–542.

Kripalani, R.H. and Kulkarni, A. (1997) Rainfall variability over south-east Asia – connections with Indian monsoon and ENSO extremes: new perspectives. *Journal of Climatology* 17, 1155–1168.

La Fage, J.P. (1987) Practical considerations of the Formosan subterranean termite in Louisiana: a 30-year-old problem. In: Tamashiro, M. and Su, N.-Y. (eds) *Biology and Control of the Formosan Subterranean Termite (Isoptera: Rhinotermitidae).* University of Hawaii College of Tropical Agriculture and Human Resources, Honolulu, Hawaii, USA, pp. 37–42.

Lainé, L.V. and Wright, D.J. (2003) The life cycle of *Reticulitermes* spp. (Isoptera: Rhinotermitidae): what do we know? *Bulletin of Entomological Research* 93, 267–378.

Lau, K.-M. and Li, M.-T. (1984) The monsoon of east Asia and its global association – a survey. *Bulletin of the American Meteorological Society* 65, 114–125.

Lax, A.R. and Osbrink, W.L.A. (2003) United States Department of Agriculture-Agriculture Research Service research on targeted management of the Formosan subterranean termite *Coptotermes formosanus* Shiraki (Isoptera: Rhinotermitidae). *Pest Management Science* 59, 788–800.

Lobry de Bruyn, L.A. and Conacher, A.J. (1990) The role of termites and ants in soil modification: a review. *Australian Journal of Soil Research* 28, 55–93.

LDAF (2005) Imposition of quarantine. Office of Agricultural and Environmental Sciences, Division of Pesticides and Environmental Programs, Louisiana Department of Agriculture and Forestry, Baton Rouge, Louisiana, USA. Available at: http://nationalplantboard.org/wp-content/uploads/docs/katrina_quarantine_3.pdf (accessed June 2016).

Messenger, M.T. and Su, N.-Y. (2005) Colony characteristics of the Formosan subterranean termite in Louis Armstrong Park, New Orleans, Louisiana. *Journal of Entomological Science* 40, 268–279.

Messenger, M.T., Su, N.-Y., and Scheffrahn, R.H. (2002) Current distribution of the Formosan subterranean termite and other termite species (Isoptera: Rhinotermitidae, Kalotermitidae) in Louisiana. *Florida Entomologist* 85, 580–587.

Mori, H. (1987) The Formosan subterranean termite in Japan: its distribution, damage, and current and potential control measures. In: Tamashiro, M. and Su, N.-Y. (eds) *Biology and Control of the Formosan Subterranean Termite (Isoptera: Rhinotermitidae).* University of Hawaii College of Tropical Agriculture and Human Resources, Honolulu, Hawaii, USA, pp. 23–26.

Nicholls, N., Gruza, G.V., Jouzel, J., Karl, T.R., Ogallo, L.A. and Parker, D.E. (1996) Observed climate variability and change. In: *Climate Change 1995: The Science of Climate Change (Intergovernmental Panel on Climate Change).* Cambridge University Press, Cambridge, UK, pp. 133–192.

Nielsen, M.G. (1997) Nesting biology of the mangrove mud-nesting ant *Polyrhachis sokolova* Forel (Hymenoptera, Formicidae) in northern Australia. *Insectes Sociaux* 44, 15–21.

Nielsen, M.G., Christian, K., Henriksen, P.G. and Birkmose, D. (2006) Respiration by mangrove ants *Camponotus anderseni* during nest submersion associated with tidal inundation in northern Australia. *Physiological Entomology* 31, 120–126.

NOAA (2005) National Oceanic and Atmospheric Administration website. Available at: http://water. weather.gov/ahps/rfc/rfc.php (accessed 9 June 2016).

Osbrink, W.L.A., Woodson, W.D. and Lax, A.R. (1999) Populations of Formosan subterranean termite, *Coptotermes formosanus* (Isoptera: Rhinotermitidae), established in living urban trees in New Orleans, Louisiana, USA. In: Robinson, W.H., Rettich, R. and Rambo, G.W. (eds) *Proceedings of the Third International Conference on Urban Pests*. Grafické Závody Hronov, Czech Republic, pp. 341–345.

Osbrink, W.L.A., Cornelius, M.L. and Lax, A.R. (2008) Effects of flooding on field populations of Formosan subterranean termites, *Coptotermes formosanus* Shiraki (Isoptera: Rhinotermitidae), in New Orleans, LA. *Journal of Economic Entomology* 101, 1367–1372.

Owens, C. (2011) *Survival and behavior of* Coptotermes formosanus *Shiraki (Isoptera: Rhinotermitidae) after flooding in New Orleans, Louisiana.* PhD thesis, University of Florida, Gainesville, Florida, USA. Available at: http:// ufdc.ufl.edu/UFE0043746/00001 (accessed 14 October 2016).

Owens, C.B., Su, N.-Y., Husseneder, C., Riegel, C. and Brown, K.S. (2012) Molecular genetic evidence of Formosan subterranean termite (Isoptera: Rhinotermitidae) colony survivorship after prolonged inundation. *Journal of Economic Entomology* 105, 518–522.

Scheffrahn, R.H., Su, N.-Y., Chase, J.A. and Forschler, B.T. (2001) New termite (Isoptera: Kalotermitidae, Rhinotermitidae) records from Georgia. *Journal of Entomological Science* 36, 109–113.

Su, N.-Y. (2003) Overview of the global distribution of the Formosan subterranean termite. *Sociobiology* 41, 7–16.

Su, N.-Y. (2005) Response of the Formosan subterranean termites (Isoptera: Rhinotermitidae) to baits or nonrepellent termiticides in extended foraging arenas. *Journal of Economic Entomology* 98, 2143–2152.

Su, N.-Y. and La Fage, J.P. (1999) Forager proportion and caste compositions of colonies of the Formosan subterranean termite (Isoptera: Rhinotermitidae) restricted to cypress trees in the Calcasieu River, Lake Charles, Louisiana. *Sociobiology* 33, 185–193.

Su, N.-Y. and Scheffrahn, R. (1990) Economically important termites in the United States and their control. *Sociobiology* 17, 77–94.

Su, N.-Y. and Tamashiro, M. (1987) An overview of the Formosan subterranean termite in the world. In: Tamashiro, M. and Su, N.-Y. (eds) *Biology and Control of the Formosan Subterranean Termite (Isoptera: Rhinotermitidae)*. University of Hawaii College of Tropical Agriculture and Human Resources, Honolulu, Hawaii, USA, pp. 3–15.

Su, N.-Y., Tamashiro, M., Yates, J.R. and Haverty, M.I. (1984) Foraging behavior of the Formosan subterranean termite (Isoptera: Rhinotermitidae). *Environmental Entomology* 13, 1466–1470.

Thorne, B.L., Traniello, J.F.A., Adams, E.S. and Bulmer, M. (1999) Reproductive dynamics and colony structure of subterranean termites of the genus *Reticulitermes* (Isoptera: Rhinotermitidae): a review of the evidence from behavioral, ecological, and genetic studies. *Ethology Ecology and Evolution* 11, 149–169.

Wheeler, G.S., Tokoro, M., Scheffrahn, R.H. and Su, N.-Y. (1996) Comparative respiration and methane production rates in Nearctic termites. *Journal of Insect Physiology* 42, 799–806.

Woodson, W.D., Wiltz, B.A. and Lax, A.R. (2001) Current distribution of the Formosan subterranean termite (Isoptera: Rhinotermitidae) in the United States. *Sociobiology* 37, 661–671.

6 Termites and a Changing Climate

Donald Ewart[1]*, Lina Nunes[2], Teresa de Troya[3] and Magdalena Kutnik[4]

[1]PO Box 1044 Research 3095, Australia; [2]LNEC, National Laboratory for Civil Engineering, Structures Department, Lisbon, Portugal; [3]Head of Wood Protection Laboratory, Forest Research Centre (CIFOR), National Institute for Agricultural and Food Research and Technology (INIA), Madrid, Spain; [4]Head of Biology Laboratory, Chair of CEN/TC 38, Technological Institute FCBA, Bordeaux, France

6.1 Introduction

The potential for rapid anthropogenic climate change to impact termite populations has concerned researchers for more than a decade (Dale, 1997; Nunes and Nobre, 2001; Erasmus *et al.*, 2002). There is much to learn from the record of climate change over geological time – information we can derive from fossils and distribution patterns. To understand the ways in which climate can impact termite populations, we must first explore the characteristics of those populations and the factors that provide them with both plasticity and firm limits.

6.2 Factors Limiting Termite Behaviour

In order to understand how termites react to changing climatic conditions, we must first explore what termites are and the parameters that control their bio-ecology. All living things have a set of requirements, abilities and interactions which may be thought of as their *ecological niche,* a multidimensional space within the boundaries of which the organism may prosper (Hutchinson, 1959;

Sillero, 2011). The niches occupied by termites are many and varied due to a wide range of *life types* (Abe, 1987; Eggleton and Tayasu, 2001). To some extent, the actions of humans have increased the size of the termite niche (Sapunov, 2008). The ecological classification of termites tends to reflect their feeding and nesting behaviours (Abe, 1987), but across this range, several limiting factors remain common to all species. All organisms require an adequate supply of food and moisture, to live within an adequate range of temperature, reproduce and disperse.

6.2.1 Food

Termites are social cockroaches (Inward *et al.*, 2007) which have evolved from ancestors that fed on decayed wood on the forest floor. While some retain the habit of eating decayed wood, others have moved on to humus, grass, apparently solid wood, and a few to the fungal mycelium itself (Abe, 1987). Most termites have a diet that requires them to digest cellulose fibres produced by plants.

Much of what is in the woody tissue of a plant is not readily digestible by termites

---

*E-mail: don_ewart@drdons.net

and some components such as oils and resins may have a protective action, causing a wide variation in the palatability of different woods (Grace *et al.*, 1996; Arango *et al.*, 2006). The type of plant and the way in which it grows tends to determine palatability, with fast-growing woody tissue generally more prone to attack by termites, for example in eucalypts (Rudman, 1964) and teak (Moya *et al.*, 2014).

6.2.2 Water

Termites live across a wide range of moisture, from drywood termites that live successfully in wood of only a few percentage points of moisture content to dampwood termites that thrive in sodden, rotted logs. Subterranean termites have distinct moisture preferences for both substrates and food and *Coptotermes formosanus* will quickly depart from a dry sand arena (Gautam, 2011). Drywood termites habitually produce dry faecal pellets, presumably a water-saving behaviour; similar pellets replace the normally wet faeces of dampwood termites when their substrate dries. These pellets are also known from some subterranean termites (Collins, 1969).

6.2.3 Temperature

The upper and lower lethal temperatures for North American pest termites were determined by Sponsler and Appel (1991) and Hu and Appel (2004). The work has been extended by others (Davis and Kamble, 1994; Woodrow and Grace, 1998; Clarke *et al.*, 2013). Well within the lethal limits for each species lies a range that is physiologically profitable. Temperature has long been recognized as an ecological resource (Magnuson *et al.*, 1979) and termites are sensitive to small fluctuations (Ettershank *et al.*, 1980). The drywood termite *Cryptotermes brevis* has been observed to prefer a lower temperature when moisture is high (Steward, 1981), suggesting an interaction of these environmental factors.

6.2.4 Reproduction

Termites are slow-moving insects that tend to nest within food or within constructions that may contain faeces, saliva, plant fibre and soil (Noirot, 1970). Living in enclosed spaces makes them vulnerable to attack by a range of parasites, predators and microorganisms that seek to live with and on them (Rosengaus *et al.*, 2010). This parasitic load is reduced by having winged reproductives that fly away from the nest, mate and start a fresh colony. Reproduction by colony fission (budding) is not commonly used (Vargo and Husseneder, 2010; Cronin *et al.*, 2013).

6.2.5 Dispersal

As well as avoiding a parasitic load, the use of winged reproductives allows for the olfactory detection of new food sources which may be distributed in dispersed patches; however, it has long been known that incipient nests may be moved to better locations (Emerson, 1938). Termites rarely initiate new colonies from flights greater than a few hundred metres, so natural dispersal is slow, even when introduced to new territory (Leniaud *et al.*, 2009; Borges *et al.*, 2014). Termites may also be dispersed over greater distances when their food/nest material is transported by flood, sea or human interference. Compared with successful tramp species like rodents and some ants, termites are less successful at rapidly reaching all suitable environments, though they probably follow similar strategies (Nobre *et al.*, 2007). The successful pest species *Coptotermes gestroi* has not yet exploited many suitable habitats in the regions where it has become established (Li *et al.*, 2013).

6.2.6 Suitable range of factors

Termites tend to have a strong preference for conditions to be stable over the short term. Nesting in soil, mounds, trees or larger structural timbers reduces the variation of temperature and moisture (Noirot, 1970;

Ewart and French, 1986) compared with surface activity. The distribution of termite species tends to follow the patterns of their temperature and moisture preferences (Emerson and Schmidt, 1955; Calaby and Gay, 1959; Haverty and Nutting, 1976).

6.3 General Warming and Polar Drift

6.3.1 Reduced species and numbers away from Equator

Termites are found between 45° N and S (Abe and Higashi, 2013). The distribution pattern is generally attributed to temperature and is mimicked where termites have been shown also to decline with elevation (Gathorne-Hardy and Eggleton, 2001). This suggests that low temperatures are generally a stronger range determinant than high temperatures. The diversity of termites generally declines with distance from the Equator, except in a few locations with Mediterranean climates (for example, see Abensperg-Traun and Steven, 1997). The situation is complicated, as climatic conditions are not symmetric about the cartographic Equator, but at approximately 3.4° N (France, 1998); the southern Mediterranean climate regions have been found to have greater diversity than those in the north, leading to a pear-shaped distribution (Eggleton, 1994). Further complications arise because of complex regional differences (Davies *et al.*, 2003).

6.3.2 Termite warming and extension in range

The principal factor limiting termites in colder climes is that they are metabolically shut down by extreme cold, to the extent that colonies may be prevented from feeding and other normal behaviours. Esenther (1969) proposed that the capacity of termites to replace their winter losses by taking advantage of an amenable summer season was an important determinant for colony survival. Lee and Chon (2011) used modelling to suggest that the seasonal differences would manifest in the form of fluctuating foraging territory. The habit of humans to heat their constructions in winter has assisted termite colonization of colder climes, because the extra heat of the buildings provides a refuge for termites to escape the severe winter cold.

6.3.2.1 Canada

The distribution of the North American native subterranean *Reticulitermes* has been extended north into Canada, where termites have dispersed with people on multiple occasions (Scaduto *et al.*, 2012) and found appropriate shelter in their dwellings, but have not dispersed away from human habitation.

6.3.2.2 Germany and Hamburg

Hamburg was infested, probably in the 1930s, by American *Reticulitermes flavipes*, which exploited a buried municipal heating system that linked the port and city (Weidner, 1952). The pipes heated the surrounding soil, thus providing the desirable subterranean warmth, and the termites established a large infestation. Control proved to be difficult (Hertel and Plarre, 2006), allowing survival of the species in the same territory.

6.3.3 Enhanced survival of founded colonies

If winter temperature minima help limit termite range, then a warming trend should see termites colonize new areas, both further away from the Equator and also at higher elevations. Newly formed termite colonies with small populations are probably most vulnerable to cold, whereas large colonies can concentrate their metabolic heat and regularly do so to maintain thermoregulation. Greaves (1967) applied dry ice to force termites to cluster in their tree nests. In the mountains of South Eastern Australia, the subterranean termite *Coptotermes lacteus* nests in trees and constructs mounds after the tree declines (Ewart, 1988). The elevated temperatures deep within these mounds are maintained in a range of less than 1°C over 24 hours (Ewart and French, 1986; Ewart, 1988). Some mounds occur at greater than 1300 m elevation, where winter snows occur. *Coptotermes lacteus* occurs from 39°C in southern Victoria to around 25°C in Queensland and probably occurred two or three degrees further south in Tasmania before the last Ice Age. It is likely that this species' habit of nesting deep in the base of the large *Eucalyptus* trees provides some early protection against cold.

6.4 Dispersal and Reaching Suitable Climes

6.4.1 Human-assisted recolonization in post Ice-Age Europe

Europe has a long and complex termite history (Martinez-Delclos and Martinell, 1995), with Eocene fossils of the tropical *Mastotermes* from about 35 million years ago (Wappler and Engel, 2006) as far north as Germany. These tropical termites died out after polar drift moved Europe into cooler climes and now only exist as a single relict species in Northern Australia (Grandcolas *et al.*, 2014). Later, indigenous European termites of the Reticulitermitidae were pushed south in the last glaciation to the Iberian refugia (Kutnik *et al.*, 2004) and to Africa. They are still in the process of returning. North African *Reticulitermes* have recolonized much of the northern Mediterranean coast, but the pattern of the species is still more related to the dispersal patterns of the peoples that brought them than it is to the African environments from which they came (Luchetti *et al.*, 2013; Dedeine *et al.*, 2016).

Europe has been colonized by *Reticulitermes flavipes*, a North American species which has been collected in Austria, France, Germany and Italy (Perdereau *et al.*, 2013). The termites were imported in timbers from the Louisiana region centred on the French port of New Orleans and have slowly dispersed from La Rochelle, most notably along railway lines, eventually reaching Paris

(Kutnik *et al.*, 2004). *Reticulitermes flavipes* has also been found on the Portuguese Azorean island of Terceira (Ferreira *et al.*, 2013) and an eradication programme based on baits has been ongoing since October 2014.

The south-west of England is warmed by the waters of the Gulf Stream, which reduces winter minima. A small population of *Reticulitermes grassei* was detected in Devon in 1994 (Verkerk and Bravery, 2001). It is not known how the termites arrived and for how long they were resident, but it was some decades later that the population was eradicated (de Bruxelles, 2010). A warming trend is likely to see much of south-west England become habitable for these termites, unless the Gulf Stream breaks down (Palter, 2015).

6.4.2 Termite dispersal

6.4.2.1 Flights

Termite reproductive flights are a successful strategy over distances of a few hundred metres, but require suitable environmental conditions, with most species requiring warmth and moisture as flight triggers. In North America, there are populations of *Reticulitermes* that typically do not fly (Esenther, 1969). Presumably these have developed some behaviours that facilitate colony fission. It is not known what factors enable this change in reproductive strategy. Certainly, what we do know of *Reticulitermes* biology suggests that there is a great deal of local variation (Lainé and Wright, 2003).

6.4.2.2 Salt, logs and vegetation islands

Logs and vegetation islands are widely known pathways for terrestrial animal dispersal across seas. While some species of termites do inhabit tropical mangrove forests and subterranean termites are found in marine timbers, it is true to say that even in those environments, termites work to limit saline exposures. The inability of termites to cope with high salinity limits their natural ability to travel in logs for short periods before the salt can penetrate. Vegetation islands, formed when river bank vegetation is eroded, are far less common than floating timbers and their contribution to termite dispersal is largely unknown. Later arrivals to the reborn Krakatau Archipelago suggest that rafting does occur (Gathorne-Hardy *et al.*, 2000), and Scheffrahn and Postle (2013) suggest that rafting, though uncommon, has long been an important means of dispersal for drywood termites.

6.4.2.3 Large timbers, railroad ties and pallets

The previously mentioned successful dispersal of North American *Reticulitermes* to Europe in large timbers has been repeated in other areas. For example, Australian *Porotermes* and *Coptotermes* have been delivered to New Zealand on many occasions in large timbers such as bridge and dock lumber, utility poles and railroad ties (or sleepers) (Miller, 1941; Pearson *et al.*, 2010). *Porotermes adamsoni* is now accepted as established in New Zealand (Pearson *et al.*, 2010).

Recycling of old railroad ties have been implicated in the spread of *Coptotermes formosanus* in the southern USA (Jenkins *et al.*, 2002), while pallets and packing crates were their presumed vehicle to colonize Lousiana (Su and Tamashiro, 1987).

6.4.2.4 Furniture and manufactured goods

Small timbers without high moisture loads and without soil contact, such as is found in furniture and manufactured goods, are most commonly exploited by drywood termites in tropical regions, with *Cryptotermes brevis* the master of small timbers (Su and Scheffrahn, 1990). For example, C. brevis is increasingly found in France, due to the transport of manufactured goods from French tropical overseas territories (FCBA, unpublished).

6.4.2.5 Boats

Apart from travelling in freight, termites may utilize boats by infesting structural and feature timbers. Drywood termites are a common pest of tropical timber fishing boats and yachts (Scheffrahn and Crowe, 2011). Subterranean termites may thrive in wooden boats where fresh water accumulates in bilges. Dhang (2014) reported on termite management in boats in the Philippines. *Coptotermes gestroi* was found to have reached Italy by yacht (Ghesini *et al.*, 2011) and the pattern of invasive *Coptotermes* around docks in Florida suggests that such transport is common (Hochmair and Scheffrahn, 2010). Scheffrahn *et al.* (2002) speculated that the arboreal nesting *Nasutitermes costalis* reached Florida either by infested boat or shipping container.

6.5 Spreading Beyond the Points of Original Introduction

While the abilities of termites to disperse is clearly limited, given enough time, some groups have dispersed widely (Bourguignon *et al.*, 2015), with continental drift being a major contributor. In the Anthropocene, termite range extensions appear to be mainly human-assisted (Scheffrahn *et al.*, 2009; Evans *et al.*, 2013). The effects of human-induced habitat changes are complex, with alterations such as the penetration of forests by roads, having an influence on termite distribution (de Sales Dambros *et al.*, 2013). It will be difficult to predict the probability rate of non-human assisted dispersal at any given location.

6.5.1 Rainfall changes

6.5.1.1 Dealing with dry

Wood-feeding termites don't occupy territories in the familiar avian or mammalian manner; rather, their occupied range exists as a series of points of interest connected by excavations, paths or constructed tunnels. The spatial pattern of this activity is greatly affected by available moisture, with termites avoiding dry areas and patches in their environment. Termites, at least the subterranean termites, are quick to reduce their effective territory when conditions are hard. This is most commonly seen where their food slowly dries and the termites eventually abandon their feeding, rather than expend more energy keeping the food damp than can be gained from eating it. Extending the subterranean gallery system to exploit new food sources is more cost-effective where soils are moist. To this end, available soil moisture is a major predictor of termite attack on structures. Drought, then, is a major limiter of termite activity, with many species seeking damp refuge, only resuming normal behaviours when water becomes more readily available. When *Coptotermes* are isolated in a building due to barrier repair or repellent soil chemical application, the survivors tend to aggregate in the dampest timbers. Where a decline in available water becomes permanent, a change in species composition may occur.

6.5.1.2 Ground water

Termites that tunnel in the soil commonly make use of ground-water resources. A good predictor of the abundance of mound-building *Coptotermes* in Australia, particularly *C. acinaciformis* and *C. lacteus*, is a geomorphology that points to subsurface drainage from nearby hills. The great success of *Coptotermes formosanus* in New Orleans has been tied to the ample presence of fresh water just below ground level. Flight and colony foundation are typically tied to times of year that are warm and have suitably damp substrates, with many termites flying when falling barometric pressure is a harbinger of more rain events (Nutting, 1969). Seasons change and soils and substrates lose moisture. Colony survival is more likely where substrates retain moisture or where the termites can gather water to keep moisture up to their required levels. Ground-water stocks may decline through changes such as reduced rainfall, increased evapotranspiration of vegetation or through human exploitation. Where climate is changing, available moisture may be as important as increasing temperature, at least over the shorter term.

6.5.1.3 Food supply

The food of termites comes from plants. Even those feeding on soil humus are essentially still eating the products of plants, with a longer time interval between production and consumption. Plants are not uniform. Plant growth varies with a host of factors such as temperature, insolation, available moisture, nutrition, soil quality, competition, wind and carbon dioxide concentration in the air. No two trees are alike: even within the trunk of a single eucalypt, CSIRO researchers found that wood palatability to termites varied vertically and horizontally (Gay *et al.*, 1955). Most termites are essentially detrivores, feeding on dead plant tissue. Consumption of live plant tissue is the exception. The dead plant tissue available to termites is not sterile, but is also subject to attack by a range of organisms (Ulyshen, 2016). Microbial life in dead plant tissue may enhance or degrade its palatability for termites, and so the direct effects of a changing climate on these commensal organisms will, at least in part, determine the overall effects on termites.

It is no coincidence that the manufacturers of termite bait products select bait wood from fast-growing tree species. Fast-grown wood tends to be more palatable than slow-grown wood. Trees that grow slowly tend to invest more heavily in the resins, oils and other metabolic products that make digestion of the cellulose more difficult. An increase in environmental heat and a decrease in the regular availability of adequate water are both likely to drive trees to produce, denser, better protected wood tissue. In arid and semi-arid Australia, droughts interrupt the production of tree growth rings to the extent that these are not as reliable as Northern Hemisphere growth rings as indicators of tree age (Brookhouse, 2006; Cook *et al.*, 2015). The production of grasses, shrubs and trees is greatly reduced during times of water stress, meaning that less cellulose is available. Termites seem to survive droughts quite well, even those lasting several years; however, population numbers must be reduced.

6.5.1.4 Extreme heat and fire

A developing feature of changing climate is a tendency for an increase in extreme weather events. High environmental heat loads result in physiological stress for both the termites and their food plants, they cause greater rates of evapotranspiration, creating a water balance issue and they increase the likelihood of fire. Jolly *et al.* (2015) investigated recent trends in maximum temperatures, minimum relative humidity, the number of rain-free days, and maximum wind speeds. They concluded that changes to these parameters meant that fire seasons, the times when ecosystem fires are likely, are becoming longer. Overgaard *et al.* (2014), working with *Drosophila*, used tolerance of extremes to map existing distributions and project future changes.

Subterranean termites are able to flee short-term heat events by retreating to the comparative stability of the soil. Termites that construct elaborate arboreal nests, such as many *Nasutitermes*, and those that live within branches, such as many *Neotermes* and *Kalotermes*, have less capacity to avoid extremes, which is likely to mean that their distributions are the first to clearly change in response to heat load variations. The partial burning of forests can result in a marked increase in termite activity (Wylie and Shanahan, 1976), resulting in economic losses, but the effects of fire can be complex and vary greatly between species and ecosystems (Davies *et al.*, 2010; Neoh *et al.*, 2015).

6.5.1.5 Flood

Kundzewicz *et al.* (2014) showed that an increasing frequency of flooding is likely. Excessive ground water limits termite activity. The viscosity of water makes walking more energetically expensive and termites under water may have difficulty obtaining sufficient oxygen. Seasonal inundation is a feature of the habitats of some Australian mound-building termites, with the well-known 'magnetic' mounds of *Amitermes meridionalis* being always found on seasonally boggy soils (Calaby and Gay, 1959; Schmidt *et al.*, 2014).

Floods carry debris downstream (Jeyasingh and Fuller, 2004) and woody debris may contain viable termite colonies. *Mastotermes darwiniensis* has an unusually southerly occurrence at Windorah, in Western Queensland (Watson and Abbey, 1993), which lies on an internal drainage line from the termites' northern distribution. Debris carried by floods may deposit in banks above normal drainage water levels and so extend the habitable spaces for termites.

Excessive periods of wet soil have been shown to cause significant reductions in the populations of ground-foraging subterranean termites (Forschler and Henderson, 1995). Osbrink *et al.* (2008) found that termites could avoid the effects of flooding where above-water-level refugia were available, while the laboratory experimental work of Cornelius and Osbrink (2010) suggested that the ability of termites to move to avoid inundation may be restricted. Using genetics to track colonies, Owens *et al.* (2012) showed that termite colonies can survive flooding. General flooding, where soil remains sodden for weeks, is likely to result in some reductions of termite population size. Increases in the frequency of such floods may result in suppressed termite populations.

6.5.1.6 Wind and storms

Trees may be damaged by high winds. Cyclonic winds can break branches and uproot trees, causing wounding of the trees, which makes more of their wood available to termites and other insects (Everham and Brokaw, 1996; Jeyasingh and Fuller, 2004). Scarred trees that become affected by wood-boring beetles and/or fungi also provide attractive colony foundation points for dispersing termites. The effect of increases in the severity or frequency of storms is likely to favour termites, at least in the short term. In urban areas, the result of storms promoting termite activity may be that the rates of termite attacks on structures are increased.

6.5.1.7 Sea-level rise

Sea levels are rising as a function of the thermal expansion of heated water (Cazenave *et al.*, 2014) and melting ice (Hansen *et al.*, 2015). In the short term of a few decades, termites will be adversely affected by saline incursion, storm surge, coastal erosion and habitat loss, particularly for mangrove termites.

6.5.1.8 Indirect change

A changing climate changes ecosystems in ways which may not be easily predicted. An example of this is the effect of an extended drought in North America, which has promoted the activity of tree-killing beetles (Bentz *et al.*, 2010), and the heat, which is permitting their range extension (Raffa *et al.*, 2013). Dead trees, riddled with beetle holes, are ideal sites for the local dampwood termites, *Zootermopsis*, to colonize and so in some areas the termite populations will expand for a few years, as they exploit this resource, and then decline. Similarly, slowly rising seawaters will kill supratidal and riparian trees (Williams *et al.*, 1999; Antonellini and Mollema, 2010), increasing the resource for some drywood termites and consequently adding pest pressure to nearby structures.

The mere presence of susceptible tree taxa in a region does not guarantee the continued success of the native termites (Ewart, 1991), as the food resource must have the required size, structure and suitable environment. The complexity of the interactions of species, especially where these species are operating at the limits of their former niche, may produce unexpected outcomes which affect tree growth, survival and hence the quality of the available timber resource (McNulty *et al.*, 2014).

6.6 Climate Change and Urban Termite Management

Climate change will alter the requirements to effectively manage urban termites across the globe. While some areas will likely see their termite risks decline, others will gain such risks for the first time. Predictive modelling and observations of similar locations will help preparations. Climate is a major determinant of the durability of in-ground timbers, such as utility poles, and changes of heat patterns and moisture levels may result in reduced service life (Wang and Wang, 2012).

Increasing sea levels inevitably mean that the gravity-dependent stormwater and sewage drains of coastal cities will require at least modification to operate at higher levels (Neumann *et al.*, 2015). Where waste waters back up, local increases in soil moisture and the dampening of timber structures will act to increase termite hazard. In the longer term, as coastal areas are inundated, mass migrations of people will create an unprecedented need for new dwelling construction. Competition for societies' economic resources will stretch budgets, creating a risk that necessary cost-cutting will result in less durable specifications and, particularly, constructions that are not adequately resistant to termites. Where resources are limited, it is important that the infrastructure responses to climate change do not increase, but rather reduce, current and future risk of losses from termite attack. It will be difficult to maintain design and construction standards in the face of immediate need; further, the risks may not be immediately apparent and the natural tendency of economists to discount the future (Gollier and Hammitt, 2014) may work against achieving adequate resilience. Unlike the rapid-breeding, easily transported global tramp pests such as the housefly (*Musca domestica*: Diptera, Muscidae), German cockroach (*Blatella germanica*: Blattodea, Blatellidae) and most stored-product insects, the human-assisted dispersal of termites is irregular, meaning that some newly suitable habitats wait many years for termites to establish.

This uncertainty may create difficulties in the communication of management preparations. A coherent strategic narrative, such as that proposed generally by Bushell *et al.* (2015), is required to ensure that the risks are not simply ignored until the opportunity for economically efficient responses has passed. Further, adaptation strategies and education are required to achieve practical outcomes (Wamsler *et al.*, 2013).

Precautionary construction requirements are common where termites are a known hazard, but these requirements need to be extended to cover all prospective construction where the termites are likely to pose a risk before the planned economic life of the structure has passed. Existing constructions, which modelling suggests will become at risk of termite attack, need to be managed in a way that minimizes future risks. This

Table 6.1. Risk reduction strategies for structural termite attacks.

Aim	Actions	Examples
Reduce or avoid	Preventative	Biosecurity actions to exclude potential pest incursions Actions to limit climate change
Resilience, vulnerability reduction	Preparative	Actions to limit the increase in termite-favouring conditions – these require knowledge of the termite species Modification/construction of infrastructure/buildings to improve termite resistance
Response to increased attacks	Management	Survey/monitoring to locate/identify populations Wide-area control actions Education
Response to incursion	Management	Attempted local eradication Biosecurity to prevent spread

Table 6.2. Methods of structural pest termite surveillance.

Method	Suitability	Advantages	Disadvantages
Visual inspection and tapping of substrates	All	Cheap and thorough	May miss early stages of attack Not all termites leave obvious surface evidence
Monitoring baited traps	Subterranean	Monitoring is quick	Bait material degrades, requiring replacement Not all species are attracted to bait Large number of monitor units required
Thermal imaging camera	Subterranean, possibly drywood and dampwood	Fast visual survey to detect heat from termite activity	Requires thermal contrast Requires significant activity to produce detectable heat Stray heat may cause confusion
Movement detecting radar	Subterranean, possibly drywood and dampwood	High degree of certainty	Slow
Adhesive alate monitoring traps with/without lights	All	High degree of certainty In most regions can be seasonally applied	Will not show early stage attacks Large number of traps required
Sniffer dog	All	Fast High degree of certainty	Dogs not welcomed in all areas Properly trained dogs are expensive
Acoustic detection	All	High degree of certainty	Slow Essentially restricted to application onto single pieces of timber as signal attenuates at joints
Methane and carbon dioxide gas detection	Subterranean, dampwood	Fast	Low certainty as gases also present with decay. Requires close application or large numbers of termites.

risk management is required in two forms: (1) restraints that prevent any maintenance actions, modification or extension that increases termite risk; and (2) as the risk becomes apparent, regular surveillance to ensure that any termite attacks are detected before damage becomes severe.

A summary of risk reduction strategies is given in Table 6.1. In order to determine when regular surveillance of structures is required, it will be necessary for municipal authorities to maintain a watching brief. The cheapest monitoring programme is likely to be one where termites are trapped during reproductive flights (see Borges *et al.*, 2014; Puckett *et al.*, 2014); however, such flights only take place when the arriving termites reach colony maturity and so may only detect termite activity several years after the first incursion. Table 6.2 summarizes potential for the commonly used detection methods. The appropriate method or mix of methods will vary between different species, locations and constructions.

6.7 Conclusion

The impacts of a changing climate on the activities of termites are complex but can be projected from our knowledge of termites around the globe. The chief factors are likely to be heat, changes to precipitation, and the frequency of extreme weather events, sea-level rise and the impact of rising CO_2 on their food resource. Termite risk will be changed in many regions and in those regions where termites are to become a risk, planning and management actions are required to avoid economic losses.

References

Abe T. (1987) Evolution of life types in termites. In: Kawano, S., Connell, J.H. and Hidaka, T. (eds) *Evolution and Coadaptation in Biotic Communities*. University of Tokyo Press, Tokyo, Japan, pp. 126–148.

Abe, T. and Higashi, M.O (2013) Isoptera. In: Levin, S.A. (ed.) *Encyclopedia of Biodiversity, Volume 4*. Academic Press, San Diego, California, USA, pp. 408–433.

Abensperg-Traun, M. and Steven, D. (1997) Latitudinal gradients in the species richness of Australian termites (Isoptera). *Australian Journal of Ecology* 22(4), 471–476.

Antonellini, M., and Mollema, P.N. (2010) Impact of groundwater salinity on vegetation species richness in the coastal pine forests and wetlands of Ravenna, Italy. *Ecological Engineering* 36(9), 1201–1211.

Arango, R.A., Green III, F., Hintz, K., Lebow, P.K. and Miller, R.B. (2006) Natural durability of tropical and native woods against termite damage by *Reticulitermes flavipes* (Kollar). *International Biodeterioration and Biodegdradation* 57, 46–150.

Bentz, B.J., Régnière, J., Fettig, C.J., Hansen, E.M., Hayes, J.L., Hicke, J.A., Kelsey, R.G., Negrón, J.F. and Seybold, S.J. (2010) Climate change and bark beetles of the western United States and Canada: direct and indirect effects. *BioScience* 60(8), 602–613.

Brookhouse, M. (2006) Eucalypt dendrochronology: past, present and potential. *Australian Journal of Botany* 54(5), 435–449.

Borges, P.A., Guerreiro, O., Ferreira, M.T., Borges, A., Ferreira, F., Bicudo, N., Nunes, L., Marcos, R.S., Arroz, A.M., Scheffrahn, R.H., Myles, T.G. (2014) *Cryptotermes brevis* (Isoptera: Kalotermitidae) in the Azores: lessons after 2 yr of monitoring in the archipelago. *Journal of Insect Science* 14(1), 172. DOI: 10.1093/jisesa/ieu034.

Bourguignon, T., Lo, N., Cameron, S.L., Šobotník, J., Hayashi, Y., Shigenobu, S., Watanabe, D., Roisin, Y., Miura, T. and Evans, T.A. (2015) The evolutionary history of termites as inferred from 66 mitochondrial genomes. *Molecular Biology and Evolution* 32(2), 406–421.

Bushell, S., Colley, T. and Workman, M. (2015) A unified narrative for climate change. *Nature Climate Change* 5(11), 971–973.

Calaby, J.H. and Gay, F.J. (1959) Aspects of the distribution and ecology of Australian termites. In: Keast, A., Crocker, R.L. and Christian, C.S. (eds) *Biogeography and Ecology in Australia*. Springer, Dordrecht, The Netherlands, pp. 211–223.

Cazenave, A., Dieng, H.B., Meyssignac, B., von Schuckmann, K., Decharme, B. and Berthier, E. (2014) The rate of sea-level rise. *Nature Climate Change* 4(5), 358–361.

Clarke, M.W., Thompson, G.J. and Sinclair, B.J. (2013) Cold tolerance of the eastern subterranean termite, *Reticulitermes flavipes* (Isoptera: Rhinotermitidae), in Ontario. *Environmental Entomology* 42(4), 805–810.

Collins, M.S. (1969) Water relations in termites. In: Krishna, K. (ed.) *Biology of Termites, Volume 1.* Academic Press, New York, USA, pp. 433–458.

Cook, E.R., Seager, R., Kushnir, Y., Briffa, K.R., Büntgen, U. *et al.* (2015) Old World mega-droughts and pluvials during the Common Era. *Science Advances* 1(10), e1500561. DOI: 10.1126/sciadv.1500561.

Cornelius, M.L. and Osbrink, W.L. (2010) Effect of flooding on the survival of Formosan subterranean termites (Isoptera: Rhinotermitidae) in laboratory tests. *Sociobiology* 56(3), 699.

Cronin, A.L., Molet, M., Doums, C., Monnin, T. and Peeters, C. (2013) Recurrent evolution of dependent colony foundation across eusocial insects. *Annual Review of Entomology* 58(1), 37–55.

Dale, V.H. (1997) The relationship between land-use change and climate change. *Ecological Applications* 7(3), 753–769.

Davies, A.B., Parr, C.L. and van Rensburg, B.J. (2010) Termites and fire: current understanding and future research directions for improved savanna conservation. *Austral Ecology* 35(4), 482–486.

Davies, R.G., Eggleton, P., Jones, D.T., Gathorne-Hardy, F.J. and Hernández, L.M. (2003) Evolution of termite functional diversity: analysis and synthesis of local ecological and regional influences on local species richness. *Journal of Biogeography* 30(6), 847–877.

Davis, R.W. and Kamble, S.T. (1994) Low temperature effects on survival of the eastern subterranean termite (Isoptera: Rhinotermitidae). *Environmental Entomology* 23(5), 1211–1214.

de Bruxelles, S. (2010) North Devon termite colony survived 12 years of attempts to kill it. *The Times,* London, 12 March. Available at: http://www.thetimes.co.uk/tto/environment/article2991.ece (accessed 5 May 2016; subscribers only).

de Sales Dambros, C., da Silva, V.N.V., Azevedo, R. and de Morais, J.W. (2013) Road-associated edge effects in Amazonia change termite community composition by modifying environmental conditions. *Journal for Nature Conservation* 21(5), 279–285.

Dedeine, F., Dupont, S., Guyot, S., Matsuura, K., Wang, C., Habibpour, B., Bagnères, A.G., Mantovani, B. and Luchetti, A. (2016) Historical biogeography of *Reticulitermes termites* (Isoptera: Rhinotermitidae) inferred from analyses of mitochondrial and nuclear loci. *Molecular Phylogenetics and Evolution* 94, 778–790.

Dhang, P. (2014) Managing subterranean termites in ships and yachts across Philippine ports. *International Pest Control* 56(5), 262–263.

Eggleton, P. (1994) Termites live in a pear-shaped world: a response to Platnick. *Journal of Natural History* 28(5), 1209–1212.

Eggleton, P. and Tayasu, I. (2001) Feeding groups, lifetypes and the global ecology of termites. *Ecological Research* 16(5), 941–960.

Emerson, A.E. (1938) Termite nests – a study of the phylogeny of behavior. *Ecological Monographs* 8(2), 247–284. DOI: 10.2307/1943251

Emerson, A.E. and Schmidt, K.P. (1955) Geographical origins and dispersions of termite genera. *Fieldiana Zoology* 37(18), 465–521. DOI: 10.5962/bhl.title.2783.

Erasmus, B.F., Van Jaarsveld, A.S., Chown, S.L., Kshatriya, M. and Wessels, K.J. (2002) Vulnerability of South African animal taxa to climate change. *Global Change Biology* 8(7), 679–693.

Esenther, G.R. (1969) Termites in Wisconsin. *Annals of the Entomological Society of America* 62(6), 1274–1284.

Ettershank, G., Ettershank, J.A. and Whitford, W.G. (1980) Location of food sources by subterranean termites. *Environmental Entomology* 9(5), 645–648.

Evans, T.A., Forschler, B.T. and Grace, J.K. (2013) Biology of invasive termites: a worldwide review. *Annual Review of Entomology* 58(1), 455–474.

Everham, E.M. and Brokaw, N.V. (1996) Forest damage and recovery from catastrophic wind. *The Botanical Review* 62(2), 113–185.

Ewart, D.M. (1988) Aspects of the ecology of the termite *Coptotermes lacteus* (Froggatt). PhD thesis, Department of Zoology, School of Biological Sciences, La Trobe University, Australia.

Ewart, D.M. (1991) Termites and forest management in Australia. In: Haverty, M.I. and Wilcox, W.W. (eds) *Proceedings of the Symposium on Current Research on Wood-destroying Organisms and Future Prospects for Protecting Wood in Use; September 13, 1989; Bend, OR.* General Technical Report PSW-GTR-128, United States Department of Agriculture, Forest Service, Pacific Southwest Research Station, Berkeley, California, USA, pp. 10–14.

Ewart, D.M. and French, J.R.J. (1986) Temperature studies on two mounds of *Coptotermes lacteus* (Isoptera). *17th IRG Meeting, Avignon, France.* Doc. No. IRG/WP/1295, International Research Group on Wood Preservation, Stockholm, Sweden. 8 pp.

Ferreira, M.T., Borges, P.A., Nunes, L., Myles, T.G., Guerreiro, O. and Scheffrahn, R.H. (2013) Termites (Isoptera) in the Azores: an overview of the four invasive species currently present in the archipelago. *Arquipélago: Life and Marine Sciences* 30, 39–55.

Forschler, B.T. and Henderson, G. (1995) Subterranean termite behavioral reaction to water and survival of inundation: implications for field populations. *Environmental Entomology* 24(6), 1592–1597.

France, R.L. (1998) Refining latitudinal gradient analyses: do we live in a climatically symmetrical world? *Global Ecology and Biogeography Letters* 7(4), 295–296.

Gathorne-Hardy, F.J. and Eggleton, P. (2001) The effects of altitude and rainfall on the composition of the termites (Isoptera) of the Leuser Ecosystem (Sumatra, Indonesia). *Journal of Tropical Ecology* 17(03), 379–393.

Gathorne-Hardy, F.J., Jones, D.T. and Mawdsley, N.A. (2000) The recolonization of the Krakatau islands by termites (Isoptera), and their biogeographical origins. *Biological Journal of the Linnaean Society* 71(2), 251–267.

Gautam, B.K. (2011) *Role of Substrate Moisture, Relative Humidity and Temperature on Survival and Foraging Behavior of Formosan Subterranean Termites.* PhD thesis, Louisiana State University, USA. Available at: http://etd.lsu.edu/docs/available/etd-10302011-181759/unrestricted/GAUTAM_diss.pdf (accessed 5 May 2016).

Gay, F.J., Greaves, T., Holdaway, F.G. and Wetherly, A.H. (1955) Standard laboratory colonies of termites for evaluating the resistance of timber, timber preservatives and other materials to termite attack. *CSIRO Bulletin* No. 277. Commonwealth Scientific and Industrial Organisation, Melbourne, Australia. 60 pp.

Ghesini, S., Puglia, G. and Marini, M. (2011) First report of *Coptotermes gestroi* in Italy and Europe. *Bulletin of Insectology* 64, 53–54.

Gollier, C. and Hammitt, J.K. (2014) The long-run discount rate controversy. *Annual Review of Resource Economics* 6(1), 273–295.

Grace, J.K., Ewart, D.M. and Tome, C.H. (1996) Termite resistance of wood species grown in Hawaii. *Forest Products Journal* 46(10), 57.

Grandcolas, P., Nattier, R. and Trewick, S. (2014) Relict species: a relict concept? *Trends in Ecology and Evolution* 29(12), 655–663.

Greaves, T. (1967) Experiments to determine the populations of tree-dwelling colonies of termites (*Coptotermes acinaciformis* (Froggatt) and *C. frenchi* Hill). Division of Entomology Technical Paper 7, CSIRO, Melbourne, Australia, pp. 19–33.

Hansen, J., Sato, M., Hearty, P., Ruedy, R., Kelley, M., Masson-Delmotte, V., Russell, G., Tselioudis, G., Cao, J., Rignot, E., Velicogna, I., Tormey, B., Donovan, B., Kandiano, E., von Schuckmann, K., Kharecha, P., Legrande, A.N., Bauer, M. and Lo, K.-W. (2015) Ice melt, sea level rise and superstorms: evidence from paleoclimate data, climate modeling, and modern observations that 2°C global warming is highly dangerous. *Atmospheric Chemistry and Physics Discussions* 15(14), pp. 20059–20179. DOI: 10.5194/acpd-15-20059-2015.

Haverty, M.I. and Nutting, W.L. (1976) Environmental factors affecting the geographical distribution of two ecologically equivalent termite species in Arizona. *American Midland Naturalist* 95(1), 20–27.

Hertel, H. and Plarre, R. (2006) Termites in Hamburg. In: Bravery, A.F. and Peek, R.-D. (eds) *COST Action E22: Environmental Optimisation of Wood Protection: Proceedings of the Final Conference 22–23 March 2004, Estoril, Portugal.* Office for Official Publications of the European Communities, Luxembourg, p. 89. Available at: http://bookshop.europa.eu/en/environmental-optimisation-of-wood-protection-pbQSNA22597.

Hochmair, H.H. and Scheffrahn, R.H. (2010) Spatial association of marine dockage with land-borne infestations of invasive termites (Isoptera: Rhinotermitidae: *Coptotermes*) in urban South Florida. *Journal of Economic Entomology* 103(4), 1338–1346.

Hu, X.P. and Appel, A.G. (2004) Seasonal variation of critical thermal limits and temperature tolerance in Formosan and eastern subterranean termites (Isoptera: Rhinotermitidae). *Environmental Entomology* 33(2), 197–205.

Hutchinson, G.E. (1959) Homage to Santa Rosalia or why are there so many kinds of animals? *The American Naturalist* 93(870), 145–159.

Inward, D., Beccaloni, G. and Eggleton, P. (2007) Death of an order: a comprehensive molecular phylogenetic study confirms that termites are eusocial cockroaches. *Biology Letters* 3(3), 331–335.

Jenkins, T.M., Dean, R.E. and Forschler, B.T. (2002) DNA technology, interstate commerce, and the likely origin of Formosan subterranean termite (Isoptera: Rhinotermitidae) infestation in Atlanta, Georgia. *Journal of Economic Entomology* 95(2), 381–389.

Jeyasingh, P.D. and Fuller, C.A. (2004) Habitat-specific life-history variation in the Caribbean termite *Nasutitermes acajutlae* (Isoptera: Termitidae). *Ecological Entomology* 29(5), 606–613.

Jolly, W.M., Cochrane, M.A., Freeborn, P.H., Holden, Z.A., Brown, T.J., Williamson, G.J. and Bowman, D.M.J.S. (2015) Climate-induced variations in global wildfire danger from 1979 to 2013. *Nature Communications* 6(7537). DOI: 10.1038/ncomms8537.

Kundzewicz, Z.W., Kanae, S., Seneviratne, S.I., Handmer, J., Nicholls, N., Peduzzi, P., Mechler, R., Bouwer, L.M., Arnell, N., Mach, K.,

Muir-Wood, R., Brakenridge, G.R., Kron, W., Benito, G., Honda, Y., Takahashi, K. and Sherstyukov, B. (2014) Flood risk and climate change: global and regional perspectives. *Hydrological Sciences Journal* 59(1), 1–28.

Kutnik, M., Uva, P., Brinkworth, L. and Bagnères, A.G. (2004) Phylogeography of two European *Reticulitermes* (Isoptera) species: the Iberian refugium. *Molecular Ecology* 13(10), 3099–3113.

Lainé, L.V. and Wright, D.J. (2003) The life cycle of *Reticulitermes* spp. (Isoptera: Rhinotermitidae): what do we know? *Bulletin of Entomological Research* 93(04), 267–278.

Lee, S.H. and Chon, T.S. (2011) Effects of climate change on subterranean termite territory size: a simulation study. *Journal of Insect Science* 11(1), 80.

Leniaud, L., Pichon, A., Uva, P. and Bagnères, A.G. (2009) Unicoloniality in *Reticulitermes urbis*: a novel feature in a potentially invasive termite species. *Bulletin of Entomological Research* 99(1), 1–10.

Li, H.F., Fujisaki, I. and Su, N.Y. (2013) Predicting habitat suitability of *Coptotermes gestroi* (Isoptera: Rhinotermitidae) with species distribution models. *Journal of Economic Entomology* 106(1), 311–321.

Luchetti, A., Scicchitano, V. and Mantovani, B. (2013) Origin and evolution of the Italian subterranean termite *Reticulitermes lucifugus* (Blattodea, Termitoidae, Rhinotermitidae). *Bulletin of Entomological Research* 103(06), 734–741.

McNulty, S.G., Boggs, J.L. and Sun, G. (2014) The rise of the mediocre forest: why chronically stressed trees may better survive extreme episodic climate variability. *New Forests* 45(3), 403–415.

Magnuson, J.J., Crowder, L.B. and Medvick, P.A. (1979) Temperature as an ecological resource. *American Zoologist* 19(1), 331–343.

Martinez-Delclos, X. and Martinell, J. (1995) The oldest known record of social insects. *Journal of Paleontology* 69(03), 594–599.

Miller, D. (1941) The species of termites in New Zealand. *New Zealand Journal of Forestry* 4, 333–334.

Moya, R., Bond, B. and Quesada, H. (2014) A review of heartwood properties of *Tectona grandis* trees from fast-growth plantations. *Wood Science and Technology* 48(2), 411–433.

Neoh, K.B., Bong, L.J., Muhammad, A., Itoh, M., Kozan, O., Takematsu, Y. and Yoshimura, T. (2015) Understanding the impact of fire on termites in degraded tropical peatlands and the mechanisms for their ecological success: current knowledge and research needs. *Ecological Research* 30(5), 759–769.

Neumann, J.E., Price, J., Chinowsky, P., Wright, L., Ludwig, L., Streeter, R., Jones, R., Smith, J.B., Perkins, W., Jantarasami, L. and Martinich, J. (2015) Climate change risks to US infrastructure: impacts on roads, bridges, coastal development, and urban drainage. *Climatic Change* 131(1), 97–109. DOI: 10.1007/s10584-013-1037-4.

Nobre, T., Nunes, L. and Bignell, D.E. (2007) Tunnel geometry of the subterranean termite *Reticulitermes grassei* (Isoptera: Rhinotermitidae) in response to sand bulk density and the presence of food. *Insect Science* 14(6), 511–518.

Noirot, C. (1970) The nests of termites. In: Krishna, K. (ed.) *Biology of Termites Volume 2*. Academic Press, New York, USA, pp. 73–125.

Nunes, L. and Nobre, T. (2001) Strategies of subterranean termite control in buildings. In: Lourenço, P.B. and Roca, P. (eds) *Historical Constructions 2001: Possibilities of Numerical and Experimental Techniques* (Proceedings of the 3rd International Seminar on Historical Constructions, held at University of Minho, 7–9 November 2001). University of Minho, Guimarães, Portugal, pp. 867–874. Available at: http://www.hms.civil.uminho.pt/events/historica2001/page%20867-874%20_135_.pdf (accessed 14 October 2016).

Nutting, W.L. (1969) Flight and colony foundation. In: Krishna, K. and Weesner, F.M. (eds) *Biology of Termites, Volume 1*. Academic Press, New York, USA, pp. 233–282

Osbrink, W.L.A., Cornelius, M.L. and Lax, A.R. (2008) Effects of flooding on field populations of Formosan subterranean termites (Isoptera: Rhinotermitidae) in New Orleans, Louisiana. *Journal of Economic Entomology* 101(4), 1367–1372.

Overgaard, J., Kearney, M.R. and Hoffmann, A.A. (2014) Sensitivity to thermal extremes in Australian *Drosophila* implies similar impacts of climate change on the distribution of widespread and tropical species. *Global Change Biology* 20(6), 1738–1750.

Owens, C.B., Su, N.Y., Husseneder, C., Riegel, C. and Brown, K.S. (2012) Molecular genetic evidence of Formosan subterranean termite (Isoptera: Rhinotermitidae) colony survivorship after prolonged inundation. *Journal of Economic Entomology* 105(2), 518–522.

Palter, J.B. (2015) The role of the Gulf Stream in European climate. *Annual Review of Marine Science* 7, 113–137.

Pearson, H.G., Bennett, S.J., Philip, B.A. and Jones, D.C. (2010) The Australian dampwood termite, *Porotermes adamsoni*, in New Zealand. *New Zealand Plant Protection* 63, 241–247.

Perdereau, E., Bagnères, A.G., Bankhead-Dronnet, S., Dupont, S., Zimmermann, M., Vargo, E.L. and Dedeine, F. (2013) Global genetic analysis reveals the putative native source of the invasive termite, *Reticulitermes flavipes*, in France. *Molecular Ecology* 22(4), 1105–1119.

Puckett, R.T., Espinoza, E.M. and Gold, R.E. (2014) Alate trap-based assessment of Formosan subterranean termite (Isoptera: Rhinotermitidae) dispersal flight phenology associated with an urbanized barrier island ecosystem. *Environmental Entomology* 43(4), 868–876.

Raffa, K.F., Powell, E.N. and Townsend, P.A. (2013) Temperature-driven range expansion of an irruptive insect heightened by weakly coevolved plant defenses. *Proceedings of the National Academy of Sciences* 110(6), 2193–2198.

Rosengaus, R.B., Traniello, J.F. and Bulmer, M.S. (2010) Ecology, behavior and evolution of disease resistance in termites. In: Bignell, D.E., Roisin, Y. and Lo, N. (eds) *Biology of Termites: A Modern Synthesis.* Springer, Dordrecht, The Netherlands, pp. 165–191.

Rudman, P. (1964) Durability in the genus *Eucalyptus. Australian Forestry* 28(4), 242–257.

Sapunov, V.B. (2008) Global dynamics of termite population: modeling, control and role in green house effect. *Earth* 1, 1–5.

Scaduto, D.A., Garner, S.R., Leach, E.L. and Thompson, G.J. (2012) Genetic evidence for multiple invasions of the eastern subterranean termite into Canada. *Environmental Entomology* 41(6), 1680–1686.

Scheffrahn, R.H. and Crowe, W. (2011) Shipborne termite (Isoptera) border interceptions in Australia and onboard infestations in Florida, 1986–2009. *Florida Entomologist* 94(1), 57–63.

Scheffrahn, R.H. and Postle, A. (2013) New termite species and newly recorded genus for Australia: *Marginitermes absitus* (Isoptera: Kalotermitidae). *Australian Journal of Entomology* 52(3), 199–205.

Scheffrahn, R.H., Cabrera, B.J., Kern Jr, W.H. and Su, N.Y. (2002) *Nasutitermes costalis* (Isoptera: Termitidae) in Florida: first record of a nonendemic establishment by a higher termite. *Florida Entomologist* 85(1), 273–275.

Scheffrahn, R.H., Kreček, J., Ripa, R. and Luppichini, P. (2009) Endemic origin and vast anthropogenic dispersal of the West Indian drywood termite. *Biological Invasions* 11(4), 787–799.

Schmidt, A.M., Jacklyn, P. and Korb, J. (2014) 'Magnetic' termite mounds: is their unique shape an adaptation to facilitate gas exchange and improve food storage? *Insectes Sociaux* 61(1), 41–49.

Sillero, N. (2011) What does ecological modelling model? A proposed classification of ecological niche models based on their underlying methods. *Ecological Modelling* 222(8), 1343–1346.

Sponsler, R.C. and Appel, A.G. (1991) Temperature tolerances of the Formosan and eastern subterranean termites (Isoptera: Rhinotermitidae). *Journal of Thermal Biology* 16(1), 41–44.

Steward, R.C. (1981) The temperature preferences and climatic adaptations of building-damaging dry-wood termites (*Cryptotermes*; Isoptera). *Journal of Thermal Biology* 6(3), 153–160.

Su, N.Y. and Scheffrahn, R.H. (1990) Economically important termites in the United States and their control. *Sociobiology* 17(1), 77–94.

Su, N.Y. and Tamashiro, M. (1987) An overview of the Formosan subterranean termite in the world. In: Tamashiro, M. and Su, N.Y. (eds) *Biology and Control of the Formosan Subterranean Termite.* College of Tropical Agriculture and Human Resources, University of Hawaii, Honolulu, Hawaii, USA, pp. 3–15.

Ulyshen, M.D. (2016) Wood decomposition as influenced by invertebrates. *Biological Reviews of the Cambridge Philosophical Society* 91(1), 70–85. DOI: 10.1111/brv.12158.

Vargo, E.L. and Husseneder, C. (2010) Genetic structure of termite colonies and populations. In: Bignell, D.E., Roisin, Y. and Lo, N. (eds) *Biology of Termites: A Modern Synthesis.* Springer, Dordrecht, The Netherlands, pp. 321–347.

Verkerk, R.H. and Bravery, A.F. (2001) The UK termite eradication programme: justification and implementation. *Sociobiology* 37(2), 351–360.

Wamsler, C., Brink, E. and Rivera, C. (2013) Planning for climate change in urban areas: from theory to practice. *Journal of Cleaner Production* 50, 68–81.

Wang, C.H. and Wang, X. (2012) Vulnerability of timber in ground contact to fungal decay under climate change. *Climatic Change* 115(3–4), 777–794.

Wappler, T. and Engel, M.S. (2006) A new record of *Mastotermes* from the Eocene of Germany (Isoptera: Mastotermitidae). *Journal of Paleontology* 80(02), 380–385.

Watson, J.A. and Abbey, H.M. (1993) *Atlas of Australian Termites.* CSIRO (Commonwealth Scientific and Industrial Research Organisation), Canberra, Australia. 155 pp.

Weidner, H. (1952) Die Ausbreitung der Termite *Reticulitermes flavipes* (Kollar) in Hamburg. *Transactions of the International Congress of Entomology* 9(1), 829.

Williams, K., Pinzon, Z.S., Stumpf, R.P. and Raabe, E.A. (1999) Sea-level rise and coastal forests on the Gulf of Mexico. *Open-File Report 99-441.*

US Geological Survey, Reston, Virginia, USA, pp. 122. Available at: http://pubs.er.usgs.gov/publication/ofr99441 (accessed 5 May 2016).

Woodrow, R.J. and Grace, J.K. (1998) Thermal tolerances of four termite species (Isoptera: Rhinotermitidae, Kalotermitidae). *Sociobiology* 32(1), 17–26.

Wylie, F.R. and Shanahan, P.J. (1976) Insect attack in fire-damaged plantation trees at Bulolo in Papua New Guinea. *Austral Entomology* 14(4), 371–382.

7 Fly Populations and Problems in a Changing Climate

Mohamed F. Sallam[1,2]*, Roberto M. Pereira[1] and Philip G. Koehler[1]

[1]*Department of Entomology and Nematology, UF/IFAS Extension, University of Florida, Gainesville, Florida, USA;* [2]*Department of Entomology, College of Science, Ain Shams University, Cairo, Egypt*

7.1 Introduction

Fly development and distribution are sub stantially varied in space and season. This variation is basically attributed to three main factors: 1) availability of food; 2) suitable breeding habitats; and 3) climate. Flies, like the rest of the insects, are ectothermic arthropods and rely on external climate to adjust their body temperatures. Their occurrence and abundance is regulated by their ecological requirements. The availability of minimum ecological requirements may sustain their existence for a particular time in a confined space. On the other hand, the accessibility of optimum conditions helps flies to expand their geographic distribution and increase their development rate. The majority of previous investigations have addressed fly biology and development rate under laboratory conditions. Few studies have also discussed the influence of climate variables on fly distribution range.

The geographic expansion and development rate of flies are climate-induced. Their importance as potential vectors of diseases and as forensic tools depends on where flies can occur in large numbers and the suitable habitats for their distribution. Studies on influential effects of climate change on flies in field conditions are few in comparison with studies on mosquito vectors of diseases, for which many ecological niche models have been established to predict their abundance and distribution. In this chapter, we summarize previous laboratory studies, generate new georeferenced data for temperature-related development of flies, and conclude with current and future prediction scenarios for fly distribution at times of natural disasters, such as tornados or tsunamis in two different temperature-gradient areas.

Our attempt to project the suitable habitats of filth flies from previous experimental studies is a potential starting point for further investigations, to generate a prediction of suitable habitats at risk for filth fly development in response to temperature changes. This study is in an early phase and requires a substantial and strong baseline data of fly distribution, which is scarce on the global scale.

7.2 Pest Fly Species

Several pest flies of present concern are possibly the same species that may be of importance as the climate changes over the next decades. The subject of filth flies has been well reviewed by Pereira *et al.* (2014). Houseflies are the most common flies globally and a species implicated in the transmission of

**E-mail: mdsallam@ufl.edu*

diseases in many areas of the world. The housefly is well adapted to many different environments and has a very short development time (Hogsette *et al.*, 1991) and high reproductive capacity. The housefly's feeding and breeding habits are important factors in the spread of dysentery, diarrhoea and other diseases (Farkas *et al.*, 1998).

Flesh flies are scavengers that feed and reach maturity on carrion and rotting meat in 7–20 days, depending on the temperature and food quality. Blow flies and bottle flies are metallic blue, and/or green and are usually the first flies to reach a carcass, where the female deposits eggs so the larvae can develop on the decaying flesh. The presence of these fly maggots is used in the determination of data used in forensic science and criminal investigations. Black soldier flies are associated with livestock waste or decaying plant material (Sheppard *et al.*, 1990, 2002) and are considered to be beneficial by reducing animal waste (Newton *et al.*, 2005) and serving as livestock feed. Drain, filter or moth flies are common in sewer systems where the larvae develop, feeding on microbial growth.

Drosophila, fruit or vinegar flies develop on rotting fruits and vegetables and the adults are able to pass through openings in most fly exclusion screens due to their small size. Eye gnats are also very small and adults carry bacteria that cause conjunctivitis when they are attracted to noses and eyes of people or animals. Phorid or humpback flies are very common throughout the world and develop on corpses and high-protein waste.

7.3 Medical Importance

Among flies, the common housefly, *Musca domestica*, has the ability to carry serious pathogens such as polio virus, rickettsia of Q fever (*Coxiella burnetii*), and bacteria such as anthrax, salmonella and cholera. The importance of the housefly in transmitting disease pathogens stems from their habit of visiting faeces and other unhygienic matter and also human food.

Flesh flies have been incriminated in accidental intestinal myiasis in humans, causing pain before larvae are expelled in the faeces. They very occasionally lay their larvae in wounds of living tissue and cause myiasis; however, they usually feed on the necrotic tissues. Adult flesh flies such as *Sarcophaga* sp. and *Wohlfahrtia* sp. are frequent visitors to wounds, excreta and decaying animal matter; therefore, they may transmit various pathogens mechanically.

7.4 Fly Bio-ecology as Related to Biotic/Abiotic Conditions

Insect survival and development is greatly influenced by climate variables, such as temperature, humidity, and other climate factors. These factors are changing due to global warming, population growth and land-use changes. Many hypotheses have been adopted to explain the dynamic of global climate changes, such as the greenhouse gas effect, Southern Oscillation, sea-surface temperature, and land-cover changes. However, these scenarios are greatly influenced by human population growth and human activities. In other words, climate change is a complex relationship between interacting subsystems, involving environmental and social variables. However, only a few studies have addressed the influence these factors have on insect development and survival, and on insect-borne disease outbreaks.

Rigorous investigation of climate change impacts on insect species and the continuous interaction between different variables included in these two complex systems will provide important information for the development of realistic distribution models for insect species (Angilletta, 2009). This is one of the major challenges for scientific understanding of the effect of climate change on insects. Therefore, it would be irrelevant to address the influence on insect development and survival caused by a single variable within a subsystem within the biotic and abiotic complex. Consideration of all variables simultaneously is thus necessary to address the complexity of the biological and abiotic relationships. In this chapter, we attempt to discuss how global climate change influences fly species and how this is being assessed, what variables are being addressed, the regions of studies, the biological impacts on fly survival, development, and distribution/range.

7.4.1 Temperature

Flies, as ectothermic arthropods, are unable to internally control their body temperature and depend on the thermal conditions of their environment. Under laboratory conditions, any slight changes in temperature significantly influence their population dynamics, survival and biological processes, which suggests that global climate change is likely to extend or limit their geographic range (Alto and Juliano, 2001). Although many studies have highlighted the response of abundance, distribution/range, interaction and phenology of different fly species to temperature, the quantitative relationships between extreme variations of temperature and the spatiotemporal distributions of these flies are still obscure. The findings of these studies are inconsistent, which may be due to data unavailability, lack of resolution of projected climate data and methodological limitations.

Survival, development and biological processes inside the insect body are influenced by changes in temperature, which may accelerate or delay these processes commensurate with the increase or decrease in temperature. Laboratory models have highlighted the thermal-dependent life stages of species such as fruit flies and mosquitoes. A small increase in temperature under laboratory conditions can increase the hatchability of many fly eggs, e.g. houseflies, mosquitoes and fruit flies (Shiravi *et al.*, 2011; Roe and Higley, 2015). Also, higher temperatures increase the rate of trans-stadial development, so the insect proceeds to the next stage in a shorter time (Shiravi *et al.*, 2011). However, the positive influence of temperature on fly development may be limited, as higher temperatures will adversely affect their survival, such as when the temperature reaches a survival threshold that can be lethal. Many fly species survive a wide range of temperature ranges, spanning temperate, tropical, subtropical, and even polar regions.

In addition, temperature impacts the geographic range of flies at local and regional scales. In the last 10 years many fly species and mosquito vectors have expanded their distribution range because of global warming and changes in the occurrence of suitable habitats in the trophic cycle (Cannon, 1998; Peñuelas *et al.*, 2002; Andrew *et al.*, 2013). The significance of the geographic expansion of flies to new areas, attributed to climate warming, causes nuisance and epidemiological situations through disease transmission or economic losses.

On the other hand, higher temperature may negatively influence the survival of flies and the development of disease pathogens inside their bodies. With regard to fly maggots that live on carcasses, the successive development of the maggots is solely dependent on the duration of the decay process, which eventually depends on climatic conditions. In summer seasons, carcasses decay at a much faster rate than in winter and spring. The increased temperature in summer speeds up the temporal succession of waves (of insects' visits to the corpses), whereas in winter the rate of development of maggots slows down. Also, the maggots' activity leads to increased temperature of the corpse, which eventually results in quick decay. *Chrysomya megacephala* and *Chrysomya rufifacies* are very good examples of calliphorid flies that are found in all the seasons of the year.

7.4.2 Precipitation

Regional climate models (CM) project a potential variability in precipitation in the near future (UK Met Office, 2011). Generally, changes in precipitation patterns in response to global climate changes vary regionally. Two points of view have been proposed for interpreting precipitation variability: 1) wet regions get wetter, meaning that the presently wet regions receive increased precipitation (Held and Soden, 2006; Chou *et al.*, 2009); and 2) warmer regions get wetter in response to increased sea-surface temperature (SST) (John *et al.*, 2009). These two mechanisms are most likely complementary rather than contradictory. The annual mean precipitation follows warmer-get-wetter; on the other hand, the seasonal mean precipitation follows the wet-get-wetter mechanism. Also, in the light of this hypothesis, we may find a strong correlation between temperature and annual precipitation, and how much the distribution range and movement of air temperature can influence the annual precipitation in warm regions. The interrelationship between temperature and precipitation is dynamic and has influences on other climate and non-climate variables including human population growth and other sources of life, such as vegetation, surface water, animal and insect distribution, bird migration, and the geographic range of floral species.

Precipitation is considered a significant variable in establishing many flies that vector diseases and other flies with forensic importance. The significance of precipitation stems from the water habitats that are created and become suitable for an abundance of plants, insect species and other animals. Generally, sufficient precipitation and local moisture are prerequisite for occurrence/abundance of many living organisms in the trophic chain. Moreover, relative humidity is a potential factor determining egg survival in insects (Juliano *et al.*, 2002).

The continuous changes and variations in precipitation, as in other climate variables, also lead to higher vector-borne disease incidences (Alsope, 2007) and wider geographic range of flies (Fischer *et al.*, 2014), but may also reduce fly-related risks in other regions (Caminade *et al.*, 2014).

7.4.3 Surface water

Surface water is defined as any water on the Earth's surface that may be replenished by precipitation and/or leakage from ground water. It is extracted by humans for livelihood, agriculture and industry, used by plants and other animals, or lost by evaporation or seepage into ground water. Surface water is one of many non-climatic variables, which sustain aquatic ecosystems with different habitats for millions of living organisms. Surface water represents, approximately, 75% of the Earth's surface, including rivers, streams, seas, oceans, lakes, etc. The significance of surface water in the ecosystem stems from the major role it plays in regulating annual and seasonal precipitation through the mechanistic cycle of water, powered by the warming of seas and oceans.

Recently, flies, as part of the biotic system in an aquatic habitat, showed high-latitudinal shift in distribution associated with regional warming (Cannon, 1998; Finlay *et al.*, 2015; Herczeg *et al.*, 2015). Any future warming of aquatic habitats is considered a significant precursor for variation and changes in biogeochemical cycles and biological productivity in terms of distribution of aquatic flies, such as hoverflies, midges and mosquitoes.

Flooding caused by surface water has the potential to cause the replacement of fly fauna of ecosystems. Eventually, newly introduced aquatic fly species, along with many other aquatic organisms, will be established in these water habitats. This drastic change in the spatial distribution of aquatic habitats will subsequently expand the geographic range of some fly species, such as hoverflies, midges and blackflies, and this may cause health problems.

7.4.4 Vegetation

Understanding the changes in the local environment in terms of vegetation is key to defining habitat requirements and the suitable ecological niches that can create or reduce insect populations. Vegetation is a very important biotic factor in the trophic chain, representing the main source of food for the entire animal kingdom, including flies. The growth of vegetation depends heavily on two climate variables: temperature and precipitation.

Flies, as non-photosynthetic organisms, derive their energy directly or indirectly from vegetation. Plants are also considered as resting and sheltering places for many flies. Since vegetation type determines the niche habitats for fly species, changes in vegetation may lead to public health problems, such as the occurrence of different fly species, including blackflies. Applications of remote sensing techniques have been intensively used since the 1970s to assess the association between vegetation indices and fly distribution/vector-borne outbreaks. The Normalized Difference Vegetation Index (NDVI) was used to monitor the abundance of mosquito vectors in Argentina (Gleiser *et al.*, 1997) and the prevalence of lymphatic filariasis in the Nile Delta, Egypt (Hassan *et al.*, 1998). The demonstrated usefulness of these indicators in predicting insect-suitable habitats and insect abundance allows implementation of insect surveillance and pest-control measures (Linthicum *et al.*, 1990; Sithiprasasna *et al.*, 2003, 2005).

During the last 20 years, severe drought and flooding have occurred with pronounced significant impacts on public health and agriculture. Climate shifting and global warming potentially affected the geographic distribution of flies and disease outbreaks (Hansen *et al.*, 2012; Blunden and Arndt, 2013). The effect of global climate warming has several manifestations on fly populations, detected as shifting vegetation habitats. Each shift results in changes in the suitability of fly habitats in these areas, which eventually may expand or constrain the distribution of flies' populations.

While dry conditions can significantly affect crop yields and pests, as well as create fire conditions, wet conditions and flooding can also cause damage to crops and wash away productive soil. With these changes, eventually other trophic systems associated with the vegetation will be impacted, leading to ecological habitat changes that may affect the development of flies in affected areas. For example, the introduced rice fields in central Kenya appeared to limit the malaria risk transmission by *Anopheles funestus* but not by *An. arabiensis* (Muturi *et al.*, 2008). Also, a recent reduction in water level in Lake Victoria has created newly emerged terrestrial habitats more suitable for *An. funestus* than for *An. gambiae* (Minakawa *et al.*, 2008).

The prolonged drought impacting some Kenyan communities resulted in changes to human food security, human and animal population movements, and grazing and water-storage practices. These changes will eventually alter the fly vector abundance/distribution dynamic and the transmission risk of diseases (Kenya Red Cross Society, 2009) in these locations. Information on the impact of agricultural activities and climate changes on flies' distribution range and disease transmission is limited, although it is increasingly important (Macintyre *et al.*, 2002).

7.4.5 Hosts

Animal, bird and human hosts play a major role in the geographic distribution of insect pests. The presence of a suitable host attracts specific pests to the vicinity of the host nest or living quarters. The continuous human population growth results in development of different forms of settlements, such as rural, urbanized and urban, which bring increasing numbers and taxa of insect pests, depending on their habitat requirements.

Overlapping habitats for hosts (human, animal, bird) and insect pests increases the complexity of potential epidemiological conditions and creates complex relationships with abiotic factors in these habitats. Urbanization significantly influences the distribution of flies. A recent study in the USA showed a greater summer temperature rise in urban areas than could be predicted by climate change alone. Also, the increase in temperatures in urbanized areas occurred faster than in adjacent rural areas (Kenward *et al.*, 2014). Moreover, the increase in temperatures in urban areas is more intense at night than during the day, with a range of 1–3°C more than in surrounding rural areas (Oke, 1997; Kenward *et al.*, 2014).

As previously discussed, temperature plays a major role in the developmental mechanisms and latitudinal distribution of fly species. However, the magnitude of this increase has not been addressed in field studies, especially in association with humans. The increase of temperature in urban areas is largely attributed to human settlements, but it does not necessarily increase the habitat suitability of flies. Some flies tolerate well the increased temperatures of human settlements and, eventually, they expand their range. Recent research suggests that urban planning, incorporating more trees, parks, a piped water system, and white roofs can reduce the effect of urban heat and eventually minimize the distribution range of urban insect pests (Kenward *et al.*, 2014).

7.5 Potential Effects of Climate Changes on Fly Species

A successful implementation of pest management programmes requires a thorough understanding of the ecology of the target pests, especially their distribution range and factors that may limit their range. Since the early 1960s, global warming has brought increased attention to the fate of insect populations in comparison with historical data. Even though local temperatures may vary and fluctuate naturally, the mean global temperature has increased. The mean global temperatures were estimated to be 1.1–6.4°C higher by the end of this century than they were at the start of the century (Oke, 1997; EPA, 2003; CDC, 2004; IPCC, 2007). Also, winter temperatures are projected to show a subtle increase with the increase of global temperatures.

Many investigations addressed the effect of climate on insect population dynamics and biology (Parmesan and Yohe, 2003; Root *et al.*, 2003; IPCC, 2007; Loarie *et al.*, 2009). For example, time required for development of fly eggs, larvae and pupae decreases with increasing temperature. Also, the increased temperatures reduce the time required for development of pathogens

inside a fly body. However, the discrepancies of spatial and temporal distribution of different insect taxa, especially flies, and the adaptability of these taxa in response to global climate change need to be investigated. Substantial evidence of changes in the geographic distribution, frequency and intensity of insect pests and diseases (for example, Bluetongue, a viral disease of ruminants, transmitted by the midge *Culicoides imicola*, *Culicoides variipennis* and other culicoids, which is moving into more temperate zones of Europe (Cannon, 1998)), has been attributed to climate change.

Currently, the projected changes in insect taxa in response to climate scenarios lack strong baseline biological records, with the exception of a few insect species that are implicated in public health issues (Woodward, 1987; Keatley *et al.*, 2002; Peňuelas *et al.*, 2002; Wilson and Maclean, 2011; Andrew *et al.*, 2013). Such climate model projections will be of increasing significance in enabling a deep understanding of the distribution limits for fly species in a changing world.

7.6 Distribution, Range Expansion and Disease Transmission

Generally, flies, as with other organisms, have minimum and maximum ecological requirements, with a range between these extremes representing their tolerance. Although the limit of tolerance varies geographically and seasonally within the same species, fast transportation methods in combination with global warming are believed to significantly increase the geographic distribution range for fly species.

With the continuous increase in global temperature and changes in precipitation, the distribution range of different fly species are drastically extending into northern regions of the globe (Alto and Juliano, 2001). A rare human umbilical cord myiasis reported in a 9-day-old infant living in a temperate climate zone (Puvabanditsin *et al.*, 2014) exemplifies the extension of fly populations into new areas. Myiasis is defined as the invasion of live mammalian tissue by the immature stage (maggots) of dipteran flies, which feed on the host's necrotic or living tissue. The rise of temperature in the northern regions of the USA will probably allow many other species of Sarcophagidae and Calliphoridae to become established and develop to later stages when their eggs are transported incidentally with animals or humans into new areas (Yüca *et al.*, 2005; Puvabanditsin *et al.*, 2014; Muntzer *et al.*, 2015). This may result in expanded distribution of myiasis, resulting in flies at northern latitudes, as in the case cited above.

In 1998, the epidemiology of *Culicoides*-borne arboviruses was dramatically changed. Northern temperate regions in Europe and North America experienced the emergence of exotic viruses, which were believed to be confined to tropical zones; eventually, this will lead to an increased global disease burden. Land use/land cover, animal trade, animal husbandry, and, on top of all these, climate change, are believed to be the driving forces in *Culicoides*-borne disease occurrences, highlighting the interaction between the insect vector population dynamics and other biotic and abiotic variables. Moreover, new *Culicoides* species were identified and incriminated along with new wild reservoir hosts in transmission (Purse *et al.*, 2015).

Although the distribution and population dynamics of horseflies were rigorously studied in many regions of Europe, little data exist on their geographic range and the factors that may impact their distribution, abundance, and activity in northern regions. In central and eastern Europe, geographic distribution of horseflies was investigated and three climatic variables – rainfall, temperature and sunshine – were found to affect tabanid fly distribution. Meanwhile, relative humidity was found to have an indirect influence on tabanid abundance through effects in the air temperature (Herczeg *et al.*, 2015). This study determined that the optimal relative humidity for these flies was ~35% whereas the optimal temperature was ~32°C.

In North Africa, human activities combined with climate warming were shown clearly to influence sand fly (*Phlebotomus*

perfiliewi and *P. perniciosus*) abundance, and the occurrence of visceral and cutaneous leishmaniasis vectors in this region (Barhoumi *et al.*, 2015). The increased irrigation, representing human activities, potentially increases the abundance of both vectors in arid bioclimatic regions. This finding showed that the abundance of both sand fly species is associated with the development of irrigated areas, suggesting that the geographic range of these species will expand to more arid areas where agricultural activities are expanded.

Another example of human-aided range expansion comes from a report where scientists and expedition staff accidentally transported flies to King George Island, off the Antarctica coast. Efforts to eradicate the newly recorded flies, such as *Trichocera maculipennis*, failed in their new habitats. Eggs of other fly species were collected from the sewage system of the research stations on the island (Powell, 2012).

The influence of climate change on black flies is also a subject of interest. Despite the decline in recorded cases of human trypanosomes, there is an increasing interest in the development of alternative control measures of black flies, and in understanding the factors that influence trypanosome transmission from the fly vector to the host, especially humans. Recently, sporadic cases of this disease were reported in southern Saudi Arabia and Yemen (Geiger *et al.*, 2015), where no cases had been previously recorded. Host movement (animal and human) and climate change are implicated in the expansion of the trypanosomiasis distribution.

In the previous decade, spatial and temporal modelling of mosquito-borne diseases using geographic information system (GIS) and satellite images allowed the prediction of the emergence of diseases such as Rift Valley Fever (RVF) 1 month before it surfaced in the Horn of Africa and the southern Arabian Peninsula (Anyamba *et al.*, 2012; Sallam *et al.*, 2013). In addition, some of the modelling investigations highlighted the interaction between climatic and nonclimatic variables in enhancing the epidemiological conditions. Mosquito-borne disease modelling is useful for the forecasting and early monitoring of arbovirus outbreaks. Moreover, this approach should be considered in addressing the geographic range of other flies that transmit diseases.

Such distribution models gained global prominence as a technology for building prediction maps. Although model results vary from conservative to optimistic, future climate change is predicted to increase air and land temperatures, with a direct effect on fly distribution and development. By the end of this century, the air temperature of central Europe is projected to reach conditions that are similar in southern Europe. The shift in climate, not only in Europe but also in other regions of the globe, may allow fly habitat shifting or expansion, leading to fly outbreaks and associate medical consequences for the human population.

7.7 Evolutionary Consequences of Climate-induced Range Shifts

Flies are greatly influenced by, and responsive to, environmental changes. A slight change in annual temperature regimes can have profound effects on basic physiological processes, development, distribution and abundance. Flies are suited to an extension of their geographic range due to their high reproductive ability and high mobility (Settele *et al.*, 2008; Ott, 2010). These characteristics enable flies to expand their geographic range within a relatively short time, once these areas become ecologically suitable.

Fittingly, fly range expansion in response to rising global temperatures and anthropogenic changes has received considerable attention in the last two decades (Parmesan, 1996; Parmesan *et al.*, 1999; Parmesan and Yohe, 2003; Crozier, 2004; Karban and Strauss, 2004; Hickling *et al.*, 2005, 2006). Although several investigations addressed such changes in general, these studies lack strong baseline data. Northward range expansions have been reported for species like lepidopterans in Europe (Parmesan *et al.*, 1999) and North America (Parmesan, 1996), and for heteropterans,

neuropterans, orthopterans and odonates in the UK (Hickling *et al.*, 2005).

The significance of the distribution expansion stems from potential economic losses and disease transmission by flies and mosquitoes. However, few studies have highlighted the evolutionary consequences that may result from these range shifts:

1. Range shifts bringing formerly allopatric species into *de novo* sympatric areas, which can eventually create new species interactions leading to population hybridization and changes in species biodiversity (Barton and Hewitt, 1985; Jiggins and Mallet, 2000; Schwarz *et al.*, 2005; Ording *et al.*, 2010).
2. Range changes causing formerly sympatric populations of the same species into allopatric areas, isolating these populations and eventually producing divergence (Rhymer and Simberloff, 1996; Wellenreuther *et al.*, 2010).

The range expansion or contraction can result in short- and long-term ecological and evolutionary consequences. So far, empirical evidence evaluating the range shifts for many insect species are scarce.

7.8 Occurrence of Climate-related Natural Disasters

The continuous change in climate may result in the increased occurrence of extreme weather phenomena in different parts of the globe, such as melting of icebergs and the development of cyclones. Extreme weather directly affects the human population, but also has the potential to severely affect other living organisms, including insects. Consequences of extreme weather are unpredictable, varying from increased biodiversity to extinction of species locally or globally. The scarcity of strong baseline data on potential effects of extreme weather phenomena on the range of flies highlights the need for further investigation.

The consistent and continuous surface-water warming and cooling of the north Pacific Ocean leads to phenomenal weather change and the resulting formation of, at certain times and places, cyclones and tornadoes. One of the major problems that may face areas subjected to weather-related natural disasters is post-strike consequences, in terms of public health and spread of epidemics due to the interruption of major sanitation services, leading to an increase in the fly population and consequent diseases.

Due to public health concerns, delineating suitable habitats for flies, as potential vectors of disease (Royle *et al.*, 2012; Fitzpatrick *et al.*, 2013) assumes greater importance. Predicted habitat suitability for fly outbreaks can be used in guiding emergency responses. Ecological niche models predict habitat suitability in response to climate variables. Climate based models, such as the maximum entropy technique (MaxEnt) (Phillips *et al.*, 2006), and the recently developed maximum likelihood technique (Max-Like) (Royle *et al.*, 2012; Fitzpatrick *et al.*, 2013) are the most frequently used prediction models for forecasting climate-induced range shifts (Lawler *et al.*, 2009).

We developed a filth fly habitat suitability model using temperature in two projected scenarios, representing current and future scenarios (Fig. 7.1) (Sallam, M.F., Pereira, R.M. and Koehler, P.G. (in preparation) Spatial projection and future scenarios of filth flies development in response to climate change in USA). Mean temperatures for New York and Miami-Dade cities were extracted and calculated from monthly minimum and maximum records from NOAA climate (Brohan *et al.*, 2006), for the current scenario (1950–2015) and from WorldClim databases (Hijmans *et al.*, 2005) for the future scenario (2050). Afterwards, temperature records were masked to both cities, converted into grid data layers and resampled to fine raster data resolution (~1 km) to enhance their accuracy using raster analysis tools in ArcGIS tool box.

The two cities were selected to represent different temperature gradients and to demonstrate the influence of latitudes on spatial/temporal distribution and range shifts of fly habitat suitability. The future scenario models were developed to illustrate how much the probability of suitable habitats for flies will be expanded in response to an increase in temperature. In our model, we did not consider any species hybridization, introgression and extinction as discussed above.

Due to the lack of strong baseline field-collected fly distribution and abundance data, and our model covering a relatively small range of temperature gradient (24.5–51.1°C), we projected the fly development rate in both space and time based on retrogressive laboratory studies (Kamal, 1958; Shiravi *et al.*, 2011; Mabika *et al.*, 2014; Roe and Higley, 2015). Then, the percentage of reduction in the development rate between 2015 and 2050 was calculated. The temperature-predicted habitats of flies can potentially produce maps to be used in emergency response by public health personnel in targeted fly-control programmes. Risk probability was categorized into five classes and the dependency of the habitat suitability risk on temperature gradient was assessed using linear regression model in R statistical packages version 3.0.3.

Surprisingly, the total areas of predicted high suitable habitats for emergency response in New York City were larger than predicted suitable habitats in Miami-Dade for both current and future scenarios. However, the development rate (hours) of flies in the latter was estimated to be smaller than those in the former because the temperature gradient in Miami-Dade is higher than in New York City. Also, the development rate is anticipated to decrease with the increase in temperatures on the temporal scale. The very low and low suitable habitats were predicted to have the largest percentage of reduction in the development rate between the current year and 2050 scenarios (Table 7.1). This relatively large reduction of the development rate for very low and low suitable habitats reflects the substantial increase in temperature for these low and very low suitable predicted habitats rather than the increase of temperature in high and very high suitable habitats.

The emergency response categories were significantly dependent on the temperature gradient in both Miami-Dade ($R^2 = 99.97$, $F_{1,4} = 1.252e^{+04}$, $P < 0.01$, for 2015; and $R^2 = 99.84$, $F_{1,4} = 1.232e^{+04}$,

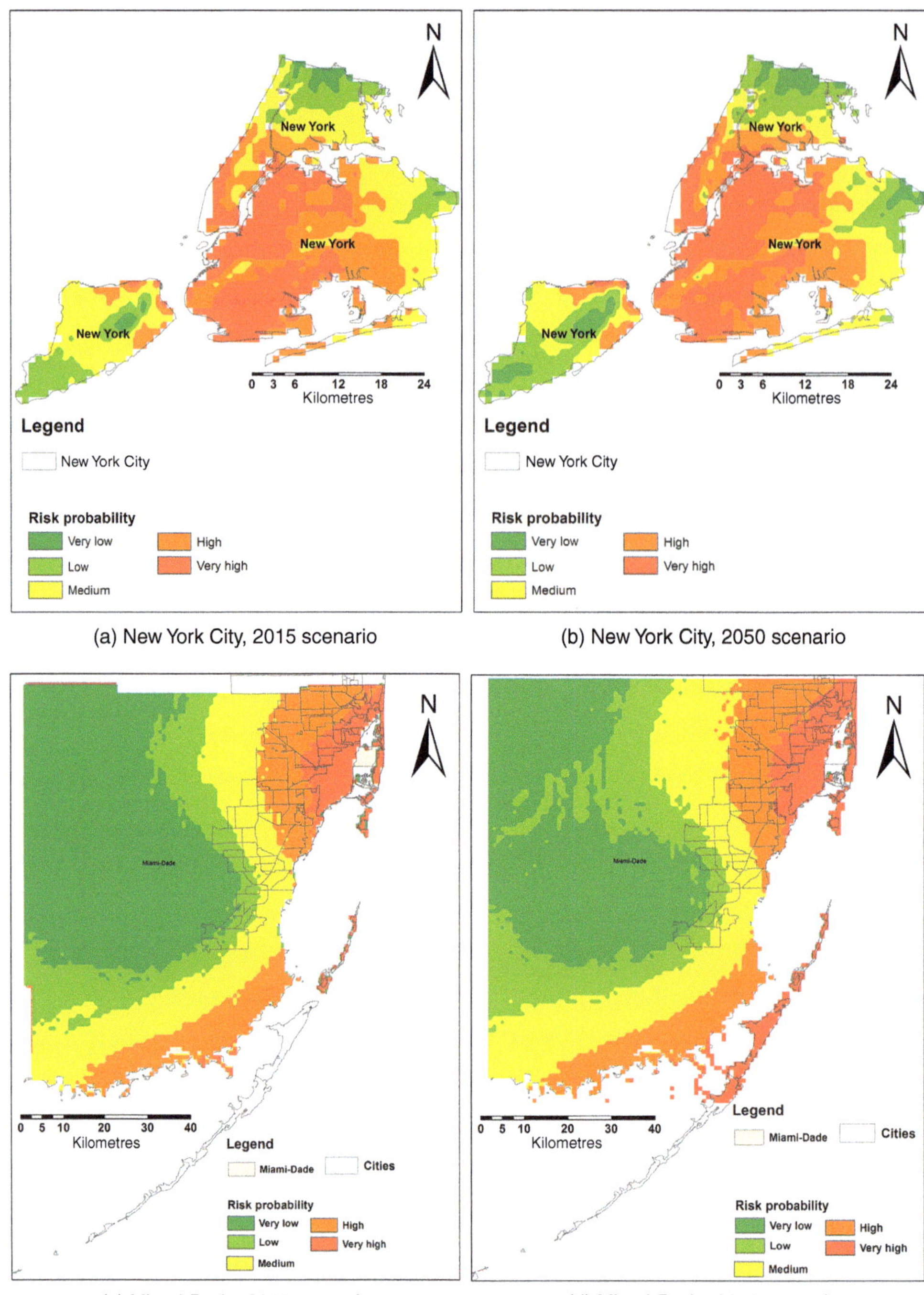

(a) New York City, 2015 scenario

(b) New York City, 2050 scenario

(c) Miami-Dade, 2015 scenario

(d) Miami-Dade, 2050 scenario

Fig. 7.1.   Predicted suitable habitats of filth flies.

Table 7.1. The percentage change of predicted suitable habitat areas and the development rate (hours) of filth breeding flies in New York and Miami-Dade cities (2015 and 2050) (Sallam, M.F., Pereira, R.M. and Koehler, P.G. (in preparation) Spatial projection and future scenarios of filth flies development in response to climate change in USA).

Risk probability	City	Temperature (Celsius)	Predicted risk area (%)			
			Current	2050	Percentage change	Percentage reduction in development rate (hours)[a]
Very high	New York	>24.5–25.0	27.59	28.69	4.00	18.30–18.33
High	New York	>24.2–24.5	24.44	24.78	1.39	18.33–18.52
Medium	New York	>23.6–24.2	33.03	22.24	−31.03	18.52–20.00
Low risk	New York	>23.1–23.6	12.18	17.94	47.29	20.00–20.07
Very low	New York	22.5–23.1	2.76	6.35	130.07	20.07–19.93
Very high	Miami-Dade	>49.2–50.0	5.02	8.41	67.53	6.20–6.31
High	Miami-Dade	>48.6–49.2	13.36	14.82	10.93	6.31–6.41
Medium	Miami-Dade	>48.1–48.6	19.64	23.50	19.65	6.41–6.47
Low risk	Miami-Dade	>47.8–48.1	13.66	19.38	41.87	6.45–6.47
Very low	Miami-Dade	47.2–47.8	48.34	33.89	−29.89	6.45–6.78

[a] The development rate was projected for egg–adult using Kamal *et al.* (1958) and Roe and Higley (2015). The reduction of the development rate was calculated from the formula (development rate during current − development rate during 2050/development rate during current × 100).

$P < 0.01$, for 2050) and New York ($R^2 = 99.94$, $F_{1,4} = 1.532e^{+04}$, $P < 0.01$, for 2015, and $R^2 = 99.96$, $F_{1,4} = 1.113e^{+04}$, $P < 0.01$, for 2050). The emergency response areas are expected to significantly expand during 2050 for the two cities. Eventually, this expansion requires appropriate substantial preparations to handle the possible breeding habitats of filth breeding flies. Moreover, our emergency response map is potentially significant, especially after catastrophic natural disasters such as tornadoes and hurricanes. Usually after a natural disaster such as a hurricane, the affected areas complain of surplus filth breeding flies impacted by the lack of appropriate handling of food wastes, disruption of sewer systems, increased landfill capacity, and decaying of dead animals.

In 1999, North Carolina encountered Hurricane Dennis, which hit the coast twice. Afterwards, Hurricane Floyd followed Dennis, leaving 12–20 inches (30–51 cm) more water. These two hurricanes, like the rest of their kind, caused the death of thousands of domestic hogs and chickens (NCRWA, 2015), and left polluted water raging down North Carolina's river.

In our model, we confirmed the findings produced from previous experimental studies (Kamal, 1958; Roe and Higley, 2015) in projecting the spatial and temporal range shifts of suitable habitats for flies in response to temperature gradients. The temporal influence of temperature on the development rate of flies has been extensively considered in previous studies; however, no study considered the role of spatial differences in temperature gradient. Our findings are potentially useful for projecting fly geographic range shifts in space and time. The use of predictive algorithms to generate distribution models (Phillips *et al.*, 2006) will potentially increase the significance of species range shifts and the associated risk, which may help in directing emergency response at times of natural disasters.

Eventually, emergency responses will have to change in the future, depending on the increased distribution range and the development rate of flies. In addition, the occurrence of natural disasters increases the areas of the emergency response. As demonstrated previously, natural disasters such as hurricanes create very suitable habitats for filth breeding flies, represented as dead hogs and chicken. Moreover, the disruption of sewage systems and human food wastes exacerbate and increase the landfill capacity. Subsequently, the increased global temperature potentially increases the suitable habitats for the development of flies and decreases the development rate.

The linear regression model confirmed the validity of our model in terms of dependency of risk probability classes on temperature gradient in both cities. The habitat suitability maps predicted from our model are potentially important in directing emergency response after catastrophic environmental disasters. However, further field investigations should consider empirical species range expansion and retraction to optimize the usefulness of models for predicting insect and disease vector outbreak risks.

7.9 Area-wide Fly Control Methods in New Climatic Conditions

Climate is in a continuous interaction with ecosystem processes and the trophic cycle at local and regional scales. Distribution, expansion and extinction of flies are dependent on many interacting variables. To understand the risk of shifts in fly distribution, regional cooperation is necessary for risk analysis, exchange of information and coordinated actions. Current surveillance and control measures should be evaluated to strengthen regional cooperation in controlling fly species.

The current knowledge on dispersal of many fly species is still ambiguous and information gaps need to be filled on both local and regional scales. Improvements are needed in surveillance and control measures related to flies and the diseases they carry, and in our understanding of the role of environmental changes in different ecological contexts, using remote sensing, geographic information systems and modelling technologies. In the light of global weather changes, there is a need to project the impacts that climate, non-climate factors and ecological processes may have on flies.

Results of these projections can guide early warning systems for active surveillance and management strategies.

Critical geographic, biological and taxonomic knowledge gaps must be resolved. In particular, there is a crucial need for better integrated surveillance techniques representing passive and active strategies, fast and cheap fly-identification methodologies, epidemiological knowledge, and better fly-control mechanisms. Recent developments in genomics and mathematical modelling may enhance ecological understanding of complex arbovirus systems. Coordinated research on climate and non-climate changes, including human activities and population growth, are needed to improve management of spatial and temporal range of invasive species. Analysis of the fly distribution on local and regional scales, based on historical and current data, and in response to different climate projections, will be essential in assessing the adaptation of these insects to the habitat shifts. Work on this global challenge requires both funding and support from local governments as well as the World Health Organization (WHO) and other international organizations.

7.10 Conclusion

Geographic distribution and range shifts of insects has come to the attention of researchers recently, in an effort to highlight the factors that may delineate their occurrence and abundance. Previous studies are rare and were usually conducted under laboratory conditions. With the continuous change in global climate, flies, like other insects, are highly affected. Although the climate is believed to be the key player in delineating the suitable habitats of flies, human and animal movements, and other environmental variables, such as surface water, vegetation, etc. play a considerable part in predicting their distribution and habitat suitability. As a matter of fact, environment and climate are in continuous interaction, and eventually this influences the habitat suitability for flies. However, few or none of the previous investigations addressed these interactions between different climate parameters, non-climate variables, and their collective influence on the distribution and development of fly species.

Flies, like other insects, have demonstrated northward movement due to global warming and human and animal movement. However, this expansion brings formerly allopatric species into the same geographic regions, which eventually creates an overlapping zone between the native and the introduced species. Also, the spread of flies is dependent on the availability of habitat requirements such as food, breeding sites and suitable climate.

The significance of investigating the geographic expansion and range shifts of flies stems from their importance in affecting human and livestock lives as a nuisance or disease vector. Studies concerned about habitat suitability and the minimum habitat requirements for fly distribution are few. This chapter addressed the consequences caused by the spread of flies in the future because of global warming, a concern emanating from increasing evidence on severe fly infestation after natural disasters. The emergency response to natural disasters and their consequences may need to be adapted, depending on the risk of fly development. The area at risk will significantly expand in the future with the continuous increase in global temperature. Emergency responses in the future need to be quick and efficient in containing larger areas at risk of fly infestation.

Modelling algorithms and niche habitat prediction methods are useful in identifying areas at risk. The significance of these modelling tools stems from their ability to develop risk maps for the habitat, covering various interactions of different climate and non-climate variables. Subsequently, these risk maps could be used in emergency responses.

References

Alsope, Z. (2007) Malaria returns to Kenya's highlands as temperatures rise. *Lancet* 370(9591), 925–926.

Alto, B.W. and Juliano, S.A. (2001) Precipitation and temperature effects on populations of *Aedes albopictus* (Diptera: Culicidae): implications for range expansions. *Journal of Medical Entomology* 38, 646–656.

Andrew, N.R., Hill, S.J., Binns, M., Bahar, M.H., Ridley, E.V., Jung, M.P., Fyfe, C., Yates, M. and Khusro, M. (2013) Assessing insect responses to climate change: What are we testing for? Where should we be heading? *PeerJ* 1(1), e11. DOI: 10.7717/peerj.11.

Angilletta, M.J. (2009) *Thermal Adaptation: A Theoretical and Empirical Synthesis*. Oxford University Press, Oxford, UK.

Anyamba, A., Linthicum, K.J., Small, J.L., Collins, K.M., Tucker, C.J., Pak, E.W., Britch, S.C., Eastman, J.R., Pinzon, J.E. and Russell, K.L. (2012) Climate teleconnections and recent patterns of human and animal disease outbreaks. *PLoS Neglected Tropical Diseases* 6(1), e1465. DOI: 10.1371/journal.pntd.0001465.

Barhoumi, W., Qualls, W.A., Archer, R.S. Fuller, D.O., Chelbi, I., Cherni, S., Derbali, M., Arheart, K.L., Zhioua, E. and Beier, J.C. (2015) Irrigation in the arid regions of Tunisia impacts the abundance and apparent density of sand fly vectors of Leishmania infantum. *Acta Tropica* 141(Pt A), 73–78. DOI: 10.1016/j.actatropica.2014.10.008.

Barton, N. and Hewitt, G. (1985) Analysis of hybrid zones. *Annual Review of Ecology and Systematics* 16, 113–148.

Blunden, J. and Arndt, D.S. (2013) State of the climate in 2012. *Bulletin of the American Meteorological Society* 94, S1–S258.

Brohan, P., Kennedy, J.J., Harris, I., Tett, S.F.B. and Jones, P.D. (2006) Uncertainty estimates in regional and global observed temperature changes: a new dataset from 1850. *Journal of Geophysical Research* 111, article ID D12106. DOI: 10.1029/2005JD006548.

Caminade, C., Ndione, J.A., Diallo, M., MacLeod, D.A., Faye, O., Ba, Y., Dia, I. and Morse, A.P. (2014) Rift Valley Fever outbreaks in Mauritania and related environmental conditions. *International Journal of Environmental Research and Public Health* 11, 903–918.

Cannon, R.J.C. (1998) The implications of predicted climate change for insect pests in the UK, with emphasis on nonindigenous species. *Global Change Biology* 4, 785–796.

CDC (2004) *Extreme Heat: A Prevention Guide to Promote Your Personal Health and Safety*. Available at: http:// http://kanehealth.com/PDFs/Emergency/Weather/ExtremeHeatTips.pdf (accessed May 2016).

Chou, C., Neelin, J., Chen, C. and Tu, J. (2009) Evaluating the 'rich-get-richer' mechanism in tropical precipitation change under global warming. *Journal of Climate* 22, 1982–2005.

Crozier, L. (2004) Warmer winters drive butterfly range expansions by increasing survivorship. *Ecology* 85, 231–241.

EPA (2003) Beating the heat: mitigating thermal impacts. *Nonpoint Source News-Notes* 72, 23–26.

Farkas, R., Hogsette, J.A. and Boerzsoenyi, L. (1998) Development of *Hydrotaea aenescens* and *Musca domestica* (Diptera: Muscidae) in poultry and pig manures of different moisture content. *Environmental Entomology* 27, 695–699.

Finlay, K., Vogt, R.J., Bogard, M.J., Wissel, B., Tutolo, B.M., Simpson, G.L. and Leavitt, P.R. (2015) Decrease in CO_2 efflux from northern hardwater lakes with increasing atmospheric warming. *Nature* 519, 215–218. DOI: 10.1038/nature14172.

Fischer, D., Thomas, S.M., Neteler, M., Tjaden, N.B. and Beierkuhnlein, C. (2014) Climatic suitability of *Aedes albopictus* in Europe referring to climate change projections: comparison of mechanistic and correlative niche modelling approaches. *Eurosurveillance* 19(6). DOI: 10.2807/1560-7917.ES2014.19.6.20696.

Fitzpatrick, M.C., Gotelli, N.J. and Ellison, A.M. (2013) MaxEnt versus MaxLike: empirical comparisons with ant species distributions. *Ecosphere* 4, 55.

Geiger, A., Ponton, F. and Simo, G. (2015) Adult blood-feeding tsetse flies, trypanosomes, microbiota and the fluctuating environment in sub-Saharan Africa. *The ISME Journal* 9, 1496–1507.

Gleiser, R.M., Gorla, D.E. and Luduena Almeida, F.F. (1997) Monitoring the abundance of *Aedes (Ochlerotatus) albifasciatus* (Macquart 1838) (Diptera: Culicidae) to the south of Mar Chiquita Lake, central Argentina, with the aid of remote sensing. *Annals of Tropical Medicine and Parasitology* 91, 917–926.

Hansen, J., Sato, M. and Ruedy, R. (2012) Perception of climate change. *Proceedings of the National Academy of Sciences USA* 109(37), 14726–14727, E12415–E12423. DOI: 10.1073/pnas.1205276109.

Hassan, A.N., Dister, S. and Beck, L. (1998) Spatial analysis of lymphatic filariasis distribution in the Nile Delta in relation to some environmental variables using geographic information system technology. *Journal of the Egyptian Society of Parasitology* 28, 119–131.

Held, M. and Soden, J. (2006) Robust responses of the hydrological cycle to global warming. *Journal of Climate* 19, 5686–5699.

Herczeg, T., Száz, D., Blahó, M., Barta, A., Gyurkovszky, M., Farkas, R. and Horváth, G. (2015) The effect of weather variables on the flight activity of horseflies (Diptera: Tabanidae) in the continental climate of Hungary. *Parasitology Research* 114(3), 1087–1097.

Hickling, R., Roy, D.B., Hill, J.K. and Thomas, C.D. (2005) A northward shift of range margins in British Odonata. *Global Change Biology* 11, 502–506.

Hickling, R., Roy, D.B., Hill, J.K., Fox, R. and Thomas, C.D. (2006) The distributions of a wide range of taxonomic groups are expanding polewards. *Global Change Biology* 12, 450–455.

Hijmans, R.J., Cameron, S.E., Parra, J.L., Jones, P.G. and Jarvis, A. (2005) Very high resolution interpolated climate surfaces for global land areas. *International Journal of Climatology* 25, 1965–1978.

Hogsette, J.A., Prichard, D.L. and Ruff, J.P. (1991) Economic effects of horn fly (Diptera: Muscidae) populations on beef cattle exposed to three pesticide treatment regimes. *Journal of Economic Entomology* 84, 1270–1274.

IPCC (2007) Summary for policymakers. In: Solomon, S., Qin, D., Manning, M., Chen, Z., Marquis, M., Averyt, K.B., Tignor, M. and Miller, H.L. (eds) *Climate Change 2007: The Physical Science Basis*. Contribution of Working Group I to the Fourth Assessment Report of the Intergovernmental Panel on Climate Change. Cambridge University Press, Cambridge, UK, pp. 1–18.

Jiggins, C.D. and Mallet, J. (2000) Bimodal hybrid zones and speciation. *Trends in Ecology and Evolution* 15, 250–255.

John, O., Allan, R.P. and Soden, J. (2009) How robust are observed and simulated precipitation responses to tropical ocean warming? *Geophysical Research Letters* 36, L14702.

Juliano, S.A., O'Meara, G.F., Morrill, J.R. and Cutwa, M.M. (2002) Desiccation and thermal tolerance of eggs and the coexistence of competing mosquitoes. *Oecologia* 130, 458–469.

Kamal, A.S. (1958) Comparative study of thirteen species of sarcosaprophagous Calliphoridae and Sarcophagidae (Diptera) I. Bionomics. *Annals of the Entomological Society of America* 51(3), 261–271. DOI: 10.1093/aesa/1051.1093.1261.

Karban, R. and Strauss, S.Y. (2004) Physiological tolerance, climate change, and a northward range shift in the spittlebug, *Philaenus spumarius*. *Ecological Entomology* 29, 251–254.

Keatley, M.R., Fletcher, T.D., Hudson, I.L. and Ades, P.K. (2002) Phenological studies in Australia: potential application in historical and future climate analysis. *International Journal of Climatology* 22, 1769–1780.

Kenward, A., Yawitz, D., Sanford, T. and Wang, R. (2014) *Summer in the City: Hot and Getting Hotter.* Climate Central, Princeton, New Jersey, USA. Available at: http://assets.climatecentral. org/pdfs/UrbanHeatIsland.pdf (accessed 15 May 2016).

Kenya Red Cross Society (2009) Kenya Drought Appeal 2009. Nairobi, Kenya. Available at: http://www.kenyaredcross.org/index.php?option= com_content&task=view&id=167&Itemid=117 (accessed 9 February 2015).

Lawler, J.J., Shafer, S.L., White, D., Kareiva, P., Maurer, E.P., Blaustein, A.R. and Bartlein, P. J. (2009) Projected climate-induced faunal change in the Western Hemisphere. *Ecology* 90, 588–597.

Linthicum, K.J., Bailey, C.L., Tucker, C.J., Mitchell, K.D., Logan, T.M., Davies, F.G., Kamau, C.W., Thande, P.C. and Wagateh, J.N. (1990) Application of polar-orbiting, meteorological satellite data to detect flooding of Rift Valley Fever virus vector mosquito habitats in Kenya. *Medical and Veterinary Entomology* 4, 433–438.

Loarie, S.R., Duffy, P.B., Hamilton, H., Asner, G.P., Field, C.B. and Ackerly, D.D. (2009) The velocity of climate change. *Nature* 462, 1052–1055.

Mabika, N., Masendu, R. and Mawera, G. (2014) An initial study of insect succession on decomposing rabbit carrions in Harare, Zimbabwe. *Asian Pacific Journal of Tropical Biomedicine* 4, 561–565. DOI: 10.12980/APJTB.4.2014C1031.

Macintyre, K., Keating, J., Sosler, S., Kibe, L., Mbogo, C.M., Githeko, A.K. and Beier, J.C. (2002) Examining the determinants of mosquito-avoidance practices in two Kenyan cities. *Malaria Journal* 15(1), 14.

Minakawa, N., Sonye, G., Dida, G.O., Futami, K. and Kaneko, S. (2008) Recent reduction in the water level of Lake Victoria has created more habitats for *Anopheles funestus. Malaria Journal* 7(1), 119.

Muntzer, A., Montagne, C., Ellse, L. and Wall, R. (2015) Temperature-dependent lipid metabolism in the blow fly *Lucilia sericata. Medical and Veterinary Entomology* 29, 305–313. DOI: 10.1111/mve.12111.

Muturi, E.J., Jacob, B.G., Kim, C.H., Mbogo, C.M. and Novak, R.J. (2008) Are coinfections of malaria and filariasis of any epidemiological significance? *Parasitology Research* 102, 175–181.

NCRWA (2015) In the aftermath of Floyd. North Carolina Riverkeepers and Waterkeeper Alliance. Available at: http://www.riverlaw.us/ hurricanefloyd.html (accessed July 2015).

Newton, L., Sheppard, C., Watson, D.W., Burtle, G. and Dove, R. (2005) *Using the Black Soldier Fly,* Hermetia illucens, *as a Value-added Tool for the Management of Swine Manure.* Waste Management Programs, North Carolina State University. Available at: http://www.cals. ncsu.edu/waste_mgt/smithfield_projects/ phase2report05/cd,web%20files/A2.pdf (accessed 14 July 2009).

Oke, T.R. (1997) Urban climates and global environmental change. In: Thompson, R.D. and Perry, A. (eds) *Applied Climatology: Principles and Practices.* Routledge, New York, USA, pp. 273–287.

Ording, G.J., Mercader, R.J., Aardema, M.L. and Scriber, J.M. (2010) Allochronic isolation and incipient hybrid speciation in tiger swallowtail butterflies. *Oecologia* 162, 523–531.

Ott, J. (2010) The big trek northwards: recent changes in the European dragonfly fauna. In: Settele, J., Penrev, L.L., Grabaum, T., Grobelink, V., Hammen, V., Klotz, S., Kotarac, M. and Kuhn, I. (eds) *Atlas of Biodiversity Risk.* Pensoft Publishers, Sofia, Bulgaria and Moscow, Russia, pp. 82–83.

Parmesan, C. and Yohe, G. (2003) A globally coherent fingerprint of climate change impacts across natural systems. *Nature* 421, 37–42.

Parmesan, C.N. (1996) Climate and species range. *Nature* 382, 765–766.

Parmesan, C.N., Ryrholm, C., Steganescu, C. and Hill, K. (1999) Poleward shifts in geographical ranges of butterfly species associated with regional warming. *Nature* 99, 579–583.

Peñuelas, J., Filella, I. and Comas, P. (2002) Changed plant and animal life cycles from 1952 to 2000 in the Mediterranean region. *Global Change Biology* 8, 531–544.

Pereira, R., Cooksey, J., Baldwin, R. and Koehler, P. (2014) Filth fly management in urban environments. In: Dhang, P. (ed.) *Urban Insect Pests: Sustainable Management Strategies.* CAB International, Wallingford, Oxfordshire, UK, pp. 43–64.

Phillips, S.J., Anderson, R.P. and Schapire, R.E. (2006) Maximum entropy modeling of species geographic distributions. *Ecological Modelling* 190, 231–259.

Powell, D. (2012) Aliens in Antarctica: visitors carry unwelcome species into a once pristine environment. *ScienceNews* May 5, 20–22.

Purse, B.V., Carpenter, S., Venter, G.J., Bellis, G. and Mullens, B.A. (2015) Bionomics of temperate and tropical *Culicoides* midges: knowledge gaps and consequences for transmission of *Culicoides*-borne viruses. *Annual Review of Entomology* 7(60), 373–392.

Puvabanditsin, S., Malikm, I., Weidner, L.M., Jadhav, S., Sanderman, J. and Mehta, R. (2014) Neonatal umbilical cord myiasis in New Jersey. *Journal of Perinatology* 34(9), 718–719.

Rhymer, J.M. and Simberloff, D. (1996) Extinction by hybridization and introgression. *Annual Review of Ecology and Systematics* 27, 83–109.

Roe, A. and Higley, L.G. (2015) Development modeling of *Lucilia sericata* (Diptera: Calliphoridae). *PeerJ* 3, e803. DOI: 10.7717/peerj.803.

Root, T.L., Price, J.T., Hall, K.R., Schneider, S.H., Rosenzweig, C. and Pounds, J.A. (2003) Fingerprints of global warming on wild animals and plants. *Nature* 421, 57–60.

Royle, J.A., Chandler, R.B., Yackulic, C. and Nichols, J.D. (2012) Likelihood analysis of species occurrence probability from presence-only data for modelling species distributions. *Methods in Ecology and Evolution* 3, 545–554.

Sallam, M.F., Ahmed, A.M.A., Abdel-Dayem, M.S. and Abdullah, M.A.R. (2013) Ecological niche modeling and land cover risk areas for Rift Valley Fever vector, *Culex tritaeniorhynchus* Giles in Jazan, Saudi Arabia. *PLoS ONE* 8(6), e65786. DOI: 10.1371/journal.pone.0065786.

Schwarz, D., Matta, B.M., Shakir-Botteri, N.L. and McPheron, B.A. (2005) Host shift to an invasive plant triggers rapid animal hybrid speciation. *Nature* 436, 546–549.

Settele, J., Kudrna, O., Harpke, A., Kuehn, I., van Swaay, C., Verovnik, R., Warren, M., Wiemers, M., Hanspach, J., Hickler, T., Kuehn, E., van Halder, I., Veling, K., Vliegenthart, A., Wynhoff, I. and Schweiger, O. (2008) Climatic risk atlas of European butterflies. *Biorisk* 1, 1–710.

Sheppard, D.C., Hinkle, N.C., Hunter, J.S. and Gaydon, D.G. (1990) Resistance in constant exposure livestock insect control systems: a partial review with some original findings on cyromazine resistance in house flies. *Florida Entomologist* 72, 360–369.

Sheppard, D.C., Tomberlin, J.K., Joyce, J.A., Kiser, B.C. and Sumner, S.M. (2002) Rearing methods for the black soldier fly (Diptera: Stratiomyidae). *Journal of Medical Entomology* 39, 695–698.

Shiravi, A.H., Mostafavi, R., Akbarzadeh, K. and Oshaghi, M.A. (2011) Temperature requirements of some common forensically important blow and flesh flies (Diptera) under laboratory conditions. *Iranian Journal of Arthropod-borne Diseases* 5(1), 54–62.

Sithiprasasna, R., Linthicum, K.J., Liu, G.J., Jones, J.W. and Singhasivanon, P. (2003) Use of GIS-based spatial modeling approach to characterize the spatial patterns of malaria mosquito vector breeding habitats in northwestern Thailand. *Southeast Asian Journal of Tropical Medicine and Public Health* 34, 517–528.

Sithiprasasna, R., Lee, W.J., Ugsang, D.M. and Linthicum, K.J. (2005) Identification and characterization of larval and adult anopheline mosquito habitats in the Republic of Korea: potential use of remotely sensed data to estimate mosquito distributions. *International Journal of Health Geographics* 4, 17.

UK Met Office (2011) Climate: observations, projections and impacts. Meteorological Office. Available at: http://www.metoffice.gov.uk/media/pdf/5/o/France.pdf.

Wellenreuther, M., Tynkkynen, K. and Svensson, E.I. (2010) Simulating range expansion: male species recognition and loss of premating isolation in damselflies. *Evolution* 64, 242–252.

Wilson, R.J. and Maclean, I.M.D. (2011) Recent evidence for the climate change threat to Lepidoptera and other insects. *Journal of Insect Conservation* 15, 259–268.

Woodward, F.I. (1987) Stomatal numbers are sensitive to increases in CO_2 from pre-industrial levels. *Nature* 327, 617–618.

Yüca, K., Caksen, H. and Sakin, Y.F. (2005) Aural myiasis in children and literature review. *Tohoku Journal of Experimental Medicine* 206, 125–130.

8 Impact of Climate Change on Medically Important Ticks in Europe and Their Control

Aleksandra Gliniewicz[1]*, Grzegorz Karbowiak[2], Ewa Mikulak[1], Marta Supergan-Marwicz[3], Agnieszka Królasik[1] and Joanna Myślewicz[4]

[1]National Institute of Public Health – National Institute of Hygiene, Warsaw, Poland; [2]W. Stefański Institute of Parasitology of Polish Academy of Sciences, Warsaw, Poland; [3]Department of General Biology and Parasitology, Medical University of Warsaw, Warsaw, Poland; [4]Warsaw University of Life Sciences, Warsaw, Poland

8.1 Introduction

Ticks are one of the most dangerous blood-sucking arthropods, attacking humans and their pets. There are about 900 species of ticks in the world (Siuda, 1993; Anderson and Magnarelli, 2008; Nowak-Chmura and Siuda, 2012), of which 44 are present in Europe. Almost half of them are of epidemiological and veterinary importance.

As ticks are external parasites and directly dependent on environmental influence, they react to climatic changes with changes in their distribution and life cycle. The most conspicuous example of such dependency among European species is *Dermacentor reticulatus* (Bullová *et al.*, 2009, Karbowiak and Kiewra, 2010; Akimov and Nebogatkin, 2011; Karbowiak, 2014), compared to other species such as *Ixodes ricinus* (Akimov and Nebogatkin 1996; Medlock *et al.*, 2013) and *Haemaphysalis punctata* (Akimov and Nebogatkin, 2012). Expansion into new areas by *Rhipicephalus sanguineus* is yet to be linked to climate change. Other species of ticks, like those from tropical countries, are also spreading in Europe, but as a result of human activity, not by the influence of climatic factors.

8.2 Occurrence of Economically Important Ticks in Europe

Ticks occupy diverse ecological zones as their preferred habitat. *Dermacentor reticulatus* inhabits moderately humid open areas or areas sparsely covered with trees. It prefers natural deciduous forests with watercourses or large reservoirs of standing water. *I. ricinus* inhabits moderately humid and shaded environments of deciduous and mixed forests with rich understories and concentrations of bushy vegetation. *H. punctata* inhabits diverse environments, such as bushes and sparse forests, forests clearings, steppes and pastures. It avoids humid forests of temperate zones (Siuda, 1993; Estrada-Peña *et al.*, 2004; Nowak-Chmura, 2013). *R. sanguineus* occurs in areas with low humidity, on steppes, forest steppes and in synanthropic environments. In addition, it might also be temporarily imported to urban

*E-mail: agliniewicz@pzh.gov.pl

environments (Siuda, 1993; Estrada-Peña *et al.*, 2004).

8.3 Tick Distribution in Europe

Dermacentor reticulatus appears in temperate climate zones throughout Eurasia, from the Atlantic Coast to the Yenisey River in Siberia. The northern border of its range runs along 56–57°N and the southern along 50°N (Siuda, 1995). According to Immler (1973), territorial scope of *D. reticulatus* is designated by the area with spring rainfall level ranging from 400 to 1000 mm and 20–22°C summer isotherms. This area is divided into two parts – the so-called western European region and the eastern region (Immler, 1973; Siuda, 1995; Siuda *et al.*, 1997). The western part encompasses populations from France (Martinod and Gilot, 1991) to the eastern part of Germany (Mačička *et al.*, 1956).

Isolated outbreaks are also observed. In the eastern area, *D. reticulatus* occurs mostly from the eastern part of Poland, through Belarus and the European part of Russia to Siberia. The northern border runs in Russia from Moscow, through Omsk to Kemerovo oblast (Ravdonikas *et al.*, 1968; Chigirik and Pleshivtseva-Eroshkina, 1969) and the southern includes the Crimean Mountains, the North Caucasus, eastern Kazakhstan and the mountains of western Altai (Mačička *et al.*, 1956). The area in Europe where *D. reticulatus* does not occur can be described as a triangle whose northern edge runs along the Baltic coast and the sides extend from central Germany, from 12–13°E to 19°E in Poland. In the south, the sides of the triangle converge, reaching to the southern border of Hungary (Karbowiak *et al.*, 2008; Karbowiak and Kiewra, 2010). The area of *D. reticulatus* incidence has not changed since first described. Since 1990 there has been a tendency of expansion into previously unoccupied areas. For example, the emergence of new *D. reticulatus* populations in eastern Germany is accompanied by new populations in western Poland in Lower Silesia and Lubuskie province (Cuber *et al.*, 2013; Kiewra *et al.*, 2015). New locations emerged in the UK, to the north of those previously known (Wall, 2012). There has also been a shift of the northern border of its range to the north to Latvia, where new populations were found (Paulauskas *et al.*, 2015). The expansion of *D. reticulatus* to the south was observed in Hungary (Farkas and Földvári, 2001) and in Ukraine (Akimov and Nebogatkin, 2011).

Over the last decade, new locations have emerged rapidly in Slovakia and the Czech Republic (Peťko *et al.*, 2008; Bullová *et al.*, 2009). Apart from changes in the area of occurrence, there has been an increase in existing tick populations (Široký *et al.*, 2011). Possible hypotheses relate to environmental conditions, climate change and human activities. There are no geographical barriers between natural forest and meadow ecosystems of the Palaearctic. The probable cause seems to be anthropogenic pressure, such as through intensive drainage and elimination of small watercourses and water reservoirs, as well as reduction in the range of occurrence of hosts (Daniel *et al.*, 1986). Since reports on the occurrence of ticks date back only from the mid-19th century (Koch, 1844), it is impossible to say when *D. reticulatus* disappeared from areas of western Germany, central Poland and Slovakia, and what factors might have been involved. The transition zone between the oceanic and continental climate stretches through central Poland. It is estimated that 52% of species found in Poland are transitional, meaning they occur both to the east and west of Polish borders. Other species on Polish territory start or end their range in this region (Matuszkiewicz, 2006) (Fig. 8.1).

The above reasons do not fully explain the gap in occurrence of *D. reticulatus*, as it ends its eastern range in Poland and reappears in the west. The same area of occurrence does not apply to the equally common tick, *Ixodes ricinus*, whose range of occurrence is uninterrupted. *I. ricinus* is found in woodlands or areas overgrown by dense bushes, whereas *D. reticulatus* prefers open areas. Open areas are characterized by greater thermal and humidity fluctuations than woodlands, where the impact of changing weather conditions is reduced by the stand and microclimate of dense forests. The surface

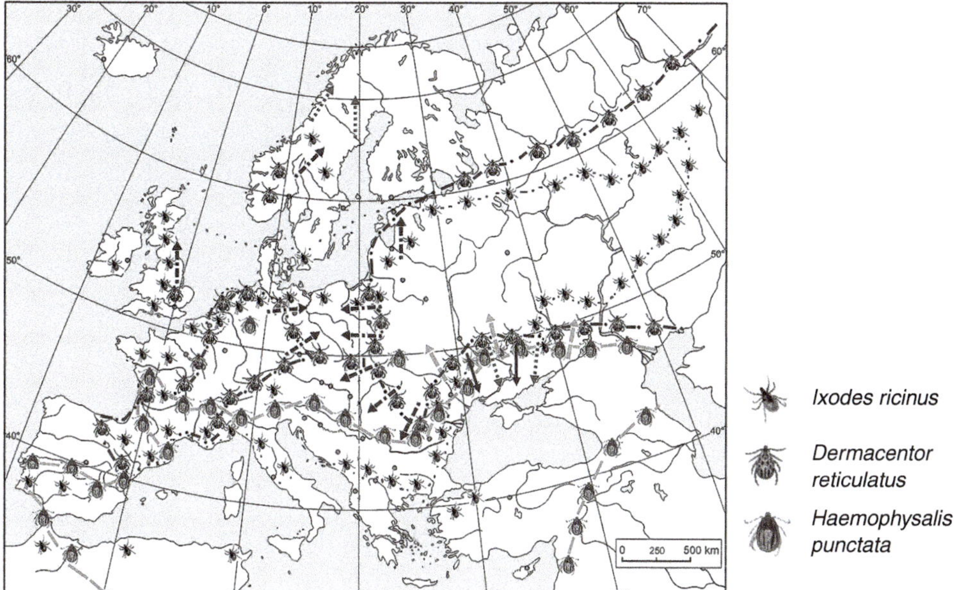

Fig. 8.1. Changes in distribution of *Ixodes ricinus*, *Dermacentor reticulatus* and *Haemophysalis punctata* ticks in Europe in the last two decades.

temperatures of soil is a factor limiting egg laying and development of eggs, which, together with required humidity levels, limits the activity period for *D. reticulatus* to 1 or 2 months a year (Gray *et al.*, 2009). In addition, juvenile forms of *D. reticulatus* are more sensitive to adverse environmental factors than *I. ricinus*, which is more flexible, helped by a broader host range and can live in open areas, while forest is often a boundary for *D. reticulatus*. This biological feature of *D. reticulatus* indicates a link between their absence in a location and local climate. Ticks spend winter under forest cover or in other secluded places in top soil, which protect wintering organisms from freezing.

The geographical range of *Ixodes ricinus* covers virtually the whole of Europe, except for Arctic regions. The northern end of its range reaches 69° N (Hvidsten *et al.*, 2015). The eastern border range runs through south-western Turkmenistan and northern Iran; the southern border is marked by the steppe zone, and simple populations are found in countries covering north-western Africa (Siuda, 1993; Estrada-Peña *et al.*, 2004; Filippova, 1977). In the mountains, the species is rare; its maximum altitude of occurrence is 700–1100 m above sea level, depending on local climatic conditions (Lindgren *et al.*, 2000; Gray *et al.*, 2009; Nowak-Chmura, 2013). *I. ricinus* is found almost everywhere in Europe; therefore, its expansion to new areas is not as spectacular as that of *D. reticulatus*. However, it has been observed in Scandinavia to the north (Lindgren *et al.*, 2000; Jaenson *et al.*, 2012); in the Ural Mountains to the east; and southern areas of western Ukraine and Crimea (Akimov and Nebogatkin, 1996). Vertical expansion was observed in Bosnia and Herzegovina, the Czech Republic and Slovakia. In the Czech Republic, the species is now observed even at altitudes above 1100 m above sea level. As in the case of *D. reticulatus*, the tick has become more common in certain areas where it had previously appeared only occasionally, as in the UK and Norway (Jaenson *et al.*, 2012; Wall, 2012).

Haemaphysalis punctata is found in the whole of Western Europe, from France to Germany and southern Sweden. The southern range includes southern Mediterranean countries; isolated populations are found in countries of North Africa and Asia Minor (Filippova, 1997; Estrada-Peña *et al.*, 2004).

The presence of the species is focal, which is why it appears to be very rare, especially in Eastern Europe. For this reason, it is difficult to ascertain its expansion into new areas; however, there was a new occurrence point found in northern Ukraine (Akimov and Nebogatkin, 2012) (Fig. 8.1).

Rhipicephalus sanguineus is associated with domestic dogs, and this host is a reason for their spread throughout the world. Africa, Mediterranean countries, the Black Sea shore and Asia Minor are considered its homeland (Nowak-Chmura, 2013). The current range is restricted to a zone of warmer subtropical climates, as it is most often found between 50°N and 35°S (Estrada-Peña *et al.*, 2004). Sometimes, however, it is imported with dogs to countries of temperate climate. With the development of tourism and trade, it is more often found in countries of Central and Northern Europe; and lately in the UK (Wall, 2012). It is still unclear whether ticks found on dogs came from a regional population or from other geographical areas. There are also no documented collections of the ticks and their juvenile form from the environment. Regardless of the potential expansion of *R. sanguineus*, there have been reports from Romania and Ukraine about the possible expansion of Russian *Rhipicephalus rossicus*, which in some places has become more numerous than *R. sanguineus* (Akimov and Nebogatkin, 2013).

8.4 Reasons for Changes in Tick Distribution Range

The reasons why ticks wander to far uninhabited areas are not fully explained. It could be through climate, by change in temperature, wind direction, rainfall frequency or socioeconomic factors related to human activities, including migration, travel and commerce (Ebert and Fleischer, 2005). Symptoms of climate change are now observed in central Europe, such as shifting of the continental climate zone to the east, a rise in average winter temperatures with a decrease in the number of frosty days, early appearance of the snow cover with shorter deposition time in lowland areas, a decrease in average temperatures during the summer with an increase in the number of hot days (Cebulak and Limanówka, 2007; Falarz, 2007). In recent years, average temperatures in Northern Europe have increased by 1.5–2.5°C. There has also been a decrease in the number of frosty days with temperatures below −12°C. These changes may affect the life cycle of ticks and result in an increase or decrease in tick activity, which influences the density and distribution of their population (Gray *et al.*, 2009). The changes may foster the development of juvenile forms, the wintering of adult forms, and extend the period of activity of host-seeking forms, and thus help their spread to areas where conditions were previously unfavourable (Lindgren *et al.*, 2000; Pet'ko *et al.*, 2008). The role of changes in snow-cover retention time and its thickness as a factor that could affect the distribution of *D. reticulatus* is indirectly indicated by observations of changes occurring in thickness of snow cover in Europe over the past 50 years. In the second half of the 20th century, there was a decline in the number of snowy days and a decrease in thickness of snow cover and its retention time (Bednorz, 2007). In northern Germany and western Poland, the area now occupied by *D. reticulatus*, there are times of entirely snowless winters, without days of frost. Areas potentially unfavourable for *D. reticulatus* can now be inhabited by this species. On the other hand, in foothill areas and in north-eastern Poland, there is a tendency towards increasing thickness of snow cover, with a decline in the number of frosty days (Falarz, 2004). These changes might also explain the vertical migration of *I. ricinus* into the mountains, observed in the Carpathian range in the Czech Republic and Slovakia.

The increase in the number of ticks is also influenced by human activities, mainly agriculture. One of the factors may be the change in the structure of agriculture. After 1990, large-scale, state-owned farms were liquidated and transformed into small ones. A portion of agricultural area thus turned into fallow land. Currently, more than 60% of farms are less than 5 ha, of which the majority are fragmented into small plots, separated by fallow land and brushwood bands (Kulikowski, 2006). The latter factor has a beneficial effect on the migration of wild animals, along with tick species, especially *D. reticulatus* and *I. ricinus*. These have been observed in Lithuania and Latvia (Paulauskas *et al.*, 2015), where deliberate conversion of large farmlands into meadows and woodland has led to an increase in protected areas to restore populations of large mammals and birds. By recreating such type of phytocenosis, humans have created suitable living conditions for ticks (Dautel *et al.*, 2006; Gray *et al.*, 2009). Moreover, the newly created populations of large mammals are not only the source of food for ticks, but also serve as a possible source of migration and colonization of new areas. Juvenile mammals usually leave their birth places and in the autumn migrate in search of new areas, thus helping to transfer parasites attached to their bodies. This especially applies to the winter months, when large numbers of *D. reticulatus* and female *I. ricinus* live on elk, bison and possibly other mammals over a period of several months (Belikova, 1956; Karbowiak *et al.*, 2003).

Factors contributing to the occurrence of small mammals, which are hosts to juvenile forms, also play a vital role in tick density and distribution. For example, pruning, thinning forests and making clearings promotes growth of rodent populations and the ticks associated with them, such as *Ixodes* and *Dermacentor* species (Kolonin, 1981). Another factor promoting ticks is the creation of green corridors through development and urbanization, which occurs in many cities. Bands of trees, undeveloped riverbeds and extensive parks with varied vegetation serve as a migration route for wild animals, travelling with ticks attached to their bodies. Compact inner-city stands create a local microclimate, characterized by lower temperatures and higher humidity than built-up areas (Szumacher, 2005), which allows some species permanent existence within cities. Tourism, creating green spaces, development of transportation, intensive exchange of goods between distant countries and spreading ticks to new areas, are classic examples of the impact of human activities on the distribution of parasites. Tourism is known as one of the factors primarily responsible for spreading *R. sanguineus* to countries outside its original range of existence. Cases of spreads originating from the Mediterranean and the countries of the tropical zone into central Europe are regularly noted (Nowak-Chmura, 2013; Siuda, 1993). *D. reticulatus* can also inhabit dogs travelling with their owners, and large animals transported for commercial or farming reasons. Cases of ticks spreading to Central Europe from remote areas have been repeatedly documented (Glaser and Gothe, 1998). Most likely, the expansion in the range of ticks is caused by a synchronous set of favourable factors that includes global changes, such as global warming, landscape and economic changes (area management and reduction of usage of pesticides and other chemicals), an increase in the number of wild animals as a result of their protection, and importation of ticks to new areas in connection with tourism and development by means of transportation (Nowak, 2010).

8.5 Tick Behaviour and Biology Influenced by Climate Change

Warming climate probably explains some recent changes in behaviour of *I. ricinus* ticks. Generally, these ticks are not active all year round. Each stage of their development has peaks at certain times of the year. Female ticks and nymphs that survive the winter are common in early spring. Females lay eggs after feeding and hatched larvae feed during the summer. In autumn, adults and nymphs dominate. For about 2 years the authors have observed changes in the annual breeding cycle of the tick: adult females appear earlier, with the disappearance of snow, and they also lay their eggs earlier, so that the larvae are active by May. These stages are also observed at the end of September, which suggests that the species lays eggs multiple times over the year (authors' observations). The authors postulate that mild winters

Table 8.1. Important pathogens transmitted by ticks in Europe.

Pathogen	Disease	Vector(s)	Area found
TBE-virus (*Flaviviridae)*	Tick-borne encephalitis (TBE)	*Ixodes ricinus*; *Ixodes persulcatus*	Whole of Europe
CCHF-virus (*Nairovirus)*	Crimean-Congo-haemorrhagic fever (CCHF)	*Hyalomma marginatum*; other *Hyalomma species*; *Ixodes ricinus*; *Dermacentor marginatus*; *Haemaphysalis punctata*	Albania; Bulgaria; France; Greece; Hungary; Kosovo; Macedonia; Moldova; Portugal; Russia; Turkey; Ukraine
Bhanja virus (*Bunyaviridae)*	Bhanja virus	*Haemaphysalis* spp.; *Dermacentor* spp; *Hyalomma* spp.; *Amblyomma* spp; *Rhipicephalus* spp.	Southern Europe; Balkans; Romania; Czech Republic; Portugal; Spain
Eyach virus (*Coltivirus)*	Eyach virus	*Ixodes ricinus*; *Ixodes ventalloi*	Germany; France; Netherlands; Czech Republic
Borrelia burgdorferi complex: *B. afzelii*, *B. burgdorferi* s.str.; *B. garinii*; *B. lusitaniae*; *B. valaisiana* *B. spielmanii*; *B. bisettii*	Lyme borreliosis (LB)	*Ixodes ricinus*; *I.* spp.; secondary vectors: *Dermacentor marginatus*; *Haemaphysalis punctata*; *Rhipicephalus sanguineus*	Whole of Europe
Borrelia spp. other than *Borrelia burgdorferi* complex	Tick-borne relapsing fever	Soft ticks (Argasidae): *Ornithodoros* spp. *Ixodes ricinus*	Greece; Cyprus; Portugal; Sweden; Germany
Rikettsia conorii	Mediterranean spotted fever (MSF)	*Rhipicephalus sanguineus*	Southern Europe: Bosnia-Herzegovina; Croatia; Greece; France; Portugal; Slovenia; Spain; Turkey. Russia; Italy
Rickettsia helvetica	Spotted fever; cardiologic problems	*Ixodes ricinus*	Switzerland; Sweden; France; Italy
Rickettsia slovaca	Febrile illness	*Dermacentor* spp.; *Dermacentor marginatus*	Armenia; Austria; Czech Republic; France; Germany; Hungary; Italy; Lithuania; Portugal;Russia; Slovakia; Spain; Switzerland
Anaplasma (*Ehrlichia) phagocytophilum*	Human granulocytic anaplasmosis (HGA)	*Ixodes ricinus*	Belgium; Bulgaria; Czech Rep; Denmark; Germany; Greece; Ireland; Italy; Netherlands; Norway; Poland; Portugal; Russia; Slovenia; Spain; Sweden; Switzerland; UK
Babesia spp.	Human babesiosis	*Ixodes* spp.; *Dermacentor* spp.; *Rhipicephalus* spp.	Ireland; France; Poland; Portugal; Russia; Switzerland; UK

probably contribute not only to the increased survival rate of ticks and their occurrence soon after daily temperatures rise above 0°C and snow cover recedes, but also to continued winter activity, which was previously unobserved. Such change in the behaviour of *D. reticulatus* was first observed in 1994, when feeding ticks were found during studies of bison (*Bison bonasus*) in Poland's Białowieża Forest during the winter. The ticks occur during other months, but rarely (Izdebska, 1998). Ticks feeding in the winter were also collected from dogs and elk (Karbowiak *et al.*, 2003). Since 2000, active adult stages have also been observed on vegetation in January and February, when for a few days there is no snow and the temperature remains above zero (Karbowiak and Szewczyk, 2014). No changes were observed in the daily activity cycle and the scope of hosts that could be linked to climate change.

8.6 Effect of Climate Change on Tick-borne Diseases

Ticks are carriers of more than 60 viruses, of which 20 are etiologic agents of human diseases and diseases of domestic animals, 11 are bacterial pathogens, including five rickettsia of spotted fever group, at least seven species of spirochetes of *Borrelia burgdorferi* sensu lato, human granulocytic anaplasmosis and human babesiosis (Gratz, 2006). Selected pathogens and diseases caused by them, as well as the vectors, are presented in Table 8.1.

The interest in the spread of tick-borne diseases in relation to climate changes has been observed since the 1980s; the number of cases of Lyme disease has gradually grown in European countries. In addition, the number of cases of tick-borne encephalitis (TBE) has also apparently increased over the past 20 years. In many European countries, cases of these two diseases have been registered for a long time, so the analysis of the impact of climate and socioeconomic factors on their prevalence is based on a number of interrelating data. In the case of other tick-borne diseases, clear records have not been kept, which is coming to light due to newly developed diagnostics. The available data are, therefore, sufficient for the analysis of prevalence of these diseases, but nothing more.

8.7 Climate Change and Lyme Disease in Europe

Ticks *Ixodes ricinus* infected with *Borrelia* spp. spirochetes are found throughout Europe. Research carried out in different regions (Gassner and Overbeek, 2007) shows that the degree of infection is very diverse. It was stated that small animals such as birds and rodents play a particularly important role in transmission of *Borrelia* spp. between vector and host. Local environmental factors and their changes may have a significant impact on populations of these animals. Changes in land cultivation, vegetation, and other human interventions can affect fluctuations in the number of reservoir organisms and, indirectly, on inhabiting them with ticks infected with *Borrelia* spp. Therefore, it is assumed that local factors cause differences in the degree of infection in different areas and create the so-called 'hot spots' or foci – places where the number of infected ticks is higher (Gassner and Overbeek, 2007). Of 33 known species of *Borrelia* spp., seven are found in Europe. These are *Borrelia afzelii, B. burgdorferi* s.s., *B. garinii, B. valaisiana, B. lusitaniae, B. spielmanii* and *B. bisettii*.

It is possible that in the future other strains may be discovered, or more virulent strains of existing species may appear through a process of horizontal gene transfer (Kurtenbach *et al.*, 2006). Species of *B. afzelii, B. burgdorferi* s.s. and *B. garinii* are primarily responsible for symptoms of human Lyme disease (Stanek and Strle, 2003). Previous reports show that certain species of *Borrelia* are associated with specific hosts. For example *B. afzelii, B. burgdorferi* s.s., *B. garinii* Tso-A 4 are associated with small rodents; *B. valaisiana* and *B. garinii* Tso-A genotypes other than Tso-A 4 are more common for birds (Kurtenbach *et al.*,

2002). In Europe, nine species of small rodents and seven medium-size mammals are also known for transmitting *Borrelia* spp. to *Ixodes ricinus* (Gern *et al.*, 1998).

The presence of small rodents, birds and mammals in an environment close to humans makes human contact with infected ticks easier. Local environmental conditions have a major impact on the survival and development cycle of ticks. However, an increasing number of ticks in the environment is not necessarily associated with a higher degree of infection, although in many instances that is the case. In European countries, there was a significant difference in the degree of infection by *Borrelia* spp. to populations of *Ixodes ricinus* from different areas. The level of infection ranged from 1.5% in Slovenia to 49.0% in Slovakia and Switzerland (Rauter and Hartung, 2005). Analysis conducted by the authors shows that *I. ricinus* infected with *Borrelia* spp. are present almost all over Europe. The dominant species proved to be *B. garinii* and *B. afzelii*, although locally in some areas other species of spirochetes dominated (e.g. *B. valaisiana* in Germany; Kampen, 2004). In Poland, the level of infected ticks varies from 2.0% to 35.0% and differs between regions (Nowak-Chmura, 2013). The differences may be significant in the case of areas in and around a single city. For example, in Warsaw and its surroundings, the prevalence of ticks infected with *Borrelia* spp. ranges from 10.7% to 20.0% (Hajdul *et al.*, 2006). Similar data of differences in tick infection come from the Netherlands, where the prevalence varies from 5% to 35% (Gassner and Overbeek, 2007).

In some European countries, it was found that infected ticks occur in urban areas or in their vicinity, in recreational areas (Gassner and Overbeek, 2007; Supergan and Karbowiak, 2009). That is why people are exposed more frequently. This may be due to expansion of people into the green areas near cities, already inhabited by infected ticks. On the other hand, creating parks and gardens from scratch means that these areas are gradually being colonized by birds and rodents bringing parasitic ticks to their new environment. Hubalek *et al.*

(2003) and Stafford *et al.* (1998) found a positive correlation between the increase in the number of ticks and the increase of cases of Lyme disease.

The impact of climate change on the increasing number of cases of Lyme disease in Europe is important; however, it is not the only reason. In southern Sweden, a link between milder winters, humidity, warm summers and increased number of Lyme disease cases was also found (Bennett *et al.*, 2006). This may be due to the fact that during warm summers, people prefer to stay outside, therefore their exposure to ticks increases. According to these authors, there will be an increase in the number of cases of the disease resulting from climate change in Europe.

8.8 Climate Change in Europe and Tick-borne Encephalitis Cases

Unlike Lyme disease, tick-borne encephalitis cases in Europe have been regularly recorded in the last 30–50 years. Since 1990, an upward trend has been observed in many countries. The exception is Austria, where vaccination coverage of around 90% of the population has led to a significant decline in prevalence since the 1990s (Kunz, 2003). It is also believed that the decline in the number of cases listed in Hungary and Croatia since 1995 is the result of changes in health care and the reduced number of laboratory tests (Randolph and Šumilo, 2007). And a significant increase in prevalence of TBE in Lithuania, Latvia, Estonia, Poland and Slovakia since the beginning of the 1990s correlates with the fall of the Soviet Union and subsequent political and sociological changes in this part of Europe.

The first risk map of vector-borne disease was created for TBE by Randolph *et al.* in 2000. Analysis of available data concerning diseases and climate showed a positive correlation between the number of cases, humidity in the summer and the increase in average temperature. In 1993 in the Baltic countries an upsurge in the number of TBE cases was noted in comparison with 1991 and 1992: 2.6-fold in Estonia; 4.5-fold in Latvia; 13.8-fold in Lithuania. This increase continued over the next 5 years and was not linked to changes in the number of *I. ricinus* and *I. persulcatus* (Šumilo *et al.*, 2007). The changes were also too substantial to be explained only by problems in the health care system in the transitional period following political change.

Humidity and temperature affect the increase in the number of ticks, including those infected with TBE. In Sweden, a relationship between the incidence of TBE and coexistence of mild winters, early springs and long autumns was observed and, it was suggested, perhaps caused a shift in the area of occurrence of *I. ricinus* further to the north (Gray *et al.*, 2009). The increase in average temperature in April, at the end of a tick's diapause and the beginning of feeding, is of particular importance. This phenomenon is important for enhancing transmission between co-feeding ticks. Similarly, Randolph *et al.* (2000) found a positive correlation between the activity of larvae and nymphs of *Ixodes ricinus*, cases of TBE and the increase in autumn temperatures. Climate change alone is not sufficient to explain observed fluctuations in the incidence of TBE in Europe, especially in Eastern Europe. Observed climate change coexists with changes in land cultivation and socioeconomic changes. After joining the European Union, Eastern European countries reduced the acreage of cultivated land for meadows, pastures and wastelands overgrown with shrub. These areas are inhabited by rodents (TBE virus transmission hosts), which are parasitized by tick larvae and nymphs.

Randolph and Šumilo (2007) drew attention to the consequences of increased recreation and use of personal transport such as cars in countries of Eastern Europe and the Baltic States. Car owners often go out of cities to green areas, where they are more exposed to contact with infected ticks. People collecting berries and mushrooms (often non-working and on a low-income) frequently go to forests and are more likely to encounter infected ticks. European analysis shows that there are countries where the incidence of TBE is decreasing. This happens

as a result of social policy of the state, which reimburses the costs of vaccinations. In the countries where patients have to cover the cost of vaccination, there has been no decrease in prevalence of disease, despite availability of a vaccine (Randolph and Šumilo, 2007).

8.9 Other Tick-borne Diseases

The increase in density of rodents caused by mild and warm winters influences the number of tick species (sometimes enabling them to survive in areas where their presence had not so far been recorded), which serve as vectors of diseases, such as babesiosis in humans and animals, and anaplasmosis. Currently, cases of the disease caused by *Anaplasma phagocytophilum*, transmitted by *Ixodes* species, are occasionally recorded, but this situation may change (Gratz, 2006). How the effect of climate change influences the increase in the range of the tick population, and consequent transmission of tick-borne diseases, can be further understood by the emergence of mathematical models based on meteorological data. The FleaTick-Risk database (http://www.fleatickrisk.com) predicts appearance of ticks in different parts of the world based on data collected over several years.

8.10 Management of the Tick Population

Control of ticks in the natural environment is extremely difficult. From the 1960s to 1990s in the United States, Canada and the Soviet Union (Russia), acaricides were used in large areas at high risks of tick-borne disease. Schulze *et al.* (2001) in their study demonstrated that one-time application of granular deltamethrin, applied along the edge of the lawn-forest in forest areas and on plots resulted in 95% reduction of *I. scapularis* nymphs and 100% reduction of *Amblyomma americanum* ticks. Similar efficacy was achieved using cypermethrin (Solberg *et al.*, 1992). Studies have shown that ticks are sensitive to all types of applied biocides (Ostfeld *et al.*, 2006), but these usually inhibit tick activity for only a single season. Furthermore, introduction of acaricides into the environment often affects a host of non-target organisms, causing reduction or elimination of other groups of arthropods, with a further possibility that ticks will develop resistance against the insecticides; the chemicals also have a devastating impact on small mammals (Schauber *et al.*, 1997). There is no selective biocidal agent which would eliminate only ticks, hence, a search for alternative methods to reduce ticks is in progress. Development of other (non-chemical) methods of controlling and reducing the number of ticks is an opportunity to reduce the spread of pathogens dangerous for humans and animals that are transmitted by ticks and arachnids. This is the reason that the use of non-specific acaricide with adverse effects on the environment was not included in the overall tick-control strategy in natural ecosystems in Sweden (Tälleklint and Jaenson, 1995). Another way to control ticks and tick-borne diseases is to reduce the number of ticks that live directly on hosts or in their habitat (e.g. rodent holes) using insecticides or acaricides. Tubes filled with permethrin or fipronil swabs can be used; the swabs are generally dragged by rodents into their burrows and nests and used as building material. Research conducted in the USA on the effectiveness of the system of traps in the area inhabited by *Peromyscus leucopus* (white-footed mouse), which is the main host for *I. scapularis*, showed a reduction in the tick population. In a study of traps carried out in Sweden, where *I. ricinus* lives on the bank vole, *Clethrionomys glareolus*, the reduction of the population was not so significant (Tälleklint and Jaenson, 1995). A satisfactory reduction in the population of ticks living off deer was obtained by using forest animal feeders impregnated with permethrin: the animals put their heads into the feeder to collect food and, through rubbing, distribute the acaricide on their neck and ears (Solberg *et al.*, 2003).

Another way to manage the tick population is through modification of their habitats, such as large forest areas, fallow lands, water bodies, as well as areas frequented by humans, such as recreational grounds within the borders of cities. These strategies rely heavily on cleaning-up work, including removing leaves and humus (which could potentially contain eggs, larvae, nymphs and adult ticks) and burning them. It has been proved that fencing and mowing are fast and effective ways to reduce the number of ticks in areas up to 1 ha (Del Fabbro, 2014). Limiting the populations of potential hosts, e.g. rodents, deer and wild boar, is also a way to exclude blood-sucking arachnids from the environment. A study by Gilbert *et al.* (2012) conducted in Scotland has shown that limiting the population of deer in the area, as well as fencing heaths and creating smaller plots available for animals, contributes to a reduction in the population of *I. ricinus* living off them. However, killing forest animals like wild boar, deer and hares solely in order to reduce the population of ticks cannot be used as the main or only way to reduce the number of these arachnids. According to the research literature, there are several species of predators, parasites and pathogenic fungi, bacteria or viruses that attack and kill ticks.

Today, however, the use of biological methods is rather a matter of research. Considering the negative effects of chemical biocides for tick control on the environment, it is important to use personal protection, such as impregnated clothes and tick repellents. Obeying a few rules can limit the number of tick attacks. The first is to avoid areas where these arthropods occur, especially during their most active periods. In such places, avoid rubbing on vegetation, going into the bushes and wild paths and resting on the grass, logs or in places with lush vegetation (Lane, 2012; Nowak-Chmura, 2013).

The use of proper clothing, such as long pants with bottoms closed with a rubber band or tucked in socks, long-sleeved blouses/shirts, and closed footwear is another way of avoiding tick attacks and is a basic form of prevention. Also the colour of clothing is important – it is easier to see a moving tick on bright fabrics (Buczek *et al.*, 2000; Nowak-Chmura, 2013). After returning home, it is important to take off and examine clothes to search for travelling ticks and then thoroughly check the entire body to search for nymphs and larvae that are often difficult to locate (Nowak-Chmura, 2013). A survey conducted in Warsaw in 2012–2013 by Gliniewicz *et al.* (2014) showed that out of 217 people surveyed, 49% used tick repellents. Responding to questions about clothing protecting against tick invasion (long sleeved-shirts, long pants, socks and closed shoes), approximately 40% stated that they put on such clothes, others (approximately 60%) did not wear appropriate clothing when walking through a forest or park. After returning from wooded areas or parks, 59% do not examine their clothes, while 41% of respondents do not check their bodies. Forty-four per cent of people questioned replied that they found ticks on their bodies, most often on the legs and torso. The question of how to remove a tick from the skin properly was answered incorrectly by 47.9% of respondents. The results of the survey show that knowledge about tick-borne diseases is limited and perhaps that is why simple preventative behaviour is not common or popular.

The basis of prevention is the use of tested and commercially available repellents – these are chemicals that repel ticks and prevent bites. The protection time varies with different brands, depending on the chemical properties of the compound, characteristics of the environment (temperature, time of day, wind speed, light and humidity), characteristics of potential host (sweating, skin condition, individual susceptibility, diet, mobility) and tick species. In tick-repelling formulations, the following active substances in different concentrations are commonly used: NN-diethyl m-toluamide (DEET); 3-(N-butyl-N-acetyl)-amino propionic acid (IR3535); KBR3023 (picaridin); and natural substances of plant origin, e.g. geraniol, citriodiol, eucalyptus oil. DEET is the active substance commonly used as an ingredient of repellents, in concentrations ranging from 7% to 31.6%. Another active substance used in repellents is IR3535. In formulations registered in Poland, it is most often found in concentrations of 7–20%. The

repellent icaridin was synthesized in 1980, and has been available on the market since 2000. The most common formulations contain icaridin in a concentration of 10–20% (Cisak *et al.*, 2012; Bissinger and Roe, 2010). Nowadays, it is believed that natural substances also contain certain insect-repellent properties. These substances include the plant oils anise, bergamot, camphor, cinnamon, clove, coconut, eucalyptus, geranium, lavender, lemon, nutmeg, orange flower, corn mint, pine, thyme oil and citronella oil from lemon eucalyptus.

Plant oils are generally less effective and provide an acceptable level of protection for a shorter period of time than chemical repellents. It is believed that phytochemicals are more friendly and harmless for users. Increasing the concentration of essential oils may increase their insect-repelling efficiency, but high concentrations can cause skin irritation. Many plant products also contain properties toxic to vertebrates, e.g. eugenol is a substance which irritates eyes and skin, and has proved to be a mutagen; citronellol is irritating for eyes, affects the reproductive and nervous systems, and is also a mutagen (Bissinger and Roe, 2010). The most commonly used tick repellents' effectiveness depends on the active ingredient and formulation. A study by Gliniewicz *et al.* (2015) on harvested ticks – males, females and nymphs of *D. reticulatus* and *I. ricinus* – examined the effectiveness of five formulations in the form of an atomizer and spray. The formulations tested differed in chemical composition and contained most of the common active ingredients: DEET (at 25% and 18%), IR3535 (in a concentration of 12% and 20%) and natural substances in the form of oils: citriodiol, citronella, lemon grass and geraniol used at various concentrations from 0.3% to 10.01%, depending on the formulation. Studies have shown that repellents containing DEET and IR3535 show 90% repelling effectiveness against ticks after a period of 1.5–2 h, whereas formulations based on plant oils were effective in 90% to about 1 h. Tested formulations containing DEET and IR3535 were more effective on the genus *Ixodes* than ticks of the genus *Dermacentor*.

In the 1970s, the first pyrethroids were synthesized. These compounds have a similar chemical structure to pyrethrum and insecticides, but with longer environmental stability. Currently, pyrethroids are used in most acaricides. An example of such a compound is permethrin, used to impregnate clothing (uniforms, clothes of forest workers, hunters' clothes) against ticks. Permethrin has a short-acting repellent and long-lasting killing effect. Studies by Faulde *et al.* (2008) showed that wearing clothing impregnated with permethrin significantly reduced the risk of tick attacks and tick bites. Out of 138 *I. ricinus* caught during the study, only six were reported in volunteers wearing permethrin-impregnated clothing, whereas 132 ticks were found on volunteers wearing non-impregnated clothing. The repelling effectiveness of clothes was set at 95.5% (Faulde *et al.*, 2008). Similar studies were conducted by Dusbábek *et al.* (1997) and Faulde *et al.* (2006).

Another important thing from the point of view of tick prevention is also proper care of pets and livestock. Pets are exposed to tick attacks during walks through vegetation. Livestock is exposed while grazing in the pastures. Therefore, proper care of animals accompanying people is necessary, i.e. checking surroundings of ears, mouth and eyes, as well as prophylaxis in the form of drops, impregnated collars, shampoos or vaccines preventing tick-borne diseases in animals and humans. Tick prevention is more effective when regular agricultural treatments are performed in the immediate surroundings. These treatments include regular grass cutting, removal of bushy undergrowth or temporary rotation of pastures for animal grazing and farm fields designated for ploughing.

8.11 Conclusion

Long-term observations of tick fauna in Europe suggest that medically important species have changed their areas of occurrence. Many authors have correlated this phenomenon with climatic changes, such as increasing temperatures in winter months, mild and long spring and autumn, and changes in snow cover. Changes in climate parameters generate changes in the environment, particularly the flora and the fauna components. Ticks can be transported by small and large mammals and birds as their hosts, enabling them to invade new territories. Expansion of ticks to new areas because of climatic changes is one cause of the observed increase in tick-borne disease cases in European countries. The coexistence of ticks and people in certain areas, the invasion of urban settlements into tick settlements, and the human tendency towards frequent outdoor activity, creates the conditions for more frequent contact between ticks and people and greater danger of tick bites. Certain socioeconomic changes also play a role when the increase of tick-borne diseases is considered.

References

Akimov, I.A. and Nebogatkin, I.V. (1996) On the southern limit of the European forest tick (*Ixodes ricinus*) range. *Vestnik zoologii* 30(6), 84–86.

Akimov, I.A. and Nebogatkin, I.V. (2011) Distribution of ticks of the genus *Dermacentor* (Acari; Ixodidae) in Ukraine. *Vestnik zoologii* 45(1), 35–40.

Akimov, I.A. and Nebogatkin, I.V. (2012) Distribution of the tick *Haemaphysalis punctata* (Acari, Ixodidae) in Ukraine. *Vestnik zoologii* 46(4), 46–51.

Akimov, I.A. and Nebogatkin, I.V. (2013) Ticks of the genus *Rhipicephalus* (Acari, Ixodidae) and their distribution in Ukraine. *Vestnik zoologii* 47(3), e-28–e-34.

Anderson, J.F. and Magnarelli, L.A. (2008) Biology of ticks. *Infectious Disease Clinics of North America* 22, 195–215.

Bednorz, E. (2007) *Zmiany występowania pokrywy śnieżnej w północnych Niemczech w latach 1950/51–1999/00* [Changes in the occurrence of snow cover in northern Germany in 1950/51–1999/00]. In: Piotrowicz, K. and Twardosz, R. (eds) *Wahania klimatu w różnych skalach przestrzennych i czasowych* [Climate change in different spatial and temporal scales]. Institute of Geography and Spatial Management, Jagiellonian University, Krakow, Poland, pp. 215–223 [in Polish].

Belikova, N.P. (1956) Materialy po zimovke iksodovykh kleshchey na zhivotnykh [Materials on wintering ticks on animals]. *Trudy Dal'nevostochnogo Filiala Akademii Nauk SSSR s. Zool.* [Proceedings of the Far Eastern branch of the Academy of Sciences of the USSR] 3(6), 265–268 [in Russian].

Bennet, L., Halling, A. and Berglund, J. (2006) Increased incidence of *Lyme borreliosis* in southern Sweden following mild winters and during warm, humid summers. *European Journal of Clinical Microbiology and Infectious Diseases* 25, 426–432.

Bissinger, B.W. and Roe, R.M. (2010) Tick repellents: past, present and future. *Pesticide Biochemistry and Physiology* 96(2), 63–79.

Buczek, A., Sadowiska, H. and Magdoń, T. (2000) Porażenie kleszczowe – przyczyna, objawy i zapobieganien [Tick paralysis – cause, symptomatic picture and prevention]. In: Buczek, A. and Błaszak, C. (eds) *Stawonogi pasożytnicze i alergogenne* [Parasitic and allergenic arthropods]. KGM, Lublin, Poland [in Polish].

Bullová, E., Lukáň, M., Stanko, M. and Peťko, B. (2009) Spatial distribution of *Dermacentor reticulatus* tick in Slovakia in the beginning of the 21st century. *Veterinary Parasitology* 165, 357–360.

Cebulak, E. and Limanówka, D. (2007) Dni z ekstremalnymi temperaturami powietrza w Polsce. In: Piotrowicz, K. and Twardosz, R. (eds) *Wahania klimatu w różnych skalach przestrzennych i czasowych.* Instytut Geografii i Gospodarki Przestrzennej, Jagiellonian University, Krakow, Poland, pp. 185–194 [in Polish].

Chigirik, E.D. and Pleshivtseva-Eroshkina, E.A. (1969) Ixodid ticks of Kemerovo Oblasti. *Meditsinskaia Parazitologiia and Parazitarnye Bolezni* 38(40), 423–426 [in Russian].

Cisak, E., Wójcik-Fatla, A., Zając, V. and Dutkiewicz, J. (2012) Repellents and acaricides as personal protection measures in the prevention of tick-borne diseases. *Annals of Agricultural and Environmental Medicine* 19, 625–630.

Cuber, P., Solarz, K., Mosialek, A., Jakubiec-Spanier, M. and Spanier, A. (2013) The first record and occurrence of the ornate cow tick *Dermacentor reticulatus* (Fabricius, 1794) in south-western Poland. *Annals of Parasitology* 59, 49–51.

Daniel, M., Černý, V. and Szymański, S. (1986) A comparison of factors influencing the distribution of *Dermacentor reticulatus* (Ixodidae) in Czechoslovakia and Poland. *Wiadomości Parazytologiczne* 32(4–6), 355–361.

Dautel, H., Dippel, C., Oehme, R., Hartelt, K. and Schettler, E. (2006) Evidence for an increased geographical distribution of *Dermacentor reticulatus* in Germany and detection of *Rickettsia* sp. RpA4. *International Journal of Medical Microbiology* 296, Suppl. 40, 149–156.

Del Fabbro, S. (2014) Fencing and mowing as effective methods for reducing tick abundance on very small, infested plots. *Ticks and Tick-borne Diseases* 6(2), 167–172. DOI: 10.1016/j.ttbdis.2014.11.009.

Dusbăbek, F., Rupeš, V., Šimek, P. and Zahradničkova, H. (1997) Enhancement of permethrin efficiacy in acaricide–attractant mixtures for control of the fowl tick *Argas persicus* (Acari: Argasidae). *Experimental and Applied Acarology* 21, 293–305.

Ebert, B. and Fleischer, B. (2005) Die globale Erwärmung und Ausbreitung von Infektionskrankheiten [Global warming and spread of infectious diseases]. *Bundesgesundheitsblatt Gesundheitsforschung Gesundheitsschutz* 48, 55–62 [in German].

Estrada-Peña, A., Bouattour, A., Camicas, J.L. and Walker, A.R. (2004) *Ticks of Domestic Animals in the Mediterranean Region. A Guide to Identification of Species.* University of Zaragoza, Zaragoza, Spain.

Falarz, M. (2004) Variability and trends in the duration and depth of snow cover in Poland in the 20th century. *International Journal of Climatology* 24(13), 1713–1727.

Falarz, M. (2007) Potencjalny okres występowania pokrywy śnieżnej w Polsce i jego zmiany w XX wieku. In: Piotrowicz, K. and Twardosz, R. (eds) *Wahania klimatu w różnych skalach przestrzennych i czasowych.* Instytut Geografii i Gospodarki Przestrzennej, Jagiellonian University, Krakow, Poland, pp. 205–213 [in Polish].

Farkas, R. and Földvári, G. (2001) A kutyák és a macskák kullancsosságának hazai vizsgálata. *Magyar Állatuosok Lapja* 123(9), 534–539.

Faulde, M., Scharninghausen, J. and Tisch, M. (2008) Preventive of perethrin-impregnated clothing to *Ixodes ricinus* ticks and associated *Borrelia burgdorferi* s.l. in Germany. *International Journal of Medical Microbiology* 298, Suppl. 1, 231–324.

Faulde, M.K., Uedelhoven, W.M., Malerius, M. and Robbins, R.G. (2006) Factory-based permethrin impregnation of uniforms: residual activity against *Aedes aegypti* and *Ixodes ricinus* in battle dress uniforms worn under field conditions, and cross-contamination during the laundering and storage process. *Military Medicine* 171, 472–477.

Filippova, N.A. (1977) Iksodovye kleshchi podsem. Ixodinae. In: *Fauna SSSR. Paukoobraznyie.* Izd-vo Nauka, Leningrad, Russia [in Russian].

Filippova, N.A. (1997) Ixodid ticks of subfamily Amblyomminae. Fauna of Russia and neighbouring countries. *Arachnoidea* 4(5), 436 [in Russian].

Gassner, F. and Overbeek, L. (2007) Lyme disease in Europe: facts and no fiction. In: Takken, W. and Knols, B.G.J. (eds) *Emerging Pests and Vector-borne Disease in Europe*. Wageningen Academic Publishers, Wageningen, The Netherlands, pp. 207–223.

Gern, L., Estrada-Peña, A., Frandsen, F., Gray, J.S., Jaenson, T.G., Jongejan, F., Kahl, O., Korenberg, E., Mehl, R. and Nuttall, P.A. (1998) European reservoir hosts of *Borrelia burgdorferi* sensu lato. *Zentralblatt für Bakteriologie* 287, 196–204.

Gilbert, L., Maffey, G.L., Ramsay, S.L. and Hester, A.J. (2012) The effect of deer management on the abundance of *Ixodes ricinus* in Scotland. *Ecological Applications* 22(2), 658–667.

Glaser, B. and Gothe, R. (1998) Importierte arthropodenübertragene Parasiten und parasitische Arthropoden beim Hund. Erregerspektrum und epidemiologische Analyse der 1995/96 diagnostizierte Fälle [Imported arthropod-borne parasites and parasitic arthropods in the dog. Species spectrum and epidemiologic analysis of the cases diagnosed in 1995/96]. *Tierärztliche Praxis* 26, 40–46 [in German].

Gliniewicz, A., Mikulak, E., Królasik, A. and Myślewicz, J. (2014) Tick-bite preventative behaviour of people on recreational areas in Warsaw. In: Müller, G., Pospischil, R. and Robinson, W.H. (eds) *Proceedings of the 8th International Conference on Urban Pests*. Executive Committee of the International Conference on Urban Pests, Zurich, Switzerland, p. 427.

Gliniewicz, A., Mikulak, E., Myślewicz, J., Królasik, A. and Supergan-Marwicz, M. (2015) Skuteczność repelentów w stosunku do komarów ni kleszczy [Effectiveness of repellents towards mosquitoes and ticks]. In: Buczek, A. and Błaszak, C. (eds) *Proceedings of the 17th International Symposium on Parasitic and Allergenic Arthropods – Medical and Sanitary Significance*, Koliber, Janowiec, Poland, p. 26 [in Polish].

Gratz, N. (2006) *Vector and Rodent-borne Diseases in Europe and North America*. World Health Organization, Geneva, Switzerland.

Gray, J.S., Dautel, H., Estrada-Peña, A., Kahl, O. and Lindgren, E. (2009) Effects of climate change on ticks and tick-borne diseases in Europe. *Interdisciplinary Perspectives on Infectious Diseases* 2009, article ID 593232.

Hajdul, M., Zaremba, M., Karbowiak, G. and Siński, E. (2006) Risk of infection with *Borrelia burgdorferi* s.l. in wooded biotops around Warsaw, Poland. In: Buczek, A. and Błaszak, C. (eds)

Proceedings of the 8th International Symposium on Parasitic and Allergenic Arthropods – Medical and Sanitary Significance. Koliber, Kazimierz Dolny, Poland, p. 74.

Hubalek, Z., Halouzka, J. and Juricová, Z. (2003): Longitudinal surveillance of the tick *Ixodes ricinus* for borreliae. *Medical and Veterinary Entomology* 17, 46–51.

Hvidsten, D., Stordal, F., Lager, M., Rognerud, B., Kristiansen, B.E., Matussek, A., Gray, J. and Stuen, S. (2015) *Borrelia burgdorferi* sensu lato-infected *Ixodes ricinus* collected from vegetation near the Arctic Circle. *Ticks and Tick-borne Diseases* 6(6) 768–773.

Immler, R.M. (1973) Untersuchungen zur Biologie and Ökologie der Zecke *Dermacentor reticulatus* (Fabricius, 1794) (Ixodidae) in einem endemischen Vorkommensgebiet [Studies on the biology and ecology of the tick *Dermacentor reticulatus* (Fabricius, 1794) (Ixodidae) in an endemic area occurrence]. *Mitteilungen der Schweizerischen Entomologischen Gesellschaft* [Notes from the Swiss Entomological Society] 46(1–2), 2–70 [in German].

Izdebska, J.N. (1998) Występowanie *Dermacentor reticulatus* (Fabricius, 1794) (Acari, Ixodidae) u żubra (*Bison bonasus*) z Puszczy Białowieskiej. *Przegląd Zoologiczny*, 42(3–4), 219–221 [in Polish].

Jaenson, T.G., Jaenson, D.G., Eisen, L., Petersson, E. and Lindgren, E. (2012) Changes in the geographical distribution and abundance of the tick *Ixodes ricinus* during the past 30 years in Sweden. *Parasites and Vectors* 5, 8.

Kampen, H.D. (2004) Substantial rise in the prevalence of *Lyme borreliosis* spirochetes in a region of western Germany over a 10-year period. *Applied and Environmental Microbiology* 70, 576.

Karbowiak, G. (2014) The occurrence of the *Dermacentor reticulatus* tick – its expansion to new areas and possible causes. *Annals of Parasitology* 60, 37–47.

Karbowiak, G. and Kiewra, D. (2010) New locations of *Dermacentor reticulatus* ticks in Western Poland: the first evidence of the merge in *D. reticulatus* occurrence areas? *Wiadomości Parazytologiczne* 56, 333–336.

Karbowiak, G. and Szewczyk, T. (2014) The duration of winter inactivity of *Dermacentor reticulatus* tick. In: *Parasitological Meeting – Parasites in the Heart of Europe. Book of Abstracts*, Vol. 4. High Tatra Mountains, Stará Lesná, Slovakia, pp. 84–85.

Karbowiak, G., Izdebska, J.N., Czaplińska, U. and Wita, I. (2003) Przypadki zimowania kleszczy z rodziny Ixodidae na żywicielach w Puszczy

Białowieskiej. In: Buczek, A. and Błaszak, C. (eds) *Stawonogi i żywiciele*. Koliber, Lublin, Poland, pp. 77–82 [in Polish].

Karbowiak, G., Bullová, E., Majláthová, V., Peťko, B., Stanko, M., Wita, I., Hapunik, J. and Czaplińska, U. (2008) The distribution and expansion of ornate dog tick *Dermacentor reticulatus*. In: *Zborník Abstraktov. Prírodne ohniskové nákazy Košice*, Vol. 21. Parazitologický Ústav SAV, Košice, Slovakia.

Kiewra, D., Szymanowski, M., Lonc, E. and Czułowska, A. (2015) *Dermacentor reticulatus* (Fabr.) in Wroclaw area-planning field measurements and analyzing spatial distribution of ticks using GIS tools. In: Buczek, A. and Błaszak, C. (eds) *Stawonogi we współczesnym świecie*. Koliber, Lublin, Poland, pp. 79–85.

Koch, C.L. (1844) Systematische Übersicht über die Ordnung der Zecker [Systematic review of the order of ticks]. *Archiv für Naturgeschichte* 10, 217–239 [in German].

Kolonin, G.V. (1981) Effect of industrial activity on ixodid tick numbers and distribution in Primor'ye region. *Zoologicheskii Zhurnal* 60(3), 363–370.

Kulikowski, R. (2006) Rolnictwo w Polsce [Agriculture in Poland]. In: Degórski, M. (ed.) Naturalne i ludzkie środowiska w Polsce. Przegląd danych geograficznych [Natural and human environment of Poland. A geographical overview]. Polish Geographical Society, Warsaw, Poland, pp. 211–231 [in Polish].

Kunz, C. (2003) TBE vaccination and the Austrian experience. *Vaccine* 21, S1/50–S51/55

Kurtenbach, K., De Michelis, S., Etti, S., Schafer, S.M., Sewell, H.-S., Brade, V. and Kraiczy, P. (2002) Host association of *Borrelia burgdorferi* sensu lato – the key role of host complement. *Trends in Microbiology* 10, 74–79.

Kurtenbach, K., Hanincová, K., Tsao, J.I., Margos, G., Fish, D. and Ogden, N.H. (2006) Fundamental processes in the evolutionary ecology of Lyme borreliosis. *Nature Reviews: Microbiology* 4, 660–669.

Lane, R.S. (2012) California forests: primary risk habitats for human exposure to *Lyme borreliosis* spirochetes. In: Skorupki, M. and Wierzbicka, A. (eds) *Vademecum wybranych chorób odzwierzęcych w środowisku leśnym* [Vademecum of selected zoonotic diseases in the forest environment]. Department of Forest Protection and Hunting, Poznan University of Life Sciences, Poznan, Poland, pp. 9–21.

Lindgren, E., Tälleklint, L. and Polfeldt, T. (2000) Impact of climatic change on the northern latitude limit and population density of the disease-transmitting European tick *Ixodes ricinus*. *Environmental Health Perspectives* 108, 119–123.

Mačička, O., Nosek, J. and Rosický, B. (1956) *Poznámky k bionómii, vývoju a hospodárskemu významu pijaka lužného* (Dermacentor pictus Herm.) *w strednej Európe*. Vydavateľstvo SAV, Bratislava, Slovakia [in Slovak].

Martinod S. and Gilot B. (1991) Epidemiology of canine babesiosis in relation to the activity of *Dermacentor reticulatus* in southern Jura (France). *Experimental and Applied Acarology* 11(2–3), 215–222.

Matuszkiewicz, J.M. (2006) Flora i fauna Polski: Naturalne i ludzkie środowiska w Polsce [Poland's flora and fauna: natural and human environment of Poland]. In: Degórski, M. (ed.) *Przegląd danych geograficznych* [A geographical overview]. Instytut Geografii i Przestrzennego Zagospodarowania im. S. Leszczyckiego [S. Leszczycki Institute of Geography and Spatial Organization], Polskie Towarzystwo Geograficzne [Polish Geographical Society], Warsaw, Poland, pp. 77–92 [in Polish].

Medlock, J.M., Hansford, K.M., Bormane, A., Derdakova, M., Estrada-Peña, A., George, J.C., Golovljova, I., Jaenson, T.G., Jensen, J.K., Jensen, P.M., Kazimirova, M., Oteo, J.A., Papa, A., Pfister, K., Plantard, O., Randolph, S.E., Rizzoli, A., Santos-Silva, M.M., Sprong, H., Vial, L., Hendrickx, G., Zeller, H. and Bortel, W.V. (2013) Driving forces for changes in geographical distribution of *Ixodes ricinus* ticks in Europe. *Parasites and Vectors* 6, 1.

Nowak-Chmura, M. (2013) *Fauna kleszczy (Ixodida) Europy środkowej*. Wydawnictwo Naukowe Uniwersytetu Pedagogicznego, Krakow, Poland [in Polish].

Nowak-Chmura, M. and Siuda, K. (2012) Ticks of Poland: review of contemporary issues and latest research. *Annals of Parasitology* 58, 125–155.

Nowak, M. (2010) The international trade in reptiles (Reptilia) the cause of the transfer of exotic ticks (Acari: Ixodida) to Poland. *Veterinary Parasitology* 169, 373–381.

Ostfeld, R.S., Price, A., Hornbostel, V.L., Benjamin, M.A. and Keesing, F. (2006) Controlling tick and tick-borne zoonoses with biological and chemical agents. *BioScience* 56, 383–394.

Paulauskas, A., Radzijevskaja, J., Mardosaitė-Busaitienė, D., Aleksandravičienė, A., Galdikas, M. and Krikštolaitis, R. (2015) New localities of *Dermacentor reticulatus* ticks in the Baltic countries. *Ticks and Tick-borne Diseases* 6, 630–635.

Peťko, B., Derdáková, M., Lenčáková, D., Majláthová, E., Bullová, E., Víchová, B., Nováková,

M., Hrkľová, G. and Lukáň, M. (2008) Epidemiologicky významné druhy kliešťov (Ixodidae) v strednej Európe v podmienkach klimatických zmien [Epidemiological important species of ticks (Ixodidae) in central Europe in the conditions of climate change]. VIII české a Slovenské Parazitologické Dny [8th Czech and Slovak Parasite Days], Sezimovo Ústí, Czech Republic, p. 78 [in Czech/Slovak].

Randolph, S.E. and Šumilo, D. (2007) Tick-borne encephalitis in Europe: dynamics of changing risk. In: Takken, W. and Knols, B.G.J. (eds) *Emerging Pests and Vector-borne Disease in Europe*. Wageningen Academic Publishers, Wageningen, The Netherlands, pp. 187–206.

Randolph, S.E., Green, R.M., Peacey, M.F. and Rogers, D.J. (2000) Seasonal synchrony: the key to tick-borne encephalitis foci identified by satellite data. *Parasitology* 121, 15–23.

Rauter, C. and Hartung, T. (2005) Prevalence of *Borrelia burgdorferi* sensu lato genospecies in *Ixodes ricinus* ticks in Europe: a metaanalysis. *Applied and Environmental Microbiology* 71, 7203–7216.

Ravdonikas, O.V., Solovey, E.A., Chumakov, M.P., Korsh, I.V. and Ivanov, D.I. (1968) Omsk hemorrhagic fever natural foci in the light of landscape distribution area properties of this disease. In: *Материалы Третьего Научного Совещания По Проблемам Медицинской Географии* [Proceedings of the third scientific meeting on problems of medical geography], November, pp. 152–154 [in Russian].

Schauber, E.M., Edge, W.D. and Wolff, J.O. (1997) Insecticide effects on small mammals: influence of vegetation structure and diet. *Ecological Applications* 7, 143–157.

Schulze, T.L., Jordan, R., Hung, R.W., Taylor, R.C., Markowski, D. and Chomsky, M.S. (2001) Efficacy of granular deltamethrin against *Ixodes scapularis* and *Amblyomma americanum* (Acari: Ixodidae) nymphs. *Journal of Medical Entomology* 38(2), 344 –346.

Široký, P., Kubelová, M., Bednář, M., Modrý, D., Hubálek, Z. and Tkadlec, E. (2011) The distribution and spreading pattern of *Dermacentor reticulatus* over its threshold area in the Czech Republic: how much is range of this vector expanding? *Veterinary Parasitology* 183, 130–135.

Siuda, K. (1993) *Kleszcze Polski (Acari: Ixodida). Część II: systematyka i rozmieszczenie.* PTP, Warsaw, Poland [in Polish].

Siuda, K. (1995) Przegląd danych dotyczących podziału Ixodida (Acari) w Polsce [Review of data on the distribution of Ixodida (Acari) in Poland.]. In: Kropczyńska, D., Boczek, J. and Tomczyk, A. (eds) *Roztocze. Fizjologiczne i ekologiczne aspekty relacji roztocze–gospodarza* [The acari. Physiological and ecological aspects of acari–host relationships]. Oficyna Dabor, Warsaw, Poland, pp. 273–280 [in Polish].

Siuda, K., Zięba, P., Bogdaszewska, Z., Stańczak, J. and Sebesta, R. (1997) Przegląd danych dotyczących rozmieszczenia *Dermacentor reticulatus* (Fabricius, 1794) (Acari: Ixodida: Ixodidae) w Polsce [Review of data of the distribution of *Dermacentor reticulatus* (Fabricius, 1794) (Acari: Ixodida: Ixodidae) in Poland], In: *Zeszyty Naukowe Ochrona środowiska*, Vol. 2. Akademia Techniczno-Rolnicza im. Jana i Jędrzeja śniadeckich, Bydgoszcz, Poland [in Polish].

Solberg, V.B., Neidhardt, K., Sardeli, M.R., Hoffmann, F.J., Stevenson, R., Boobar, L.R. and Harlan, H.J. (1992) Field evaluation of two formulations of cyfluthrin for control of *Ixodes dammini* and *Amblyomma americanum* (Acari: Ixodidae). *Journal of Medical Entomology* 29, 634–638.

Solberg, V.B., Miller, J.A., Hadfield, T. Burge, R. Schech, J.M. and Pound, J.M. (2003) Control of *Ixodes scapularis* (Acari: Ixodidae) with topical self-application of permethrin by white-tailed deer inhabiting NASA, Beltsville, Maryland. *Journal of Vector Ecology* 28(1), 117–134.

Stafford, K.C., Carter, M.L., Magnarelli, L.A., Ertel S.-H. and Mshar, P.A. (1998) Temporal correlations between tick abundance and prevalence of ticks infected with *Borrelia burgdorferi* and increasing incidence of Lyme disease. *Journal of Clinical Microbiology* 36, 1240–1244.

Stanek, G. and Strle, F. (2003) Lyme borreliosis. *Lancet* 362, 1639–1647.

Šumilo, D., Askoliene, L., Bormane, A., Vasilenko, V., Golovjova, I. and Randolph, S.E. (2007) Climate change cannot explain the upsurge of tick-borne encephalitis in the Baltics. *PLoS ONE* 2(6), article ID e500. DOI: 10.1371/journal.pone.0000500.

Supergan, M. and Karbowiak, G. (2009) The estimation scale of endangerment with tick attacks on recreational towns areas. *Przegląd Epidemiologiczny* 63, 67–71.

Szumacher, I. (2005) Funkcje ekologiczne parków miejskich [Ecological functions of urban parks]. *Prace i studia geograficzne* 36, 107–120 [in Polish].

Tälleklint, L. and Jaenson, T.G.T. (1995) Control of *Lyme borreliosis* in Sweden by reduction of tick vector: an impossible task? *International Journal of Angiology* 4, 34–37.

Wall, R. (2012) A ticking clock for tickborne disease? *Veterinary Record* 170, 326–328.

9 Climate Change and its Effect on Urban Mosquitoes in South America

Tamara Nunes de Lima-Camara[1]* and Nildimar Alves Honório[2,3]

[1]Laboratório de Entomologia em Saúde Pública/Culicídeos, Departamento de Epidemiologia, Faculdade de Saúde Pública da USP, São Paulo, Brazil; [2]Laboratório de Mosquitos Transmissores de Hematozoários, Instituto Oswaldo Cruz, Fundação Oswaldo Cruz, Manguinhos, Rio de Janeiro, Brazil; [3]Núcleo Operacional Sentinela de Mosquitos Vetores-NOSMOVE/FIOCRUZ, Manguinhos, Rio de Janeiro, Brazil

9.1 Introduction

9.1.1 Urban mosquitoes in South America

South America is an extensive continent with 13 countries and three dependencies that exhibits diverse patterns of weather and climate, including tropical, subtropical and extratropical features (Garreaud *et al.*, 2009). Because of this heterogeneity in weather and in other abiotic conditions, as well as a complex geological history, South America is well known for its biodiversity and regions with numerous endemic species (Sigrist and Carvalho, 2009). This biodiversity harbours a great range of disease vectors that are sensitive to climate changes. Indeed, many important vector-borne diseases are endemic in South American countries, like leishmaniasis, Chagas disease, malaria and dengue fever. All these vector-borne diseases are highly influenced by seasonal weather variations, socioeconomic status of humans, vector control programmes, drug resistance and environmental changes (Githeko *et al.*, 2000).

Several vector-borne diseases that are influenced by seasonal weather are vectored by mosquitoes. Mosquitoes belong to the phylum Arthropoda, class Insecta, order Diptera, suborder Nematocera and family Culicidae. In common with all members of the family Culicidae, mosquitoes show a holometabolous development, going through a complete metamorphosis, with an egg, aquatic larva (with four instars), aquatic pupa and adult stages. Adults are epidemiologically important because most mosquito females need a blood meal to develop eggs. During the blood-feeding, mosquito females can either become infected with a pathogen or transmit the pathogen to a susceptible human host (Consoli and Lourenço-de-Oliveira, 1994; Eldridge, 2008).

Natural mosquito habitats are abundant in many areas of the world, regardless of the landscape modifications caused by humans. Nevertheless, human activities, such as clearing land for subsistence agriculture or dam construction for hydroelectric power or recreational purposes, may exacerbate existing mosquito-associated problems by expanding habitats, creating new habitats, or altering habitats in such a way that once-limited mosquito populations may

*E-mail: limacamara@usp.br

now explode with the availability of new habitats. The effects of change in land use by humans have long been recognized as an important factor for the adaptation of mosquitoes to urban areas and the exacerbation of several mosquito-borne diseases (Norris, 2004). Over time, several mosquito species have become truly urban, and in this chapter we will focus on urban mosquitoes that serve as a vector to humans.

9.1.2 *Aedes* vectors and arbovirosis

In South America, *Aedes (Stegomyia) aegypti* is considered an important vector of several arboviruses, like dengue (DENV), chikungunya (CHIKV), Yellow fever (YFV) and Zika (ZIKV), with a broad distribution in this continent (Consoli and Lourenço-de-Oliveira, 1994; Lourenço-de-Oliveira *et al.*, 2013). *Ae. aegypti* frequently shows endophilic and anthropophilic behaviours, resting inside houses (Lima-Camara *et al.*, 2006) and blood-feeding on human hosts (Scott *et al.*, 1993; Hoeck *et al.*, 2003). It is considered an urban mosquito with a close association to humans, which increases the contact between vector and host and the chances of arbovirus transmission (Consoli and Lourenço-de-Olivera, 1994).

Aedes (Stegomyia) albopictus is another important mosquito species in South America. This species was first described in India and even more so than *Ae. aegypti*, *Ae. albopictus* has a broad distribution in tropical as well as temperate countries (Consoli and Lourenço-de-Oliveira, 1994). *Ae. albopictus* has been incriminated as a dengue vector in some areas of Colombia and Costa Rica (Méndez *et al.*, 2006; Calderón-Arguedas *et al.*, 2015), and studies in Brazil and Mexico have demonstrated the occurrence of vertical dengue transmission in this species (Ibáñez-Bernal *et al.*, 1997; Martins *et al.*, 2012). In addition, Brazilian populations of *Ae. albopictus*, for example, have shown notorious competence to become infected and transmit DENV (Miller and Ballinger, 1988; Castro *et al.*, 2004), CHIKV (Vega-Rúa *et al.*, 2014) and ZIKV (Chouin-Carneiro *et al.*, 2016) under laboratory conditions. Therefore, it is extremely important to maintain entomological surveillance and control strategies of *Ae. albopictus* in South American countries (Pancetti *et al.*, 2015). *Ae. albopictus* males and females can be frequently found resting outside houses, which makes this mosquito a more exophilic species than *Ae. aegypti* (Thavara *et al.*, 2001; Gomes *et al.*, 2005; Lima-Camara *et al.*, 2006).

Climate variables, like rising temperatures, changes in rainfall and relative humidity, directly influence the seasonal dynamics of *Ae. aegypti* and *Ae. albopictus*. These conditions favour an increase in the number of natural or artificial breeding sites available and consequently the development of the vector (Viana and Ignotti, 2013). In addition, higher temperatures seem to favour the development and expansion of *Ae. aegypti* and *Ae. albopictus* (Glasser and Gomes, 2002; Beserra *et al.*, 2006).

Dengue is an arbovirus disease, ubiquitous throughout the tropics and subtropics and strongly influenced by temperature, rainfall and levels of urbanization. Recently, a study estimated 96 million apparent infections and 293 million unapparent infections of dengue cases in the world per year (Bhatt *et al.*, 2013). The etiologic agent of dengue fever is a virus from the family *Flaviviridae* and genus *Flavivirus*, with four closely related, but antigenically distinct serotypes: DENV-1, DENV-2, DENV-3 and DENV-4 (Halstead, 1988). In the last few years, South America has reported more than 1 million clinical cases of dengue annually, which makes this arbovirosis one of the most important mosquito-borne diseases in the region. From 1998 to 2012, almost 80% of the total dengue fever cases reported in South America occurred in the countries of the Southern Cone, especially Brazil (Lourenço-de-Oliveira *et al.*, 2013).

Chikungunya fever is an acute viral disease caused by the RNA chikungunya arbovirus, which belongs to the *Togaviridae* family, genus *Alphavirus*. CHIKV was first isolated in Tanzania, Africa, in 1952, and phylogenetic analysis indicates three genotypes: West African, East/Central/South Africa (ECSA) and Asian (Sam *et al.*, 2015). It is transmitted to humans by both *Ae. aegypti* and *Ae. albopictus* infected females (Lahariya and Pradhan, 2006). The symptoms of chikungunya fever are similar to dengue, but arthritis (joint pain) is much more common and severe in this case. The characteristic severe arthralgia gives the name of this disease, which means 'that which bends up' in the Makonde language spoken in some areas of Tanzania and northern Mozambique (Kucharz and Cebula-Byrska, 2012). From 2004 to 2007, CHIKV emerged as a global pathogen, with outbreaks in Africa, Asia, Europe and the Indian and Pacific islands (Reiskind *et al.*, 2008; CDC, 2014). In late 2013, the first autochthonous transmission of CHIKV in the Americas was identified in Caribbean countries and at the end of 2014, countries of South America, such as French Guiana, Venezuela, Colombia, Suriname, Brazil, Paraguay and Ecuador reported autochthonous transmission of CHIKV (CDC, 2014). A recent study showed high vector competence for CHIKV of *Ae. aegypti* and *Ae. albopictus* populations from several countries of South America, including Venezuela, Peru, Brazil, Paraguay, Uruguay and Argentina (Vega-Rúa *et al.*, 2014).

Zika is an arbovirosis caused by a virus of *Flaviviridae* family and genus *Flavivirus*. ZIKV was first isolated in a rhesus monkey in 1947, in the Zika forest in Uganda, but the first isolation in humans occurred in Nigeria. Serological evidence of human infection with this arbovirus were also reported in other African countries, such as Egypt, Tanzania, Gabon and Sierra Leone, as well as Asian countries such as India, Malaysia, Thailand and Indonesia (Hayes, 2009). In 2007, an outbreak of Zika was reported in Yap Island, Micronesia, in the Pacific Ocean, while the French Polynesia, also in Oceania, recorded a major outbreak of the disease in October 2013. Thus, the circulation and transmission of ZIKV out of the African and Asian continents were confirmed (Hayes, 2009). In 2015, Brazil recorded the first autochthonous human cases of Zika, confirming the recent entry of this arbovirus in South America. Other countries of this continent have reported the circulation of ZIKV as well, like Venezuela and Paraguay. Also

transmitted through the bite of infected *Aedes* females, especially *Ae. aegypti*, the symptoms of Zika include fever, arthralgia, myalgia, headache, conjunctivitis and maculopapular rash (Duffy *et al.*, 2009).

9.1.3 *Culex quinquefasciatus* and West Nile virus

The transmission of West Nile virus (WNV) involves some species of birds and mosquitoes, whereas human and horses are considered accidental hosts (Gubler, 2007). In the USA, approximately 65 different mosquito species have been found naturally infected with WNV between 1999 and 2012. Nevertheless, just a few *Culex* mosquito species are responsible for the virus transmission in nature (bird–*Culex*–bird) and subsequent spread to humans (Petersen *et al.*, 2013). In the USA, *Cx. pipiens* is considered the most effective vector in the northern half of this country and *Cx. quinquefasciatus* plays an important role as a WNV vector in the southern states (Ciota and Kramer, 2013; Petersen *et al.*, 2013).

Cx. quinquefasciatus is widely distributed in South America. This species shows a close association with humans and it is frequently found inside houses of urban and suburban areas of South America (Forattini, 2002). This species presents several ecological and biological characteristics that indicate this mosquito as a potential WNV vector in South America. As mentioned before, it is considered an urban mosquito, with a close association with humans. Moreover, studies on host preferences of *Cx. quinquefasciatus* females from South America indicate this mosquito is an opportunistic species, blood-feeding on different vertebrates, such as birds, dogs, man and oxen (Gomes *et al.*, 2003; Lorosa *et al.*, 2010; Carvalho *et al.*, 2014). The first isolation of WNV was in a febrile woman at Omogo in the West Nile district of Uganda, in 1937 (Smithburn *et al.*, 1940). Since then, WNV has always been associated with small and sporadic outbreaks of a self-limiting febrile illness without complications and with few cases of neurological disorders, known as West Nile fever (Couissinier-Paris, 2006). Nevertheless, WNV showed a large geographic distribution, with reports of viral circulation in different countries of Africa (Hurlbut *et al.*, 1956), Europe (Ernek *et al.*, 1971), Asia (Pavri and Singh, 1965) and Australia.

In the Americas, the first outbreak of WNV occurred in New York City in the USA in the summer of 1999, with deaths reported in humans, horses and birds (CDC, 1999). In South America, the WNV circulation was not described until 2005, when evidence of seropositivity to WNV in horses in Colombia was identified (Mattar *et al.*, 2005). Serological evidence of WNV circulation in South American countries has been reported in birds and horses in Colombia, Venezuela and Argentina (Mattar *et al.*, 2005; Bosch *et al.*, 2007; Diaz *et al.*, 2008; Tauro *et al.*, 2012). Moreover, encephalitis cases suggesting human clinical infection with WNV have also been reported in Argentina (Morales *et al.*, 2006; Artsob *et al.*, 2009). The first serological evidence of WNV circulation in Brazil was in 2009, in horses from Pantanal, Mato Grosso do Sul state (Pauvolid-Correa *et al.*, 2011). Further evidence of WNV circulation in this country was also identified in horses of the Midwest of Brazil (Melandri *et al.*, 2012; Ometto *et al.*, 2013). In December 2014, the first human case of WNV was reported in Piauí, Brazil (WHO, 2014).

9.1.4 *Anopheles darlingi* and malaria

Malaria is considered a public health concern in tropical and subtropical regions in sub-Saharan Africa, Central and South America, the Middle East, Southeast Asia and Oceania (Bloland, 2001). The etiological agent of malaria is a protozoan of family Plasmodiidae and genus *Plasmodium*.

The transmission of *Plasmodium* to man occurs when an infected mosquito female of the genus *Anopheles* blood-feeds. *Anopheles darlingi* belongs to subgenus *Nyssorhynchus* and has a wide geographic distribution in South and Central America, extending from south Mexico to north Argentina, and from the east of the Andes chain to the coast of the Atlantic Ocean, being the main vector of malaria in South America (Deane, 1948; Forattini, 1987; Hiwat and Bretas, 2011). This anopheline shows an anthropophilic and endophilic behaviour, invading residences during the evening hours to blood-feed on man. The breeding sites of *An. darlingi* are represented by large water collections, fish tanks, rivers and streams, shaded or partially shaded and poor in salts (Forattini, 2002).

The frequency of *An. darlingi* seems to be closely related to the annual cycle of rainfall, although the relationship between the abundance of this species and rainfall seems to differ at different localities or regions. Forattini (1987) considered that the occurrence of heavy rains could possibly flood breeding sites and create flood currents that carry away immature forms, which was also found by Pajot *et al.* (1977) in French Guiana, who observed that heavy rains are followed by a decrease or a total absence of *An. darlingi*. This phenomenon has been referred to as the 'flush-out theory' (Barros *et al.*, 2011). Larvae of *An. darlingi* seem to require stable chemical and physical conditions in the breeding sites, which is generally found in large, deep and clear water bodies, such as lakes, swamps or large rivers (Deane *et al.*, 1948; Rachou, 1958; Rozendaal, 1992). Adequate larval conditions may depend on depth of the water, temperature, pH, chemical stability and light/shade proportions (Rachou, 1958). Singer and Castro (2001) considered the forest margins to be the principal breeding sites for *An. darlingi* in the Amazon. This was experimentally confirmed by Barros and Honório (2015): based on clustering of larvae at deforestation transitions of a temporary river, a 'forest fringe model' was proposed. The model predicts that three factors that co-occur at forest fringes contribute to increased breeding of *An. darlingi*: availability of human blood meals; stable surface water; and low luminance at water perimeters.

Historically, malaria was controlled in many parts of the world because of socioeconomic developments in many developed countries. In addition, specific control programmes that used insecticides such as DDT were implemented in the 1960s and 1970s and helped in eradicating or controlling malaria in many areas in Asia and Latin America, and eradicated it from Europe. Nevertheless, the development of widespread resistance to insecticides and the drugs used to treat the disease have also contributed to the current global situation (Martens *et al.*, 1999). About 3.2 billion people – almost half of the world's population – live in areas at risk of malaria. Nevertheless, between 2000 and 2015, malaria incidence and malaria death rates respectively fell by 37% and by 60% globally. Sub-Saharan Africa carries the largest number of malaria cases in the world: in 2015, the region was responsible for 89% of malaria cases and 91% of malaria deaths (WHO, 2015).

Although malaria is typically considered mainly a problem of rural and poor areas, this disease has been a concern in urban settings (Tadei *et al.*, 1988; Tatem and Hay 2004; Cabral *et al.*, 2010). The process of urbanization includes landscape modification and transformation of environs through demand for resources and improved communications (Tatem *et al.*, 2013). In spite of being considered a riverine, forest-dwelling species, *An. darlingi* is becoming more abundant in peri-urban areas, increasing malaria risk. This has been associated with human-driven environmental changes such as deforestation.

9.2 Climate Change

The climate is a complex and interactive system consisting of the atmosphere, land surface, snow, oceans and living organisms. According to the IPCC (Intergovernmental Panel on Climate Change), the climate system can be influenced by its own internal dynamics or by external forces, such as volcanic eruptions, solar variations and human-induced changes in atmosphere composition (Hegerl *et al.*, 2007).

In 1750, the world watched the Industrial Revolution transforming hand tools and handmade items into machine-manufactured and mass-produced goods. When the Industrial Revolution took hold, society's attention turned from the rural home to the urban factory and from human power to mechanical power. Indeed, this change generally helped life, but also hindered it as well. In addition to the notorious decline of work conditions and the increased number of women and children working long hours for low wages in dangerous conditions, pollution, such as carbon dioxide levels in the atmosphere, started to continually rise.

It is generally thought that greenhouse gases are harmful for living organisms. On the contrary, our planet Earth is habitable because of the natural greenhouse effect in its atmosphere. Gases that contribute to the natural greenhouse effect on Earth's atmosphere, such as water vapour, carbon dioxide and ozone, act as a partial blanket for the longwave radiation coming from the surface, warming the Earth and allowing life to exist. However, in the last century, human activities have intensified this blanketing effect through the continual release of greenhouse gases (Hegerl *et al.*, 2007), making the Earth warmer.

The Earth's surface has been warmer in the last three decades than in any preceding decade since 1850. From 1880 up to 2012, it is estimated that the global temperature – associated with land and ocean surface temperatures – increased by approximately 1°C, mainly because of anthropogenic greenhouse gas emissions from the pre-industrial era onwards. It can be estimated with a high degree of certainty that about half of the anthropogenic CO_2 emissions in the past two centuries occurred during the last 40 years (IPCC, 2007). If greenhouse gas emissions continue at the same current rate or above, we might have larger changes in the global climate system during the 21st century than those observed during the 20th century. For the end of the 21st century (2090–2099), several models evaluated different scenarios for global average surface warming in relation to 1980–1999 and the best case predicted an increase of 1.8°C (range 1.1°C–2.9°C), whereas the worst case predicted an increase of 4.0°C (range 2.4°C–6.4°C) (IPCC, 2007). Climate change is a great challenge for public health, because it can have direct and indirect effects. The former is related to extreme climate conditions, like heatwaves and natural disasters (hurricanes, floods and droughts), whereas the latter is related to vector-borne and water-transmission diseases.

For South America, climate changes over the past century have included changes in precipitation rates and intensities in several countries, such as southern Brazil, Paraguay and Argentina. Climate change could affect South American ecosystems and has the potential to endanger subsistence farmers and pastoral peoples, which represent a large portion of the rural populations of the Andean plateaus and tropical and subtropical forest areas. Moreover, environmental deterioration resulting from the misuse of land might be aggravated by the impacts of climate change on water availability and agricultural lands, because of coastal inundation stemming from sea-level rise and flooding (IPCC, 2007).

Infectious and vector-borne diseases are among the most important causes of morbidity and mortality in South America and they are much more common in tropical and subtropical areas than in temperate areas. Global warming could extend the coverage area of these diseases and increase the chances of outbreaks in new areas. When cholera was introduced into Peru in 1993, for example, it produced an outbreak that spread to other countries of South America, such as Argentina. Cholera and other diarrhoeas are associated with the distribution and quality of surface water, as well as with flooding and water shortages. Vector-borne diseases could also be impacted by climate change, since they could expand to new areas with favourable conditions for pathogens, reservoirs and vectors (IPCC, 2007). Vulnerabilities vary with geography, age, gender, race, ethnicity and socioeconomic status, and potentially the rise in large cities.

9.3 Climate Change and Urban Mosquitoes

It is recognized that climate change will affect vector-borne diseases dynamics, like dengue, chikungunya, West Nile and malaria. Mosquito vectors are currently spreading into temperate areas, including Europe, exposing vulnerable human populations to pathogens that they have never had contact with previously. The disease transmission dynamics englobes key elements that need to be well integrated: the vector (in this case, mosquito), the pathogen, the vertebrate host and the environment. Understanding how climate change acts on this multidisciplinary topic will help us to improve not only reliable and robust projections for future global and regional vector-borne disease burdens, but also vector- and disease-control strategies to reduce the impact of these disease outbreaks in established regions and prevent them in vulnerable populations (Parham *et al.*, 2015).

Urban mosquitoes have the remarkable characteristic of being well adapted to urban and suburban environments, with a strong association with humans. The distribution and geographic expansion of urban mosquitoes are highly influenced by environmental and social aspects, including the climate, population density and economic activity (Patz *et al.*, 2005). Climate change can directly affect disease transmission by shifting the vector's geographic range or by changing some biological, ecological and physiological aspects of urban mosquitoes (Patz *et al.*, 1996). Several studies have explored the influence of climatic change in the occurrence of infectious diseases, such as dengue and malaria. Some of these studies documented this influence based on mathematical models; others have reported the effects of abiotic variations, such as temperature, on the vector's life cycle under laboratory conditions.

9.3.1 Abiotic variations and mosquito vectors

Abiotic variations, especially temperature, humidity and precipitation, affect several biological aspects of mosquito vectors, such as survival, reproduction, distribution and abundance (Paz, 2015). Increasing temperature due to global climate change may cause a decrease in the time needed for parasite development; changes in mosquito abundance and survivorship; changes in the gonotrophic cycle; and changes in larval development and pupation rates (Patz and Olson, 2006).

Under natural conditions, the density of DENV, CHIKV and ZIKV vectors – *Ae. aegypti* and *Ae. albopictus* – is affected by temperature variation (Consoli and Lourenço-de-Oliveira, 1994; Kuno, 1995; Honório *et al.*, 2009; Juliano *et al.*, 2002). Moreover, several studies have evaluated the effects of temperature on biological aspects of *Ae. aegypti*, *Ae. albopictus* and the West Nile vector, *Cx. quinquefasciatus*, under laboratory conditions. At 25°C, *Cx. quinquefasciatus* and *Ae. aegypti* required 8.72 and 8.86 days, respectively, to complete the total development cycle from egg hatch to adult emergence. Nevertheless, at 27°C, this total development time decreased to 7.79 and 7.30 days for *Cx. quinquefasciatus* and *Ae. aegypti*, respectively (Rueda *et al.*, 1990). Ciota *et al.* (2014) also reported that temperature increases are associated with decreased development time of *Cx. quinquefasciatus*. Males and females of *Ae. albopictus* exposed to cyclic temperatures of 20°C/27°C showed a higher development rate than those exposed to cyclic temperatures of 18°C/25°C (Lowenberg-Neto and Navarro-Silva, 2004). These data indicate that the temperature increase can decrease the development time of *Aedes* and *Culex* mosquitoes, enhancing population growth.

Calado and Navarro-Silva (2002) evaluated the effects of temperature on the blood-feeding and oviposition behaviours of *Ae. albopictus* females under 12 hours of light and 12 hours of dark (LD 12:12), at 20°C, 25°C and 30°C. At 25°C, the mean number of blood-feedings of *Ae. albopictus* females was significantly higher than the mean number of blood-feedings of females exposed to 20°C. Nevertheless, no significant difference was observed between the mean number of blood-feedings of females exposed to 25°C and 30°C. In addition, the mean number of eggs/day/females was not significantly different at 25°C and 30°C, but it was significantly higher than the mean number of eggs/day/females at 20°C. Thus, the increase of 5°C (25°C–30°C) did not negatively affect the frequency of blood-feeding nor the oviposition activity of *Ae. albopictus* females.

Extrinsic incubation period (EIP) is defined as the viral incubation period from the time when a female blood-feeds on a viremic human host up to the time she becomes infectious. Thus, the virus present in the blood meal invades several tissues of the mosquito, and reaches the salivary glands, when the EIP is completed and the female is considered infectious (Chan and Johansson, 2012). EIP is an important component of virus transmission dynamics and it can be affected by temperature (McLean *et al.*, 1974; Watts *et al.*, 1987). Rohani *et al.* (2009) evaluated the effects of temperature on the EIP of DENV-2 and DENV-4 in *Ae. aegypti* females exposed to 26°C, 28°C and 30°C. The viruses were first detected in the salivary glands on Day 9 at 26°C and 28°C and on Day 5 at 30°C. These data indicate that the incubation period of the viruses decreases when the extrinsic incubation temperature increases. Thus, an increase of global temperature could enhance dengue virus transmission by *Ae. aegypti* females.

Nevertheless, it is important to point out that temperature can negatively affect the survival of *Aedes*, *Culex* and *Anopheles* adults as well (Calado and Navarro-Silva, 2002; Ciota *et al.*, 2014; Christiansen-Jucht *et al.*, 2015). For example, the longevity of males and females of *Aedes albopictus* exposed to 30°C was significantly lower than those exposed to 25°C (Calado and Navarro-Silva, 2002). Similarly, Ciota *et al.* (2014) observed a significant decrease in median survival of *Cx. quinquefasciatus* when the temperature increased from 28°C to 32°C, whereas adult mortality of *Anopheles gambiae* – a malaria vector in Africa – was positively associated with a 4°C increase in temperature (from 23°C to 27°C) (Christiansen-Jucht *et al.*, 2015).

9.4 Climate Predictions and Mosquito Vectors

Maps of the geographical distribution of mosquito vectors have always been available, but only recently have these data been analysed in order to extract information on the environmental constraints keeping a species within its characteristic range. Data from weather stations and satellites can be allied to determine which predictor variable is most useful for describing vector distributions or for foretelling alterations in distribution with climate change (Rogers and Packer, 1993). The choice of climate variables is an important issue to consider when modelling the geographical expansion of urban mosquito vectors to new areas. Several climate variables have already been considered in vector-borne disease modelling studies, such as minimum, maximal and mean temperature, humidity, rainfall and ENSO (El-Niño–Southern Oscillation). Temperature and relative humidity seem to be the best significant predictors of dengue transmission, for example (Naish *et al.*, 2014).

Using an empirical model, Hales *et al.* (2002) estimated changes in the geographical limits of dengue fever transmission and the size of population at risk between the 2050s and 2080s, based on future climate change situations defined by the IPCC. The authors defined humidity as the best individual predictor of dengue fever distribution and confirmed that the geographical limits of dengue fever transmission are influenced by climate. Data from 1990 indicate that almost 30% of the world's population lived in areas of dengue transmission. With climate and population projections for 2085, this will increase to 50–60% (Hales *et al.*, 2002). Climate change associated with higher temperatures or higher rainfall can also affect the spatial and temporal distribution of dengue and chikungunya, increasing the transmission. Nevertheless, several studies have indicated that drought is an important factor that needs to be considered as well, since water storage increases the number of suitable breeding sites for mosquitoes (Barclay, 2008; Measron and Paterson, 2014).

Campbell *et al.* (2015) assessed the potential global distribution of *Ae. aegypti* and *Ae. albopictus* in the future using correlative ecological niche modelling approaches. For *Ae. aegypti*, a potential expansion in eastern North America, South Asia and east and southward in Africa and Asia was observed, as well as a potential broadening distribution in interior South America. For *Ae. albopictus*, a potential expansion in eastern North America and East Africa as well as eastern and southern South America was indicated. Nevertheless, the reorganization of the distribution of *Ae. aegypti* and *Ae. albopictus* strongly depends on their particular ecological niche profile and their climate tolerance (Campbell *et al.*, 2015).

In addition to climate change, it is important to point out that social, political and economic aspects, such as human migration, deforestation and poverty in urban areas, can intensify the transmission of dengue and other arbovirosis, increasing the availability of breeding sites for vectors, for example (Githeko *et al.*, 2000). Rapid urbanization and international travel are important factors that can explain the expansion of dengue, chikungunya and Zika in the Americas. Highly urbanized communities show a great concentration of people living in relatively small geographic spaces and the greater the population density, the greater the risk of mosquito-transmitted disease. In addition, the increased mobility of people within and between countries, as well as from rural areas to cities, also contributes to the spread of arbovirosis. Thus, although there is strong evidence of a correlation between climate variability and arbovirosis incidence, living conditions are also crucially important (Barclay, 2008; Barcellos *et al.*, 2009).

Githeko *et al.* (2000) indicate that the increase of global temperature will positively affect the spread of several vector-borne diseases in new areas, but this effect will be influenced by human settlement patterns in different regions. Thus, the relationship between global climate change and the epidemiology of vector-borne diseases should be assessed at a continental level. The total number of dengue cases, for instance, has increased in the last decades in South

America, where most people live in urban settlements. With a rise of 4°C in 2100, we could expect an increase of dengue transmission intensity in most of South America (Githeko *et al.*, 2000).

The epidemiology of West Nile fever is affected by several climatic variables, and changes in these parameters may alter the dynamics of this arbovirosis transmission. Temperature, for example, correlates positively with viral replication rates, growth rates of *Culex* mosquito populations, viral transmission efficiency to birds, and geographical variation in human case incidence. Floods caused by precipitation lead to higher mosquito abundance and potential disease outbreaks in humans, whereas droughts caused by lack of precipitation can facilitate population outbreaks of some mosquito species and increase the number of standing water containers that attract several species of mosquitoes and birds. Relative humidity is positively correlated with vector population dynamics and morbidity in humans (Paz, 2015).

Few studies have evaluated the impact of climate change on West Nile virus epidemiology in South America probably because very few human cases with WNV were reported. Nevertheless, the IPCC has reported an overall increase of temperature and precipitation in South America which could facilitate the spread of *Cx. quinquefasciatus* and WNV in this continent (Paz, 2015).

Climate–malaria models have also shown an increase in the malaria transmission season in several areas like eastern Africa, Central America and southern Brazil for the next few years (Caminade *et al.*, 2014). Both *Anopheles* and *Plasmodium* are sensitive to temperature. Thus, climate change could affect malaria transmission by increasing its distribution where it is currently limited by low temperature; decreasing its distribution in areas where it becomes too dry for mosquitoes; increasing or decreasing the months of transmission in areas where it is endemic; and increasing the risk of outbreaks in areas where the disease has been eradicated but vectors are still present, such as in Europe or the United States (Kovats *et al.*, 2001).

9.5 Conclusions

The Earth has been undergoing global climate change since the Industrial Revolution, when human activities started to emit significant amounts of carbon dioxide into the atmosphere. The increasing urbanization of cities further increased the emission of this gas, as well as other gases that contribute to the greenhouse phenomenon. Thus, currently, most scientists agree that the world is undergoing global warming, with serious consequences to man. Among these consequences, natural disasters, such as hurricanes and floods, long periods of drought, heavy precipitation and increasing average temperatures can be observed. In addition, the emerging and re-emerging of infectious diseases are affected by several factors, including transport globalization, urbanization and global climate changes. Climate change would directly affect disease transmission by shifting the vectors' biology and geographic range.

South America is the fourth-largest continent in the world, and it is located mostly in the Southern Hemisphere, showing diverse patterns of weather and climate, including tropical, subtropical and extratropical features. Moreover, this continent is considered endemic to several mosquitoborne diseases that are sensitive to climate, including dengue and other arbovirosis. In South America, the most important mosquito vectors are *Ae. aegypti* and *Ae. albopictus*, which transmit DENV, CHIKV and ZIKV, *Cx. quinquefasciatus*, considered a potential vector of WNV, and *An. darlingi*, the main malaria vector.

Temperature is one of the most important abiotic factors that can affect the biology of these urban mosquito vectors. The total development time of these mosquito species (egg, larva, pupa and adult) depends on climatic variables, especially temperature. The development time of mosquito immatures is generally shortened by warm temperatures, which increases the population of the vectors. Moreover, higher temperatures reduce the extrinsic incubation period (EIP), resulting in a higher proportion of infected mosquitoes.

Higher temperatures could also contribute to the expansion of the pathogens transmitted by *Ae. aegypti*, *Ae. albopictus*, *Cx. quinquefasciatus* and *An. darlingi* and several models have shown that the distribution of vectors and vector-borne diseases can be highly affected by temperature and humidity. Nevertheless, social, political and economic characteristics of each country are important vulnerabilities that will influence the climate change impacts in a stronger or weaker way. Thus, we may assume that mosquito vectors currently confined to the tropics may spread into temperate regions if global warming occurs, but it is not certain that the diseases they are associated with will be as prevalent in the newly invaded areas as elsewhere.

Since climate change is already underway, it is important to focus on research that aims to minimize the impacts of this phenomenon. The research team should include entomologists, climatologists and epidemiologists, and the greatest challenge for these professionals is to share expertise to maintain an active and continuous surveillance on mosquito vectors, as well as a rapid diagnostic of arboviruses suspected in human cases, in order to reduce the impacts of current climate variability on urban mosquito vectors and to minimize possible future epidemics.

References

Artsob, H., Gubler, D.J., Enria, D.A., Morales, M.A., Pupo, M., Bunning, M.L. and Dudley, J.P. (2009) West Nile Virus in the New World: trends in the spread and proliferation of West Nile Virus in the Western Hemisphere. *Zoonoses and Public Health* 56, 357–369.

Barcellos, C., Monteiro, A.M.V., Corvalán, C., Gurgel, H.C., Carvalho, M.S., Artaxo, P., Hacon, S. and Ragoni, V. (2009) Mudanças climáticas e ambientais e as doenças infecciosas: cenários e incertezas para o Brasil [Climatic and environmental changes and their effect on infectious diseases: scenarios and uncertainties for Brazil]. *Epidemiologia e Serviços de Saúde* 18, 285–304.

Barclay, E. (2008) Is climate change affecting dengue in the Americas? *Lancet* 371, 973–974.

Barros, F.S. and Honório, N.A. (2015) Deforestation and malaria on the Amazon frontier: larval clustering of *Anopheles darlingi* (Diptera: Culicidae) determines focal distribution of malaria. *The American Journal of Tropical Medicine and Hygiene* 93, 939–953.

Barros, F.S.M., Honório, N.A. and Arruda, M.E. (2011) Survivorship of *Anopheles darlingi* (Diptera: Culicidae) in relation with malaria incidence in the Brazilian Amazon. *PLoS ONE* 6(8), e22388.

Beserra, E.B., Junior, F.P.C., Santos, J.W., Santos, T.S., Fernandes, C.R.M. (2006) Biologia e exigências térmicas de *Aedes aegypti* (L.) (Diptera: Culicidae) provenientes de quatro regiões bioclimáticas da Paraíba. *Neotropical Entomology* 35, 853–860.

Bhatt, S.P., Gething, W., Brady, O.J., Messina, J.P., Farlow, A.W., Moyes, C.L., Drake, J.M., Brownstein, J.S., Hoen, A.G., Sankoh, O., Myers, M.F., George, D.B., Jaenisch, T., Wint, G.R., Simmons, C.P., Scott, T.W., Farrar, J.J. and Hay, S.I. (2013) The global distribution and burden of dengue. *Nature* 496, 504–507.

Bloland, P.B. (2001) Drug resistance in malaria. WHO/CDS/CSR/DRS, Switzerland. Available at: http://www.who.int/csr/resources/publications/drugresist/malaria.pdf (accessed December 2015).

Bosch, I., Herrera, F., Navarro, J., Lentino, M., Dupuis, A., Maffei, J., Jones, M., Fernández, E., Pérez, N., Pérez-Emán, J., Guimarães, A.E., Barrera, R., Valero, N., Ruiz, J., Velásquez, G., Martinez, J., Comach, G., Komar, N., Spielman, A. and Kramer, L. (2007) West Nile virus, Venezuela. *Emerging Infectious Diseases* 13, 651–653.

Cabral, A.C., Fé, N.F., Suárez-Mutis, M.C., Bóia, M.N. and Carvalho-Costa, F.A. (2010) Increasing incidence of malaria in the Negro River basin, Brazilian Amazon. *Transactions of the Royal Society of Tropical Medicine and Hygiene* 104, 556–562.

Calado, D.C. and Navarro-Silva, M.A. (2002) Evaluation of the temperature influence on the development of *Aedes albopictus*. *Revista de Saúde Pública* 36, 173–179.

Calderón-Arguedas, O., Troyo, A., Moreira-Soto, R.D., Marín, R. and Taylor, L. (2015) Dengue viruses in *Aedes albopictus* Skuse from a pineapple plantation in Costa Rica. *Journal of Vector Ecology* 40(1), 184–186.

Caminade, C., Kovats, S., Rocklov, J., Tompkins, A.M., Morse, A.P., Colón-González, F.J., Stenlund, H., Martens, P. and Lloyd, S.J. (2014) Impact of climate change on global malaria distribution. *Proceedings of the National Academy of Sciences* 111, 3286–3291.

Campbell, L.P., Luther, C., Moo-Llanes, D., Ramsey, J.M., Danis-Lozano, R. and Peterson, A.T. (2015) Climate change influences on global distributions of dengue and chikungunya virus vectors. *Philosophical Transactions of the Royal Society B* 370, article ID 20140135. DOI: 10.1098/rstb.2014.0135.

Carvalho, G.C., Malafronte, R.S., Miti Izumisawa, C., Souza Teixeira, R., Natal, L. and Marrelli, M.T. (2014) Blood meal sources of mosquitoes captured in municipal parks in São Paulo, Brazil. *Journal of Vector Ecology* 39, 146–152.

Castro, M.G., Nogueira, R.M.R., Schatzmayr, H.G., Miagostovich, M.P. and Lourenço-de-Oliveira, R. (2004) Dengue virus detection by using reverse transcription-polymerase chain reaction in saliva and progeny of experimentally infected *Aedes albopictus* from Brazil. *Memórias Instituto Oswaldo Cruz* 99, 809–814.

CDC (1999) Outbreak of West Nile-like viral encephalitis – New York, 1999 (Centers for Disease Control and Prevention). Available at: http://www.cdc.gov/mmwr/preview/mmwrhtml/mm4838a1.htm (accessed January 2015).

CDC (2014) Chikungunya virus: symptoms, diagnosis, & treatment (Centers for Disease Control and Prevention). Available at: http://www.cdc.gov/chikungunya/symptoms/index.html (accessed December 2014).

Chan, M. and Johansson, M.A. (2012) The incubation periods of dengue viruses. *PLoS ONE* 7(11), article ID e50972. DOI: 10.1371/journal.pone.0050972.

Chouin-Carneiro, T., Vega-Rua, A., Vazeille, M., Yebakima, A., Girod, R., Goindin, D., Dupont-Rouzeyrol, M., Lourenço-de-Oliveira, R. and Failloux, A.B. (2016) Differential susceptibilities of *Aedes aegypti* and *Aedes albopictus* from the Americas to Zika virus. *PLoS Neglected Tropical Diseases* 10(3), article ID e0004543.

Christiansen-Jucht, C.D., Parham, P.E., Saddler, A., Koella, J.C. and Basáñez, M.G. (2015) Larval and adult environmental temperatures influence the adult reproductive traits of *Anopheles gambiae* s.s. *Parasites and Vectors* 8, 456.

Ciota, A.T. and Kramer, L.D. (2013) Vector-virus interactions and transmission dynamics of West Nile virus. *Viruses* 5, 3021–3047.

Ciota, A.T., Matacchiero, A.C., Kilpatrick, A.M. and Kramer, L.D. (2014) The effect of temperature on life history traits of *Culex* mosquitoes. *Journal of Medical Entomology* 51, 55–62.

Consoli, R.A.G.B. and Lourenço-de-Oliveira, R. (1994) Principais mosquitos de importância sanitária no Brasil [Main mosquitoes' health importance in Brazil]. Fiocruz, Rio de Janeiro, Brazil.

Couissinier-Paris, P. (2006) West Nile virus in Europe and Africa: still minor pathogen, or potential threat to public health? *Bulletin de la Société de Pathologie Exotique* 99, 348–354.

Deane, L.M., Causey, O.R. and Deane, M.P. (1948): Notas sobre a distribuição e a biologia dos anofelinos das regiões nordestina e amazônica do Brasil [Notes on the distribution and biology of *Anopheles* of the northeast and Amazon regions of Brazil]. *Revista Serviço Especial de Saúde Pública* 1, 827–965.

Diaz, L.A., Komar, N., Visintin, A., Júri, M.J.D., Stein, M., Allende, R.L., Spinsanti, L., Konigheim, B., Aguilar, J., Laurito, M., Almirón, W. and Contigiani, M. (2008) West Nile Virus in birds, Argentina. *Emerging Infectious Diseases* 14, 689–691.

Duffy, M.R., Chen, T.H., Hancock, W.T., Powers, A.M., Kool, J.L., Lanciotti, R.S., Pretrick, M., Marfel, M., Holzbauer, S., Dubray, C., Guillaumot, L., Griggs, A., Bel, M., Lambert, A.J., Laven, J., Kosoy, O., Panella, A., Biggerstaff, B.J., Fischer, M. and Hayes, E.B. (2009) Zika virus outbreak on Yap Island, Federated States of Micronesia. *New England Journal of Medicine* 360, 2536–2543.

Eldridge, B.F. (2008) *The Biology and Control of Mosquitoes in California*. Training manual prepared in collaboration with Vector-Borne Disease Section, Center for Infectious Diseases, California Department of Public Health, Davis, California, USA. Available at: https://www.cdph.ca.gov/certlic/occupations/Documents/ManualBiologyandControlofMosquitoesinCA.pdf (accessed June 2016).

Ernek, E., Kozuch, O., Sekeyová, M., Hudec, K. and Folk, C. (1971) Antibodies to arboviruses in birds in Czechoslovakia. *Acta Virologica* 15, 335.

Forattini, O.P. (1987) Exophilic behavior of *Anopheles darlingi* root in a southern region of Brazil. *Revista Saúde Pública* 21, 291–304.

Forattini, O.P. (2002) *Culicidologia Médica*, Vol. 2: *Identificação, biologia, epidemiologia*. Editora da Universidade de São Paulo – EDUSP, São Paulo, Brazil.

Garreaud, R.D., Vuille, M. Compagnucci, R. and Marengo, J. (2009) Present-day South American climate. *Paleogeography, Palaeoclimatology, Palaeoecology* 281, 180–195.

Githeko, A.K., Lindsay, S.W., Confalonieri, U.E. and Patz, J.A. (2000) Climate change and vector-borne diseases: a regional analysis. *Bulletin World Health Organization* 78, 1136–1147.

Glasser, C.M. and Gomes, A.C. (2002) Climate and the superimposed distribution of *Aedes aegypti* and *Aedes albopictus* on infestation of São Paulo State, Brazil. *Revista Saúde Pública* 36, 166–172.

Gomes, A.C., Silva, N.N., Marques, G.R. and Brito, M. (2003) Host-feeding patterns of potential human disease vectors in the Paraíba Valley region, State of São Paulo, Brazil. *Journal of Vector Ecology* 28, 74–78.

Gomes, A.C., de Souza, J.M., Bergamaschi, D.P., Dos Santos, J.L.F., Andrade, V.R., Leite, O.F., Rangel, O., de Souza, S.S., Guimarães, N.S. and de Lima, V.L. (2005) Anthropophilic activity of *Aedes aegypti* and of *Aedes albopictus* in area under control and surveillance. *Revista Saúde Pública* 39, 206–210.

Gubler, D.J. (2007) Emerging infections: the continuing spread of West Nile Virus in the Western Hemisphere. *Clinical Infectious Diseases* 45, 1039–1046.

Hales, S., De Wet, N., Maindonald, J. and Woodward, A. (2002) Potential effect of population and climate changes on global distribution of dengue fever: an empirical model. *Lancet* 360, 830–834.

Halstead, S.B. (1988) Pathogenesis of dengue: challenges to molecular biology. *Science* 239, 479–481.

Hayes, E.B. (2009) Zika virus outside Africa. *Emerging Infectious Diseases* 15, 1347–1350.

Hegerl, G.C., Zwiers, F.W., Braconnot, P., Gillett, N.P., Luo, Y., Marengo Orsini, J.A., Nicholls, N., Penner, J.E. and Stott, P.A. (2007) Understanding and attributing climate change. In: Solomon, S., Qin, D., Manning, M., Chen, Z., Marquis, M., Averyt, K.B., Tignor, M. and Miller, H.L. (eds) *Climate Change 2007: The Physical Science Basis*. Contribution of Working Group I to the Fourth Assessment Report of the Intergovernmental Panel on Climate Change. Cambridge University Press, Cambridge, UK and New York, USA, pp. 663–745.

Hiwat, H. and Bretas, G. (2011) Ecology of *Anopheles darlingi* Root with respect to vector importance: a review. *Parasites and Vectors* 4, 177.

Hoeck. P.A.E., Ramberg, F.B., Merrill, S.A., Moll, C. and Hagerdorn, H.H. (2003) Population and parity levels of *Aedes aegypti* collected in Tucson. *Journal of Vector Ecology* 28, 65–73.

Honório, N.A, Codeço, C.T., Alves, F.C., Magalhães, M.A. and Lourenço-de-Oliveira, R. (2009) Temporal distribution of Aedes aegypti in different districts of Rio de Janeiro, Brazil, measured by two types of traps. *Journal of Medical Entomology* 46, 1001–1014.

Hurlbut, H.S., Riizk, F., Taylor, R.M. and Work, T.H. (1956) A study of the ecology of West Nile virus in Egypt. *American Journal of Tropical Medicine and Hygiene* 5(4), 579–620.

Ibáñez-Bernal, S., Briseño, B., Mutebi, J.P., Argot, E., Rodríguez, G., Martínez-Campos, C., Paz, R., de la Fuente-San Román, P., Tapia-Conyer, R. and Flisser, A. (1997) First record in America of *Aedes albopictus* naturally infected with dengue virus during the 1995 outbreak at Reynosa, Mexico. *Medical Veterinary Entomology* 11(4), 305–309.

IPCC (2007) The regional impacts of climate change. Intergovernmental Panel on Climate Change. Available at: http://www.ipcc.ch/ipccreports/sres/regional/144.htm#indirec (accessed November 2015).

Juliano, S.A., O'Meara, G.F., Morrill, J.R. and Cutwa, M.M. (2002) Desiccation and thermal tolerance of eggs and the coexistence of competing mosquitoes. *Oecologia* 130, 458–469.

Kovats, R.S., Campbell-Lendrum, D., McMichael, A.J., Woodward, A. and Cox, J. (2001) Early effects of climate change: do they include changes in vector borne diseases? *Philosophical Transactions of the Royal Society of London, Series B: Biological Sciences* 356, 1057–1068.

Kuno, G. (1995) Review of the factors modulating dengue transmission. *Epidemiology Review* 17, 321–335.

Kucharz, E.J. and Cebula-Byrska, I. (2012) Chikungunya fever. *European Journal of Internal Medicine* 23, 325–329.

Lahariya, C. and Pradhan, S.K. (2006) Emergence of chikungunya virus in Indian subcontinent after 32 years: a review. *Journal of Vector Borne Diseases* 43, 151–160.

Lima-Camara, T.N., Honório, N.A. and Lourenço-de-Oliveira, R. (2006) Freqüência e distribuição espacial de *Aedes aegypti* e *Aedes albopictus* (Diptera: Culicidae) no Rio de Janeiro, Brasil [Frequency and spatial distribution of *Aedes aegypti* and *Aedes albopictus* (Diptera: Culicidae) in Rio de Janeiro, Brazil]. *Cadernos de Saúde Pública* 22, 2079–2084.

Lorosa, E.S., Faria, M.S., de Oliveira, L.C., Alencar, J. and Marcondes, C.B. (2010) Blood meal identification of selected mosquitoes in Rio de Janeiro, Brazil. *Journal of American Mosquito Control Association* 26, 18–23.

Lourenço-de-Oliveira, R., Vega-Rua, A., Vezzani, D., Willat, G., Vazeille, M., Mousson, L. and Failloux, A.B. (2013) *Aedes aegypti* from temperate regions of South America are highly competent to transmit dengue virus. *BMC Infectious Diseases* 13, 610–617.

Lowenberg-Neto, P. and Navarro-Silva, M.A. (2004) Development, longevity, gonotrophic cycle and oviposition of *Aedes albopictus* Skuse (Diptera: Culicidae) under cyclic temperatures. *Neotropical Entomology* 33, 29–33.

Martens, P., Kovats, R.S., Nijhof, S., de Vires, P., Livermore, M.T., Bradley, D.J., Cox, J. and

McMichael, A.J. (1999) Climate change and future populations at risk from malaria. *Global Environmental Change* 9, S89–S107.

Martins, V.E., Alencar, C.H., Kamimura, M.T., de Carvalho Araújo, F.M., De Simone, S.G., Dutra, R.F. and Guedes, M.I. (2012) Occurrence of natural vertical transmission of dengue-2 and dengue-3 viruses in *Aedes aegypti* and *Aedes albopictus* in Fortaleza, Ceará, Brazil. *PLoS ONE* 7(7), article ID e41386. DOI: 10.1371/journal.pone.0041386.

Mattar, S., Edwards, E., Laguado, J., González, M., Alvarez, J. and Komar, N. (2005) West Nile virus antibodies in Colombian horses. *Emerging Infectious Diseases (Letters)* 11, 1497–1498.

McLean, D.M., Clarke, A.M., Coleman, J.C., Montalbetti, C.A., Skidmore, A.G., Walters, T.E. and Wise, R. (1974) Vector capability of *Aedes aegypti* mosquitoes for California encephalitis and dengue viruses at various temperatures. *Canadian Journal of Microbiology* 20, 255–262.

Meason, B. and Paterson, R. (2014) Chikungunya, climate change, and human rights. *Health Human Rights* 16, 105–112.

Melandri, V., Guimarães, A.E., Komar, N., Nogueira, M.L., Mondini, A., Fernandez-Sesma, A., Alencar, J. and Bosch, I. (2012) Serological detection of West Nile virus in horses and chicken from Pantanal, Brazil. *Memórias do Instituto Oswaldo Cruz* 107(8), 1073–1075.

Méndez, F., Barreto, M., Arias, J.F., Rengifo, G., Muñoz, J., Burbano, M.E. and Parra, B. (2006) Human and mosquito infections by dengue viruses during and after epidemics in a dengue-endemic region of Colombia. *American Journal of Tropical Medicine and Hygiene* 74(4), 678–683.

Miller, B.R. and Ballinger, M.E. (1988) *Aedes albopictus* mosquitoes introduced into Brazil: vector competence for yellow fever and dengue viruses. *Transactions of the Royal Society of Tropical Medicine and Hygiene* 82, 476–477.

Morales, M.A., Barrandeguy, M., Fabbri, C., Garcia, J.B., Vissani, A., Trono, K., Gutierrez, G., Pigretti, S., Menchaca, H., Garrido, N., Taylor, N., Fernandez, F., Levis, S. and Enria, D. (2006) West Nile virus isolation from equines in Argentina, 2006. *Emerging Infectious Diseases* 12, 1559–1561.

Naish, S., Dale, P., Mackenzie, J.S., McBride, J., Mengersen, K. and Tong, S. (2014) Climate change and dengue: a critical and systematic review of quantitative modelling approaches. *BMC Infectious Diseases* 14, 167.

Norris, D.E. (2004) Mosquito-borne diseases as a consequence of land use change. *EcoHealth* 1, 19–24.

Ometto, T., Durigon, E.L., Araujo, J., Aprelon, R., Aguiar, D.M., Cavalcante, G.T., Melo, R.M., Levi, J.E., Azevedo Jr, S.M., Petry, M.V., Neto, I.S., Serafini, P., Villalobos, E., Cunha, E.M., Lara, M.C.C.S.H., Nava, A.F.D., Nardi, M.S., Hurtado, R., Rodrigues, R., Sherer, A.L., Sherer, J.F.M., Geraldi, M.P., Seixas, M.M.M., Peterka, C., Bandeira, D.S., Pradel, J., Vachiery, N., Labruna, M.B., Camargo, L.M., Lanciotti, R. and Lefrancois, T. (2013) West Nile virus surveillance, Brazil, 2008–2010. *Transactions of the Royal Society of Tropical Medicine and Hygiene* 107, 723–730.

Pajot, F.X., Le Pont, F., Molez, J.F. and Degallier, N. (1977) Aggressivité d'*Anopheles* (*Nyssorhynchus*) *darlingi* Root, 1926 (Diptera, Culicidae) en Guyane française. *Cahiers Orstom: Série Entomologie Médicale et Parasitologie* 15, 15–22.

Pancetti, F.G., Honório, N.A., Urbinatti, P.R., Lima-Camara, T.N. (2015) Twenty-eight years of *Aedes albopictus* in Brazil: a rationale to maintain active entomological and epidemiological surveillance. *Revista da Sociedade Brasileira de Medicina Tropical* 48(1), 87–89.

Parham, P.E., Waldock, J., Christophides, G.K. and Michael, E. (2015) Climate change and vector-borne diseases of humans. *Philosophical Transactions of the Royal Society of London, Series B: Biological Sciences* 370, 1665.

Patz, J.A. and Olson, S.H. (2006) Malaria risk and temperature: Influences from global climate change and local land use practices. *Proceedings of the National Academy of Sciences* 103, 5635–5636.

Patz, J.A., Epstein, P.R., Burke, T.A. and Balbus, J.M. (1996) Global climate change and emerging infectious diseases. *Journal of the American Medical Association* 275, 217–223.

Patz, J.A., Campbell-Lendrum, D., Holloway, T. and Foley, J.A. (2005) Impact of regional climate change on human health. *Nature* 438, 310–317.

Pauvolid-Correa, A., Morales, M.A., Levis, S., Figueiredo, L.T.M., Couto-Lima, D., Campos, Z., Nogueira, M.F., Silva, E.E., Nogueira, R.M. and Schatzmayr, H.G. (2011) Neutralising antibodies for West Nile virus in horses from Brazilian Pantanal. *Memórias do Instituto Oswaldo Cruz* 1064, 467–474.

Pavri, K.M. and Singh, K.R.P. (1965) Isolation of West Nile virus from Culex fatigans mosquitoes from western India. *Indian Journal of Medical Research* 53, 501–505.

Paz, S. (2015) Climate change impacts on West Nile virus transmission in a global context. *Philosophical Transactions of the Royal Society of London, Series B: Biological Sciences* 370(1665), article ID 20130561. DOI: 10.1098/rstb.2013.0561.

Petersen, L.R., Brault, A.C. and Nasci, R.S. (2013) West Nile virus: review of the literature. *Journal of the American Medical Association* 310, 308–315.

Rachou, R.G. (1958) Anofelinos do Brasil. Comportamento das espécies vetoras de malária [Anopheles Brazil. Behaviour of the vector species of malaria]. *Revista Brasileira de Malariologia e Doenças Tropicais* 10, 145–181.

Reiskind, M.H., Pesko, K., Westbrook, C.J. and Mores, C.N. (2008) Susceptibility of Florida mosquitoes to infection with chikungunya virus. *American Journal of Tropical Medicine and Hygiene* 78, 422–425.

Rogers, D.J. and Packer, M.J. (1993) Vector-borne diseases, models, and global change. *Lancet* 342, 1282–1284.

Rohani, A., Wong, Y.C., Zamre, I., Lee, H.L. and Zurainee, M.N. (2009) The effect of extrinsic incubation temperature on development of dengue serotype 2 and 4 viruses in *Aedes aegypti* (L.). *Southeast Asian Journal of Tropical Medicine and Public Health* 40, 942–950.

Rozendaal, J.A. (1992) Relations between *Anopheles darlingi* breeding habitats, rainfall, river level and malaria transmission rates in the rain forest of Suriname. *Medical Veterinary Entomology* 6, 16–22.

Rueda, L.M., Patel, K.J., Axtell, R.C. and Stinner, R.E. (1990) Temperature-dependent development and survival rates of *Culex quinquefasciatus* and *Aedes aegypti* (Diptera: Culicidae). *Journal of Medical Entomology* 27, 892–898.

Sam, I.C., Kümmerer, B.M., Chan, Y.F., Roques, P., Drosten, C. and AbuBakar, S. (2015) Updates on chikungunya epidemiology, clinical disease, and diagnostics. *Vector Borne and Zoonotic Diseases* 15, 223–230.

Scott, T.W., Clark, G.G., Lorenz, L.H., Amerasinghe, P.H., Reiter, P. and Edman, J.D. (1993) Detection of multiple blood feeding in *Aedes aegypti* (Diptera: Culicidae) during a single gonotrophic cycle using a histologic technique. *Journal of Medical Entomology* 30, 94–99.

Smithburn, J.S., Hughes, T.P., Burke, A.W. and Paul, J.H. (1940) A neurotropic virus isolated from the blood of a native of Uganda. *American Journal of Tropical Medicine and Hygiene* 20, 471–492.

Sigrist, M.S. and Carvalho, C.J.B. (2009) Historical relationships among areas of endemism in the tropical South America using Brooks Parsimony Analysis (BPA). *Biota Neotropica* 9, 79–90.

Singer, B.H. and Castro, M.C. (2001) Agricultural colonization and malaria on the Amazon Frontier. *Annals of the New York Academy of Sciences* 954, 184–222.

Tadei, W.P., dos Santos, J.M., Costa, W.L. and Scarpassa, V.M. (1988) Biology of Amazonian *Anopheles*. XII. Occurrence of *Anopheles* species, transmission dynamics and malaria control in the urban area of Ariquemes (Rondônia). *Revista Instituto Medicina Tropical Sao Paulo* 30, 221–251.

Tatem, A.J. and Hay, S.I. (2004) Measuring urbanization pattern and extent for malaria research: a review of remote sensing approaches. *Journal of Urban Health* 81, 363–376.

Tatem, A.J., Gething, P.W., Smith, D.L. and Hay, S.I. (2013) Urbanization and the global malaria recession. *Malaria Journal* 12, 1–10.

Tauro, L., Marino, B., Diaz, L.A., Lucca, E., Gallozo, D., Spinsanti, L. and Contigiani, M. (2012) Serological detection of St Louis encephalitis virus and West Nile virus in equines from Santa Fe, Argentina. *Memórias do Instituto Oswaldo Cruz* 107, 553–556.

Thavara, U., Tawatsin, A., Chansang, C., Kongngamsuk, W., Paosriwong, S., Boon-Long, J., Rongsriyam, Y. and Komalamisra, N. (2001) Larval occurrence, oviposition behavior and biting activity of potential mosquito vectors of dengue on Samui Island, Thailand. *Journal of Vector Ecology* 26, 172–180.

Vega-Rúa, A., Zouache, K., Girod, R., Failloux, A.B. and Lourenço-de-Oliveira, R. (2014) High level of vector competence of *Aedes aegypti* and *Aedes albopictus* from ten American countries as a crucial factor in the spread of Chikungunya virus. *Journal of Virology* 88, 6294–6306.

Viana, D.V. and Ignotti, E. (2013) The ocurrence of dengue and weather changes in Brazil: a systematic review. *Revista Brasileira Epidemiologia* 16, 240–256.

Watts, D.M., Burke, D.S., Harrison, B.A., Whitmire, R.E. and Nisalak, A. (1987) Effect of temperature on the vector efficiency of *Aedes aegypti* for dengue 2 virus. *American Journal of Tropical Medicine and Hygiene* 36, 143–152.

WHO (2014) West Nile virus – Brazil (World Health Organization). Available at: http://www.who.int/csr/don/15-december-2014-wnv/en (accessed January 2015).

WHO (2015) Neglected tropical diseases (World Health Organization). Available at: http://www.who.int/neglected_diseases/diseases/en (accessed June 2015).

10 Urbanization, Climate Change and Malaria Transmission in Sub-Saharan Africa

Eliningaya J. Kweka[1,2,*], Humphrey D. Mazigo[2], Yousif E. Himeidan[3], Domenica Morona[2] and Stephen Munga[4]

[1]Division of Livestock and Human Diseases Vector Control, Tropical Pesticides Research Institute, Arusha, Tanzania; [2]Department of Medical Parasitology and Entomology, Catholic University of Health and Allied Sciences, Mwanza, Tanzania; [3]Africa Technical Research Centre, Vector Health International, Arusha, Tanzania; [4]Centre for Global Health Research, Kenya Medical Research Institute, Kisumu, Kenya

10.1 Introduction

In recent years, malaria transmission has increasingly been reported within urban areas across sub-Saharan Africa (Ntonga et al., 2015). Malaria cannot be considered a rural disease any longer. With the growing human population in urban areas, the chances of urban transmission may be higher than rural (Staedke et al., 2003; Donnelly et al., 2005; Hay et al., 2005). One of the major aspects of sub-Saharan Africa's development which has had a significant impact on malaria transmission is urbanization (Tatem et al., 2013). Due to rapid growth of the urban population in sub-Saharan Africa, there is great change in the physical landscape and transformation of the environment due to increased demand of human settlements. Urbanization has created opportunities which have led to improvement of the socioeconomic status, resulting in better health and housing conditions for the population (Dye, 2008; Alirol et al., 2011). Although the impact of rapid urbanization within sub-Saharan Africa has been recognized, its impact on the environment, such as climate change and transmission of malaria, has not been critically explored.

10.2 Factors Leading to Urban Migration in Sub-Saharan Africa

Urban migration in sub-Saharan Africa has rapidly increased in the 21st century (UN-DESA, 2012). By 2009, half of the world's population lived in cities, with much of the urbanization process occurring in low-income developing countries (UN-DESA, 2012). Migration of population into urban areas in sub-Saharan Africa has increased from 11% in 1950 to over 35% in 2010 and is expected to rise to 50% by 2040 (UN-DESA, 2012). The main cause of migration from rural areas to urban areas is the search for employment and the accessibility to basic needs (UN-DESA, 2012). The

*E-mail: pat.kweka@gmail.com

141

urbanization growth in low-income countries often results in the development and expansion of unplanned settlements with poor-quality housing, inadequate water supply and sanitation facilities, and overcrowding (UN-HABITAT, 2008). The combination of a high population and poor urban planning has had a profound effect on malaria and other diseases by increasing the possibility of transmission of the parasites that cause them (Snow *et al.*, 1999; Tatem *et al.*, 2008). These living conditions are particularly favourable to the transmission of malaria and other protozoan diseases in urban settings (Utzinger and Keiser, 2006; Alirol *et al.*, 2011).

10.3 Climate and Vector Behaviour Changes in Relation to Malaria in Africa

Among other human vector-borne diseases, malaria distribution varies from low transmission to epidemic distribution depending on location, time and variations of rainfall and temperature (Lindsay and Birley, 1996; Martens *et al.*, 1999; Afrane *et al.*, 2005, 2006, 2012b; Oyewole and Awolola, 2006; Mills *et al.*, 2010). The change of environment is partly caused by deforestation for various land-use purposes, such as construction of buildings to accommodate the urban growing population. These changes might bring about vector species distribution-related changes and subsequently malaria spread, which may lead to a high level of malaria transmission and infection (Oyewole and Awolola, 2006).

Temperature, a major climatic variation, is estimated to change and by the year 2100 is estimated to increase by 1.0 to 3.5°C, which will facilitate vector-borne disease transmission, particularly malaria (Watson *et al.*, 1996). The spatial and temporal changes in temperature, humidity and precipitation which are anticipated to take place will affect the vector's ecology and biology, and increase the risk of disease transmission (Githeko *et al.*, 2000). The elevated risk of vector-borne diseases increases with climate

change, on which a critical component of the vector's life cycle depends, including transmission efficiency (Lindsay and Birley, 1996). For most malaria vectors, the minimum temperature is reported to range between 14°C and 18°C, while the maximum is between 35°C and 40°C. At a temperature range of 30–32°C, there is likelihood of increased vectoral capacity, such as reduction of the incubation period for parasites, although survival rates of vectors might be decreased. Main malaria vectors such as *Anopheles gambiae* s.s., *An. arabiensis* and *An. funestus* are sensitive to temperature changes in the immature stage of the life cycle in aquatic stages and in terrestrial as adults (Rueda *et al.*, 1990; Bayoh and Lindsay, 2004; Afrane *et al.*, 2006; Lyons *et al.*, 2013a, 2014; Walker *et al.*, 2013). The increase in water surface temperature further shortens the larval developmental time, while for adults it increases the chance of parasite development and transmission (Rueda *et al.*, 1990; Bayoh and Lindsay, 2004; Afrane *et al.*, 2006). The blood meal digestion time is reduced with the increase in temperature, which subsequently influences the parasite transmission rate (Gillies, 1953; Afrane *et al.*, 2006, 2012a). Each of the above described phenomenon has a significant impact on malaria transmission.

Apart from temperature changes, vector biology and parasite development are also affected by precipitation, which can have both short- and long-term effects on vector habitat stability, a critical component of vector biology. An increase in precipitation influences the quality and availability of breeding habitats for malaria vectors (Githeko *et al.*, 2000). The increase in density of vegetation due to precipitation affects the availability of resting sites. African urban areas have a high diversity of vector species with the potential to redistribute to a new habitat if driven to it by climate change, and this could lead to a new disease pattern. *Anopheles* mosquito species differ in their responses to precipitation and temperature, which make their study complex and extremely risky (Christiansen-Jucht *et al.*, 2015). There is experimental evidence that climate change and an increase in global temperature affects the distribution of mosquito-borne diseases, mostly malaria, by accelerating vector ecology changes which influence the life cycle (Mills *et al.*, 2010; Lyons *et al.*, 2013b) and disease transmission efficiency (Mills *et al.*, 2010; Christiansen-Jucht *et al.*, 2014; Moller-Jacobs *et al.*, 2014), which all depend on climatic variables, such as rainfall, temperature and desiccation.

Climate variation, in particular, is expected to influence vector survivorship (Afrane *et al.*, 2006; Christiansen-Jucht *et al.*, 2014), and also feeding pattern, reproduction, behaviour and development rates (Afrane *et al.*, 2006; Gage *et al.*, 2008). Due to climate change by increase in temperature, there are areas where the number of malaria cases has been found to increase; in urban Kericho, the increase of temperature by 0.2°C has led to increased malaria-related cases (Omumbo *et al.*, 2011) and new vector species have been found – *An. arabiensis*, colonizing new areas of Mt Kenya highland (Chen *et al.*, 2006) and in western Kenya highlands (Kweka *et al.*, 2011).

10.3.1 Factors contributing to the emergence of urban malaria in sub-Saharan Africa

It is understood that urbanization results in improved infrastructure, through high-quality housing and sanitation, thus reducing the availability of breeding sites (Chang *et al.*, 2014). Despite these encouraging factors, malaria transmission has persisted in some urban Africa settlements at a higher level than in the surrounding rural areas, with an entomological inoculation rate (EIR) of more than 80 infective bites per person per year, as illustrated in Libreville in Gabon (Matthys *et al.*, 2006; Mourou *et al.*, 2012). Studies point to a number of factors in sub-Saharan African cities, which in combination may contribute to continued malaria transmission. These include urban agricultural practices, livestock keeping, urban outdoor lifestyle (such as retiring late under a bed net), abandoned water bodies and the lack of land-use planning (Afrane *et al.*, 2004, 2012a; Sattler *et al.*, 2005; Matthys *et al.*, 2006; Castro *et al.*, 2009, 2010; Dongus *et al.*, 2009; Stoler *et al.*, 2009; Maheu-Giroux and Castro, 2013a).

In addition, the unplanned growth of cities throughout sub-Saharan Africa has led to increased malaria transmission, helped by continued rural practices, such as growing small gardens, and creating potential breeding sites for *Anopheles gambiae* sensu lato (Afrane *et al.*, 2004; Castro *et al.*, 2009, 2010). The practice of urban agriculture, combined with poor drainage systems, puts the population living in urban areas, including those living in the outskirts, at risk of contracting malaria. It is notable that the city outskirts are characterized by poor housing and unpaved roads, allowing water to accumulate (Mwangangi *et al.*, 2012).

Reviews of urbanization and incidence of malaria in sub-Saharan Africa have shown three patterns of transmission zones: 1) urban; 2) peri-urban; and 3) rural (Robert *et al.*, 2003; Girardin *et al.*, 2004; Keiser *et al.*, 2004; Hay *et al.*, 2005) and have reported their EIR to be 19, 64 and 126 infectious bites per person per year respectively (Hay *et al.*, 2005). The intensity of malaria transmission varies with factors such as the geographical parameters of the location, such as altitude, proximity to sea, presence of a river and flood plain, the year-round climate, land use, human movement characteristics, local vector species, breeding sites, waste management and local malaria intervention programmes (Epstein, 2010).

10.3.2 Human urbanization and climate change

In Africa, human settlements and their activities have negative implications for land use, land cover, topography and forest (Minakawa *et al.*, 2005; Mushinzimana *et al.*, 2006; Zhou *et al.*, 2007). All of these activities are the result of expansion of agriculture and the need for food security, the need for an energy source (firewood) and demand for construction materials (Tuno *et al.*, 2005). Since the 1980s, climate changes have been witnessed in different parts of Africa, in addition to increases in the urban human population. In the western Kenya urban area of Kakamega, the forest loss due to urbanization shows a large increase (Tuno *et al.*, 2005). This has led to the creation of potential breeding sites for *An. gambiae* s.l. (Minakawa *et al.*, 2005; Munga *et al.*, 2006; Zhou *et al.*, 2007). The rise in temperature in the region has further increased both adult mosquito survivorship and sporogony life cycle development in the mosquito (Afrane *et al.*, 2005, 2006, 2012a). In the Kericho area of central Kenya, human activities such as tea estate expansion and deforestation have subsequently led to a temperature increase of 2°C (Omumbo *et al.*, 2011). The rise in temperature has led to an increase in vector population in areas above 1500 m above sea level, with efficient transmittance of malaria. These are the areas which were formerly known to be malaria-free in Africa (Chen *et al.*, 2006). In Tanzania, the western Usambara mountains (more than 800 m above sea level), previously covered by forest, have now been converted to human settlements and various forms of agricultural practices, leading to the development of *Anopheles* breeding sites (Lindsay *et al.*, 2000; Balls *et al.*, 2004). This change has led to the establishment of a malaria vector population in a previously malaria-free zone. The land-use and topography changes have also led to a rise in temperature, enabling the survival of *Anopheles gambiae* s.s., *An. arabiensis* and *An. funestus* (Lindsay *et al.*, 2000; Balls *et al.*, 2004; Zhou *et al.*, 2007; Kweka *et al.*, 2012b).

Urbanization, therefore, leads to climate change by modifying topography, land cover and deforestation (Tuno *et al.*, 2005). In the African scenario, part of the climate change has been caused by deforestation and rapid unplanned urbanization, which, in turn, has increased the number of urban malaria cases (Keiser *et al.*, 2004; Kasili *et al.*, 2009).

10.4 Urban Malaria Transmission and Human Settlements

In urban areas in Africa, malaria transmission shows a clear gradient in the number of cases from the urban centre to the outskirts of the cities (Robert *et al.*, 2003; Keiser *et al.*, 2004; Hay *et al.*, 2005). In a survey of urban areas, the *Plasmodium falciparum* parasitic rate was found to be 24% in the urban centre and 38.6% in the peri-urban area of Ouagadougou in Burkina Faso, and other urban areas in Africa show a similar trend (Wang *et al.*, 2005a,b, 2006a,b,c). This is a common observation for African cities, which grow outwards from a town centre, with poor and unplanned housing (Byrne, 2007). The populations migrating from rural African villages to urban areas tend to bring their practices with them, thus creating potential vector breeding sites on the outskirts of the city (Adiamah *et al.*, 1993; Afrane *et al.*, 2004, 2012a; Wang *et al.*, 2005a; Castro *et al.*, 2009, 2010).

In contrast, in Gabon, malaria transmission has been found to peak more often in urban centres than in peri-urban areas, with an EIR of 87.8 and 13.3 infectious bites per person per year respectively, as a consequence of poor housing and unplanned settlements surrounding the city centre (Mourou *et al.*, 2012). In Benin, the highest malaria parasite prevalence was found in an intermediate zone (9%) between the urban centre (2.6%) and the urban outskirts (2.5%). This can be explained by the practice of urban agriculture in the intermediate zone and the presence of salty lagoons in the periphery, which limits the breeding of the malaria vector *An. gambiae* (Wang *et al.*, 2005a). A similar scenario has been reported in Kumasi in Ghana and Dar es Salaam in Tanzania, where urban gardens and small-scale farming have created potential mosquito-breeding sites, thus augmenting parasite transmission in the city periphery (Afrane *et al.*, 2004, 2012a; Castro *et al.*, 2009, 2010; Dongus *et al.*, 2009). In growing cities in Africa, the quality of housing and planning of plots can also play a major role in urban malaria transmission control (Geissbuhler *et al.*, 2007; Ogoma *et al.*, 2009, 2010).

10.4.1 Environmental and human-made factors leading to urban malaria transmission

In Africa it can be postulated that human activities have created potential vector-breeding sites, allowing the breeding of malaria vectors throughout the year. These habitats are considered artificial rather than natural breeding sites and provide abundant sources of mosquito larvae in urban Africa (Afrane *et al.*, 2004, 2012a; Carlson *et al.*, 2004; Klinkenberg *et al.*, 2008; Chaki *et al.*, 2009; Stoler *et al.*, 2009; Castro *et al.*, 2010; Siri *et al.*, 2010 ; Alemu *et al.*, 2011). In urban environments, the major breeding sites found throughout Africa are human-made habitats, mostly preferable to natural habitats for *An. gambiae* s.l., hence increasing the risk of malaria transmission in urban areas. All of these major and minor potential habitats have been found to harbour all aquatic stages of *An. gambiae* s.l. (Afrane *et al.*, 2004, 2012a; Impoinvil *et al.*, 2008a,b; Klinkenberg *et al.*, 2008; Castro *et al.*, 2009, 2010; Dongus *et al.*, 2009; Stoler *et al.*, 2009). The most common artificial habitats are drains or gutters, ditches, truck tyres, leaking water pipes, foliage and sites created by urban agriculture. Also considered as minor potential breeding sites in urban settings are water tanks, construction sites and swimming pools.

10.4.2 Role of urban agriculture

Failing agriculture triggered by drought, poor productivity, pests and other peripheral factors such as transportation has encouraged the practice of urban agriculture. Cities provide adequate infrastructure as well as resources such as fresh water to grow crops. This form of agriculture has expanded recently in urban areas across sub-Saharan Africa (Afrane *et al.*, 2004, 2012a; Klinkenberg *et al.*, 2008; Baragatti *et al.*, 2009; Dongus *et al.*, 2009; Stoler *et al.*, 2009). Urban food supply demand, mostly of vegetables, has fuelled urban agriculture practices. These practices have an impact on food security, on employment and on the fight against malnutrition and poverty, but provide a favourable environment for the breeding and survival of the aquatic stages of the malaria vectors, and subsequently, maintain malaria transmission within urban areas (Castro *et al.*, 2009; Dongus *et al.*, 2009; Yadouleton *et al.*, 2009, 2010; Afrane *et al.*, 2012a). Drainage ditches in agricultural fields create potential breeding sites due to the accumulation of very shallow slow-moving water, which favours gravid *An. gambiae* mosquito females for oviposition. These ditches have been surveyed in Abidjan, Cote d'Ivoire, Dar es Salaam, Tanzania and Accra, Ghana, and have been found to be potential breeding sites (Afrane *et al.*, 2004, 2012a; Klinkenberg *et al.*, 2008; Castro *et al.*, 2009; Dongus *et al.*, 2009; Stoler *et al.*, 2009; Matthys *et al.*, 2010). In Abidjan, Cote d'Ivoire, anopheline larvae were found to occur in all urban rice fields surveyed, during wet and dry seasons (Matthys *et al.*, 2006, 2010). Other breeding habitats found in urban environments include irrigation wells and ditches for furrow systems (Dongus *et al.*, 2009; Machault *et al.*, 2009, 2010). Large water bodies are likely to be more productive for breeding than smaller ones due to fewer disturbances by irrigation activities. These urban agricultural irrigated fields have created a malaria risk for their surrounding human population (Fournet *et al.*, 2010; Yadouleton *et al.*, 2010). In Maputo, Mozambique, the malaria parasite density was found to be higher among people who lived and worked in urban agricultural sites throughout the city, compared to those living in rural areas (Macedo de Oliveira *et al.*, 2011).

10.5 The Contribution of Human Activity to Urban Malaria Transmission

Human activity, both in rural and urban areas, has been intensively associated with malaria (Minakawa *et al.*, 2005; Tuno *et al.*, 2005; Munga *et al.*, 2006). Human factors contributing to malaria transmission have been subdivided into four groups discussed hereafter.

10.5.1 Human movements

In general due to lower rates of malaria prevalence in urban areas, the exposure to infection is consequently low among urban populations, which, in turn, reduces their immunity to malaria and increases the risk of getting infected when coming into contact with infected mosquitoes (Carme, 1994). This was observed in Burkina Faso, Côte d'Ivoire and Zambia, where infection was found to be correlated with exposure to high infection risk (Ng'andu *et al.*, 1989; Wang *et al.*, 2005b; Baragatti *et al.*, 2009). This was also noted in some West African cities during the month of October, at a time when the urban population usually returns from vacation in their rural dwellings and when rural youths migrate to urban areas to seek employment after an agricultural season in the rural areas (Ceesay *et al.*, 2012). Other studies from Kenyan and Ethiopian towns have shown a similar increase in malaria transmission risk when people travel from malaria-free to malaria-endemic zones due to low immunity (Dennis *et al.*, 2004; Shanks *et al.*, 2005; Alemu *et al.*, 2014).

10.5.2 Housing structure

In urban Africa, house structure has improved comparatively to rural houses (Sachs and Malaney, 2002; Sachs, 2002; Gallup and Sachs, 2001). The improved quality of the housing reduces entry routes for mosquitoes (Ogoma *et al.*, 2009, 2010; Kampango *et al.*, 2013). A study conducted in the Gambia showed that houses with mud walls, open eaves and without ceilings had higher malaria infections rates among children, similarly to observations from south-central Lao PDR and Kenya (Lindsay and Snow, 1988; Adiamah *et al.*, 1993; Kirby *et al.*, 2008a,b; Atieli *et al.*, 2009; Hiscox *et al.*, 2013). In Dar es Salaam, Tanzania, Ogoma *et al.* showed that the number of mosquitoes

sampled indoors had decreased considerably in urban modernized houses compared to local rural houses (Ogoma *et al.*, 2009, 2010). Kirby *et al.* (2008a,b) reported that the presence of a ceiling and screening of the windows and doors, as in most urban houses, had a high impact on the number of indoor host-seeking mosquitoes. Elevated house walls have been found to decrease malaria incidence and mosquito densities indoors (Charlwood *et al.*, 2003; Lindsay *et al.*, 2003). A study carried out in a peri-urban area of Burkina Faso showed that houses using electricity as a source of light had higher infection rates than houses in a similar environment using biomass for cooking, which produces smoke that repels mosquitoes and deters them from going inside the house (Yamamoto *et al.*, 2009).

10.5.3 Community factors

Poor sanitation and waste disposal, as well as poor hygiene, play key roles in malaria transmission in urban environments. Sanitation is determinant in removing the liquid waste which accumulates in pools of stagnant water and forms breeding sites. Having waste disposal facilities such as toilets in Accra, Ghana reduced the risk of both malaria and diarrhoea (Fobil *et al.*, 2011, 2012). Toilets in some urban areas in Libreville have been found to be potential malaria vector breeding sites (Mourou *et al.*, 2012). The bricks and soil digging holes left around houses after their construction in Kisumu city in western Kenya were witnessed to play a role in malaria transmission as they served as habitats for vectors' breeding (Carlson *et al.*, 2004).

10.5.4 Household socioeconomic status

In urban areas, a higher socioeconomic status is closely associated with factors such as better education, access to television and radio broadcasting, affordability of personal protection tools, prompt treatment and better sanitation and hygiene environments

(Mireji *et al.*, 2008; Ogoma *et al.*, 2010; Imbahale *et al.*, 2011a,b). These factors influence the basic knowledge of vector-breeding sites, malaria transmission and control among individuals, so urban groups with a higher socioeconomic status have a notably reduced risk of malaria infection (De Beaudrap *et al.*, 2011). In Dar es Salaam, Tanzania, a higher socioeconomic status was associated with a high-quality house with screened windows, a door and blocked areas reducing or completely preventing entry of mosquitoes (Ogoma *et al.*, 2009, 2010). Slum-like areas of urban Africa, with low socioeconomic status, have greatly contributed to an increased malaria disease risk as seen in cities such as Nairobi, Kenya and Libreville, Gabon (Kasili *et al.*, 2009; Mourou *et al.*, 2012).

10.6 Contribution of Human Lifestyle to Urban Malaria

Human lifestyle in the use of mosquito-control tools varies drastically from rural to urban areas. Urban areas have more exposure and access to tools than rural areas. The wide and intensive implementation of various malaria-control tools in urban areas, including long-lasting insecticidal-treated nets (LLINs) and indoor residual spray (IRS) has triggered a change in vector behaviour. In Dar es Salaam in Tanzania, a wide coverage of control tools (IRS and LLINs) caused the behaviour of the vector *An. gambiae* s.s. to change from endophilic and endophagic to exophilic and exophagic (Russell *et al.*, 2011). This adaptation increases the survivorship of mosquitoes and the possibility to escape from insecticide-treated material and surfaces. The urban population tends to wake up early in the morning (04.00 h) and retire around midnight (00.00 h), a period which falls within the active biting cycle of *An. gambiae*. This trend influences outdoor malaria transmission, as a substantial proportion of the population is unprotected outdoors (Russell *et al.*, 2011). This could explain the maintenance of malaria foci in urban areas with the movements of the vectors restricted by building structures. The outdoor human population should therefore use personal protection tools, such as plant-based repellents (Kweka *et al.*, 2008a,b) or synthetic repellents such as DEET (Kweka *et al.*, 2012a; Chen-Hussey *et al.*, 2014; Deressa *et al.*, 2014) during the active biting cycles.

10.7 Adaptation of Malaria Vectors to Urban Climate

The adaptation of vector species to the changing climate as well as urbanization can play a vital role in disease persistence, as well as transmission. In the early 2000s, the malaria vector *An. gambiae* s.l. has adapted to breeding in polluted urban water bodies (Klinkenberg *et al.*, 2008; Mireji *et al.*, 2008, 2010a,b). *An. gambiae* s.l. has been found breeding in polluted water bodies in Kenya (Mireji *et al.*, 2010a), Côte d'Ivoire (Matthys *et al.*, 2006), Cameroon (Antonio-Nkondjio *et al.*, 2011) and Ghana (Afrane *et al.*, 2004, 2012a). In Kisumu and Malindi, Kenya, and Lagos in Nigeria, *An gambiae* s.s. larvae have been found to adapt to water polluted with a high concentration of heavy metals such as lead, copper and iron, and other water containing human faeces and petroleum products (Awolola *et al.*, 2007; Mireji *et al.*, 2008, 2010a,b). Other species such as *An. arabiensis* and *An. funestus* were found to be less tolerant to pollutants such as heavy metals, but highly tolerant to water turbidity (Awolola *et al.*, 2007; Kasili *et al.*, 2009). This suggests that *An. funestus* and *An. arabiensis* are not as well adapted to habitats with urban pollutants as *An. gambiae* s.s., implying that the latter is becoming a prominent urban malaria vector, as witnessed in Dar es Salaam, Tanzania (Sattler *et al.*, 2005; Russell *et al.*, 2011), Kisumu, Nairobi, and Malindi in Kenya (Mireji *et al.*, 2008, 2010a,b; Kasili *et al.*, 2009), in Benin (Yadouleton *et al.*, 2009, 2010), in Lagos, Nigeria (Awolola *et al.*, 2007), Kumasi and Accra, Ghana (Afrane *et al.*, 2004, 2012a) and Yaoundé, Cameroon (Antonio-Nkondjio *et al.*, 2011).

10.8 Conclusion

Parts of Africa which were formerly free of malaria have now witnessed the establishment of the disease, due to various factors, including climate change. Various aspects of vector biology, including the life cycle, are influenced by climate parameters such as temperature, precipitation, vegetation density and mosquito management programmes. Urbanization is also a component linked to climate change, and inadequate urban planning policies for a sustainable monitoring of waste disposal, drainage of channels, cleaning and maintenance of swimming pools, and regulation of urban agricultural practices could encourage malaria vectors. Housing improvements also have a role to play in reducing the contact between the human population and vectors.

References

Adiamah, J.H., Koram, K.A., Thomson, M.C., Lindsay, S.W., Todd, J. and Greenwood, B.M. (1993) Entomological risk factors for severe malaria in a peri-urban area of the Gambia. *Annals of Tropical Medicine Parasitology* 87, 491–500.

Afrane, Y.A., Klinkenberg, E., Drechsel, P., Owusu-Daaku, K., Garms, R. and Kruppa, T. (2004) Does irrigated urban agriculture influence the transmission of malaria in the city of Kumasi, Ghana? *Acta Tropica* 89, 125–134.

Afrane, Y., Lawson B., Githeko A. and Yan, G. (2005) Effects of microclimatic changes caused by land use and land cover on duration of gonotrophic cycles of *Anopheles gambiae* (Diptera: Culicidae) in western Kenya highlands. *Journal of Medical Entomology* 42, 974–980.

Afrane, Y.A., Zhou, G., Lawson, B.W., Githeko, A.K. and Yan, G. (2006) Effects of microclimatic changes caused by deforestation on the survivorship and reproductive fitness of *Anopheles gambiae* in western Kenya highlands. *American Journal of Tropical Medicine and Hygiene* 74, 772–778.

Afrane, Y., Lawson, B., Brenya, R., Kruppa, T. and Yan, G. (2012a) The ecology of mosquitoes in an irrigated vegetable farm in Kumasi, Ghana: abundance, productivity and survivorship. *Parasites and Vectors* 5, 233.

Afrane, Y.A., Githeko, A.K. and Yan, G. (2012b) The ecology of *Anopheles* mosquitoes under climate change: case studies from the effects of deforestation in East African highlands. *Annals of the New York Academy of Sciences* 1249, 204–210.

Alemu, A., Tsegaye, W., Golassa, L. and Abebe, G. (2011) Urban malaria and associated risk factors in Jimma Town, south-west Ethiopia. *Malaria Journal* 10, 173.

Alemu, K., Worku, A., Berhane, Y. and Kumie, A. (2014) Men traveling away from home are more likely to bring malaria into high altitude villages, Northwest Ethiopia. *PLoS ONE* 9(4), article ID e95341.

Alirol, E., Getaz, L., Stoll, B., Chappuis, F. and Loutan, L. (2011) Urbanisation and infectious diseases in a globalised world. *Lancet Infectious Diseases* 11, 131–141.

Antonio-Nkondjio, C., Fossog, B., Ndo, C., Djantio, B., Togouet, S., Awono-Ambene, P., Costantini, C., Wondji, C. and Ranson, H. (2011) *Anopheles gambiae* distribution and insecticide resistance in the cities of Douala and Yaounde (Cameroon): influence of urban agriculture and pollution. *Malaria Journal* 10, 154.

Atieli, H., Menya, D., Githeko, A. and Scott, T. (2009) House design modifications reduce indoor resting malaria vector densities in rice irrigation scheme area in western Kenya. *Malaria Journal* 8, 108.

Awolola, T.S., Oduola, A.O., Obansa, J.B., Chukwurar, N.J. and Unyimadu, J.P. (2007) *Anopheles gambiae* s.s. breeding in polluted water bodies in urban Lagos, southwestern Nigeria. *Journal of Vector Borne Diseases* 44, 241–244.

Balls, M.J., Bødker, R., Thomas, C.J., Kisinza, W., Msangeni, H.A. and Lindsay, S.W. (2004) Effect of topography on the risk of malaria infection in the Usambara Mountains, Tanzania. *Transactions of the Royal Society of Tropical Medicine and Hygiene* 98, 400–408.

Baragatti, M., Fournet, F., Henry, M.-C., Assi, S., Ouedraogo, H., Rogier, C. and Salem, G. (2009) Social and environmental malaria risk factors in urban areas of Ouagadougou, Burkina Faso. *Malaria Journal* 8, 13.

Bayoh, M.N. and Lindsay, S.W. (2004) Temperature-related duration of aquatic stages of the afro-tropical malaria vector mosquito *Anopheles gambiae* in the laboratory. *Medical and Veterinary Entomology* 18, 174–179.

Byrne, N. (2007) Urban malaria risk in sub-Saharan Africa: where is the evidence? *Travel Medicine and Infectious Disease* 5, 135–137.

Carlson, J., Byrd, B. and Omlin, F. (2004) Field assessments In western Kenya link malaria vectors to environmentally disturbed habitats during the dry season. *BMC Public Health* 4, 33.

Carme, B. (1994) Reducing the risk of malaria acquisition by urban dwellers of sub-Saharan Africa during travel in malaria-endemic areas. *Journal of Infectious Diseases* 170, 257–258.

Castro, M., Tsuruta, A., Kanamori, S., Kannady, K. and Mkude, S. (2009) Community-based environmental management for malaria control: evidence from a small-scale intervention in Dar es Salaam, Tanzania. *Malaria Journal* 8, 57.

Castro, M.C., Kanamori, S., Kannady, K., Mkude, S., Killeen, G.F. and Fillinger, U. (2010) The importance of drains for the larval development of lymphatic filariasis and malaria vectors in Dar es Salaam, United Republic of Tanzania. *PloS Neglected Tropical Diseases* 4, article ID e693.

Ceesay, S.J., Bojang, K.A., Nwakanma, D., Conway, D.J., Koita, O.A., Doumbia, S.O., Ndiaye, D., Coulibaly, T.F., Diakité, M., Traoré, S.F., Coulibaly, M., Ndiaye, J.-L., Sarr, O., Gaye, O., Konaté, L., Sy, N., Faye, B., Faye, O., Sogoba, N., Jawara, M., Dao, A., Poudiougou, B., Diawara, S., Okebe, J., Sangaré, L., Abubakar, I., Sissako, A., Diarra, A., Kéita, M., Kandeh, B., Long, C.A., Fairhurst, R.M., Duraisingh, M., Perry, R., Muskavitch, M.A.T., Valim, C., Volkman, S.K., Wirth, D.F. and Krogstad, D.J. (2012) Sahel, savana, riverine and urban malaria in West Africa: similar control policies with different outcomes. *Acta Tropica* 121, 166–174.

Chaki, P., Govella, N., Shoo, B., Hemed, A., Tanner, M., Fillinger, U. and Killeen, G. (2009) Achieving high coverage of larval-stage mosquito surveillance: challenges for a community-based mosquito control programme in urban Dar es Salaam, Tanzania. *Malaria Journal* 8, 311.

Chang, A.Y., Fuller, D.O., Carrasquillo, O. and Beier, J.C. (2014). Social justice, climate change, and dengue. *Health and Human Rights* 16, 93–104.

Charlwood, J., Pinto, J., Ferrara, P., Sousa, C., Ferreira, C., Gil, V. and Do Rosario, V. (2003) Raised houses reduce mosquito bites. *Malaria Journal* 2, 45.

Chen, H., Githeko, A.K., Zhou, G., Githure, J.I. and Yan, G. (2006) New records of *Anopheles arabiensis* breeding on the Mount Kenya Highlands indicate indigenous malaria transmission. *Malaria Journal* 5, 17.

Chen-Hussey, V., Behrens, R. and Logan, J. (2014) Assessment of methods used to determine the safety of the topical insect repellent N,N-diethyl-M-toluamide (DEET). *Parasites and Vectors* 7, 173.

Christiansen-Jucht, C., Parham, P.E., Saddler, A., Koella, J.C. and Basáñez, M.-G. (2014) Temperature during larval development and adult maintenance influences the survival of *Anopheles gambiae* s.s. *Parasites and Vectors* 7, 489.

Christiansen-Jucht, C., Erguler, K., Shek, C., Basáñez, M.-G. and Parham, P. (2015) Modelling *Anopheles gambiae* s.s. population dynamics with temperature- and age-dependent survival. *International Journal of Environmental Research and Public Health* 12, 5975.

De Beaudrap, P., Nabasumba, C., Grandesso, F., Turyakira, E., Schramm, B., Boum, Y. and Etard, J.-F. (2011) Heterogeneous decrease in malaria prevalence in children over a six-year period in south-western Uganda. *Malaria Journal* 10, 132.

Dennis, G., Kimutai, B. and Maguire, J. (2004) Travel as a risk factor for malaria requiring hospitalization on a highland tea plantation in western Kenya. *Journal of Travel Medicine* 11, 354–358.

Deressa, W., Yihdego, Y., Kebede, Z., Batisso, E., Tekalegne, A. and Dagne, G. (2014) Effect of combining mosquito repellent and insecticide treated net on malaria prevalence in southern Ethiopia: a cluster-randomised trial. *Parasites and Vectors* 7, 132.

Dongus, S., Nyika, D., Kannady, K., Mtasiwa, D., Mshinda, H., Gosoniu, L., Drescher, A., Fillinger, U., Tanner, M., Killeen, G. and Castro, M.C. (2009) Urban agriculture and *Anopheles* habitats in Dar es Salaam, Tanzania. *Geospatial Health* 3, 189–210.

Donnelly, M.J., McCall, P.J., Lengeler, C., Bates, I., D'Alessandro, U., Barnish, G., Konradsen, F., Klinkenberg, E., Townson, H., Trape, J.F., Hastings, I.M. and Mutero, C. (2005) Malaria and urbanization in sub-Saharan Africa. *Malaria Journal* 4, 12.

Dye, C. (2008) Health and urban living. *Science* 319, 766–769.

Epstein, P. (2010) The ecology of climate change and infectious diseases: comment. *Ecology* 91(3), 925–928. DOI: 10.1890/09-0761.1.

Fobil, J.N., Kraemer, A., Meyer, C.G. and May, J. (2011) Neighborhood urban environmental quality conditions are likely to drive malaria and diarrhea mortality in Accra, Ghana. *Journal of Environmental and Public Health* 2011, article ID 484010. 10 pp.

Fobil, J., Levers, C., Lakes, T., Loag, W., Kraemer, A. and May, J. (2012) Mapping urban malaria and diarrhea mortality in Accra, Ghana: evidence of vulnerabilities and implications for urban health policy. *Journal of Urban Health* 89, 977–991.

Fournet, F., Cussac, M., Ouari, A., Meyer, P.-E., Toe, H., Gouagna, L.-C. and Dabire, R. (2010) Diversity in anopheline larval habitats and adult composition during the dry and wet seasons in Ouagadougou (Burkina Faso). *Malaria Journal* 9, 78.

Gage, K.L., Burkot, T.R., Eisen, R.J. and Hayes, E.B. (2008) Climate and vectorborne diseases. *American Journal of Preventive Medicine* 35, 436–450.

Gallup, J.L. and Sachs, J.D. (2001) The economic burden of malaria. *American Journal of Tropical Medicine Hygiene* 64, 85–96.

Geissbuhler, Y., Chaki, P., Emidi, B., Govella, N., Shirima, R., Mayagaya, V., Mtasiwa, D., Mshinda, H., Fillinger, U., Lindsay, S., Kannady, K., De Castro, M., Tanner, M. and Killeen, G. (2007) Interdependence of domestic malaria prevention measures and mosquito-human interactions in urban Dar es Salaam, Tanzania. *Malaria Journal* 6, 126.

Gillies, M. (1953) The duration of the gonotrophic cycle in *Anopheles gambiae* and *Anopheles funestus*, with a note on the efficiency of hand catching. *East African Medical Journal* 30(4), 129–135.

Girardin, O., Dao, D., Koudou, B.G., Essé, C., Cissé, G., Yao, T., N'goran, E.K., Tschannen, A.B., Bordmann, G., Lehmann, B., Nsabimana, C., Keiser, J., Killeen, G.F., Singer, B.H., Tanner, M. and Utzinger, J. (2004) Opportunities and limiting factors of intensive vegetable farming in malaria endemic Côte d'Ivoire. *Acta Tropica* 89, 109–123.

Githeko, A.K., Lindsay, S.W., Confalonieri, U.E. and Patz, J.A. (2000) Climate change and vector-borne diseases: a regional analysis. *Bulletin of the World Health Organization* 78, 1136–1147.

Hay, S.I., Guerra, C.A., Tatem, A.J., Atkinson, P.M. and Snow, R.W. (2005) Urbanization, malaria transmission and disease burden in Africa. *Nature Review Microbiology* 3, 81–90.

Hiscox, A., Khammanithong, P., Kaul, S., Sananikhom, P., Luthi, R., Hill, N., Brey, P.T. and Lindsay, S.W. (2013) Risk factors for mosquito house entry in the Lao PDR. *PLoS ONE* 8(5), e62769.

Imbahale, S., Mweresa, C., Takken, W. and Mukabana, W. (2011a) Development of environmental tools for anopheline larval control. *Parasites and Vectors* 4, 130.

Imbahale, S., Paaijmans, K., Mukabana, W., Van Lammeren, R., Githeko, A. and Takken, W. (2011b) A longitudinal study on *Anopheles* mosquito larval abundance in distinct geographical and environmental settings in western Kenya. *Malaria Journal* 10, 81.

Impoinvil, D.E., Mbogo, C.M., Keating, J. and Beier, J.C. (2008a) The role of unused swimming pools as a habitat for *Anopheles* immature stages in urban Malindi, Kenya. *Journal of American Mosquito Control Association* 24, 457–459.

Impoinvil, D.E., Keating, J., Mbogo, C.M., Potts, M.D., Chowdhury, R.R. and Beier, J.C. (2008b) Abundance of immature *Anopheles* and *Culicines* (Diptera: Culicidae) in different water body types in the urban environment of Malindi, Kenya. *Journal of Vector Borne Diseases* 33, 107–116.

Kampango, A., Braganca, M., Sousa, B.D. and Charlwood, J.D. (2013) Netting barriers to prevent mosquito entry into houses in southern Mozambique: a pilot study. *Malaria Journal* 12, 99.

Kasili, S., Odemba, N., Ngere, F.G., Kamanza, J.B., Muema, A.M. and Kutima, H.L. (2009) Entomological assessment of the potential for malaria transmission in Kibera slum of Nairobi, Kenya. *Journal of Vector Borne Diseases* 46, 273–279.

Keiser, J., Utzinger, J., Caldas de Castro, M., Smith, T.A., Tanner, M. and Singer, B.H. (2004) Urbanization in sub-Saharan Africa and implication for malaria control. *American Journal of Tropical Medicine and Hygiene* 71, 118–127.

Kirby, M., Green, C., Milligan, P., Sismanidis, C., Jasseh, M., Conway, D. and Lindsay, S. (2008a) Risk factors for house-entry by malaria vectors in a rural town and satellite villages in the Gambia. *Malaria Journal* 7, 2.

Kirby, M., Milligan, P., Conway, D. and Lindsay, S. (2008b) Study protocol for a three-armed randomized controlled trial to assess whether house screening can reduce exposure to malaria vectors and reduce malaria transmission in the Gambia. *Trials* 9, 33. DOI: 10.1186/1745-6215-9-33.

Klinkenberg, E., McCall, P., Wilson, M., Amerasinghe, F. and Donnelly, M. (2008) Impact of urban agriculture on malaria vectors in Accra, Ghana. *Malaria Journal* 7, 151.

Kweka, E., Mosha, F., Lowassa, A., Mahande, A., Kitau, J., Matowo, J., Mahande, M., Massenga, C., Tenu, F., Feston, E., Lyatuu, E., Mboya, M., Mndeme, R., Chuwa, G. and Temu, E. (2008a) Ethnobotanical study of some of mosquito repellent plants in north-eastern Tanzania. *Malaria Journal* 7, 152.

Kweka, E., Mosha, F., Lowassa, A., Mahande, A., Mahande, M., Massenga, C., Tenu, F., Lyatuu, E., Mboya, M. and Temu, E. (2008b) Longitudinal evaluation of *Ocimum* and other plants effects on the feeding behavioral response of mosquitoes (Diptera: Culicidae) in the field in Tanzania. *Parasites and Vectors* 1, 42.

Kweka, E., Zhou, G., Lee, M.-C., Gilbreath, T., Mosha, F., Munga, S., Githeko, A. and Yan, G. (2011) Evaluation of two methods of estimating larval habitat productivity in western Kenya highlands. *Parasites and Vectors* 4, 110.

Kweka, E., Munga, S., Mahande, A., Msangi, S., Mazigo, H., Adrias, A. and Matias, J. (2012a)

Protective efficacy of menthol propylene glycol carbonate compared to N, N-diethyl-methylbenzamide against mosquito bites in northern Tanzania. *Parasites and Vectors* 5, 189.

Kweka, E.J., Zhou, G., Munga, S., Lee, M.-C., Atieli, H.E., Nyindo, M., Githeko, A.K. and Yan, G. (2012b) Anopheline larval habitats seasonality and species distribution: a prerequisite for effective targeted larval habitats control programmes. *PLoS ONE* 7(12), e52084.

Lindsay, S. and Birley, M. (1996) Climate change and malaria transmission. *Annals of Tropical Medicine and Parasitology* 90, 573–588.

Lindsay, S.W. and Snow, R.W. (1988) The trouble with eaves; house entry by vectors of malaria. *Transactions of the Royal Society of Tropical Medicine and Hygiene* 82, 645–646.

Lindsay, S.W., Bødker, R., Malima, R., Msangeni, H.A. and Kisinza, W. (2000) Effect of 1997–98 El Niño on highland malaria in Tanzania. *Lancet* 355, 989–990.

Lindsay, S.W., Jawara, M., Paine, K., Pinder, M., Walraven, G.E.L. and Emerson, P.M. (2003) Changes in house design reduce exposure to malaria mosquitoes. *Tropical Medicine and International Health* 8, 512–517.

Lyons, C., Coetzee, M. and Chown, S. (2013a) Stable and fluctuating temperature effects on the development rate and survival of two malaria vectors, *Anopheles arabiensis* and *Anopheles funestus*. *Parasites and Vectors* 6, 104.

Lyons, C.L., Coetzee, M. and Chown, S.L. (2013b) Stable and fluctuating temperature effects on the development rate and survival of two malaria vectors, *Anopheles arabiensis* and *Anopheles funestus*. *Parasite and Vectors* 6, 104.

Lyons, C.L., Coetzee, M., Terblanche, J.S. and Chown, S.L. (2014) Desiccation tolerance as a function of age, sex, humidity and temperature in adults of the African malaria vectors *Anopheles arabiensis* and *Anopheles funestus*. *Journal of Experimental Biology* 217, 3823–3833.

Macedo de Oliveira, A., Mutemba, R., Morgan, J., Streat, E., Roberts, J., Menon, M. and Mabunda, S. (2011) Prevalence of malaria among patients attending public health facilities in Maputo City, Mozambique. *American Journal of Tropical Medicine and Hygiene* 85, 1002–1007.

Machault, V., Gadiaga, L., Vignolles, C., Jarjaval, F., Bouzid, S., Sokhna, C., Lacaux, J.-P., Trape, J.-F., Rogier, C. and Pages, F. (2009) Highly focused anopheline breeding sites and malaria transmission in Dakar. *Malaria Journal* 8, 138.

Machault, V., Vignolles, C., Pages, F., Gadiaga, L., Gaye, A., Sokhna, C., Trape, J.-F., Lacaux, J.-P. and Rogier, C. (2010) Spatial heterogeneity and temporal evolution of malaria transmission risk in Dakar, Senegal, according to remotely sensed environmental data. *Malaria Journal* 9, 252.

Maheu-Giroux, M. and Castro, M. (2013a) Do malaria vector control measures impact disease-related behaviour and knowledge? Evidence from a large-scale larviciding intervention in Tanzania. *Malaria Journal* 12, 422.

Martens, P., Kovats, R., Nijhof, S., De Vries, P., Livermore, M., Bradley, D., Cox, J. and McMichael, A. (1999) Climate change and future populations at risk of malaria. *Global Environmental Change* 9, S89–S107.

Matthys, B., N'Goran, E.K., Koné, M., Koudou, B.G., Vounatsou, P., Cissé, G., Tschannen, A.B., Tanner, M. and Utzinger, J. (2006) Urban agricultural land use and characterization of mosquito larval habitats in a medium-sized town of Cote d'Ivoire. *Journal of Vector Ecology* 31, 319–333.

Matthys, B., Koudou, B.G., N'Goran, E.K., Vounatsou, P., Gosoniu, L., Koné, M., Gissé, G. and Utzinger, J. (2010) Spatial dispersion and characterisation of mosquito breeding habitats in urban vegetable-production areas of Abidjan, Côte d'Ivoire. *Annals of Tropical Medicine and Parasitology* 104, 649–666.

Mills, J.N., Gage, K.L. and Khan, A.S. (2010) Potential influence of climate change on vector-borne and zoonotic diseases: a review and proposed research plan. *Environmental Health Perspectives* 118, 1507–1514.

Minakawa, N., Munga, S., Atieli, F., Mushinzimana, E., Zhou, G., Githeko, A.K. and Yan, G. (2005) Spatial distribution of anopheline larval habitats in western Kenyan highlands: effects of land cover types and topography. *American Journal of Tropical Medicine and Hygiene* 73, 157–165.

Mireji, P.O., Keating, J., Hassanali, A., Mbogo, C.M., Nyambaka, H., Kahindi, S. and Beier, J.C. (2008) Heavy metals in mosquito larval habitats in urban Kisumu and Malindi, Kenya, and their impact. *Ecotoxicology and Environmental Safety* 70, 147–153.

Mireji, P.O., Keating, J., Hassanali, A., Impoinvil, D.E., Mbogo, C.M., Muturi, M.N., Nyambaka, H., Kenya, E.U., Githure, J.I. and Beier, J.C. (2010a) Expression of metallothionein and α-tubulin in heavy metal-tolerant *Anopheles gambiae* sensu stricto (Diptera: Culicidae). *Ecotoxicology and Environmental Safety* 73, 46–50.

Mireji, P.O., Keating, J., Hassanali, A., Mbogo, C.M., Muturi, M.N., Githure, J.I. and Beier, J.C. (2010b) Biological cost of tolerance to heavy metals in the mosquito *Anopheles gambiae*. *Medical and Veterinary Entomology* 24, 101–107.

Moller-Jacobs, L.L., Murdock, C.C. and Thomas, M.B. (2014) Capacity of mosquitoes to transmit malaria depends on larval environment. *Parasites and Vectors* 7, 593.

Mourou, J.-R., Coffinet, T., Jarjaval, F., Cotteaux, C., Pradines, E., Godefroy, L., Kombila, M. and Pages, F. (2012) Malaria transmission in Libreville: results of a one year survey. *Malaria Journal* 11, 40.

Munga, S., Minakawa, N., Zhou, G., Mushinzimana, E., Barrack, O.-O.J., Githeko, A.K. and Yan, G. (2006) Association between land cover and habitat productivity of malaria vectors in western Kenyan highlands. *American Journal of Tropical Medicine and Hygiene* 74, 69–75.

Mushinzimana, E., Munga, S., Minakawa, N., Li, L., Feng, C.C., Bian, L., Kitron, U., Schmidt, C., Beck, L., Zhou, G., Githeko, A.K. and Yan, G. (2006) Landscape determinants and remote sensing of anopheline mosquito larval habitats in the western Kenya highlands. *Malaria Journal* 5, 13.

Mwangangi, J., Midega, J., Kahindi, S., Njoroge, L., Nzovu, J., Githure, J., Mbogo, C. and Beier, J. (2012) Mosquito species abundance and diversity in Malindi, Kenya and their potential implication in pathogen transmission. *Parasitology Research* 110, 61–71.

Ng'andu, N., Watts, T.E., Wray, J.R., Chela, C. and Zulu, B. (1989) Some risk factors for transmission of malaria in a population where control measures were applied in Zambia. *East Africa Medical Journal* 66, 728–37.

Ntonga, P., Mbida, J., Tonga, C., Belong, P., Ngo Hondt, O., Magne, G., Peka, M. and Lehman, L. (2015) Impact of vegetable crop agriculture on anopheline agressivity and malaria transmission in urban and less urbanized settings of the south region of Cameroon. *Parasites and Vectors* 8, 293.

Ogoma, S., Kannady, K., Sikulu, M., Chaki, P., Govella, N., Mukabana, W. and Killeen, G. (2009) Window screening, ceilings and closed eaves as sustainable ways to control malaria in Dar es Salaam, Tanzania. *Malaria Journal* 8, 221.

Ogoma, S.B., Lweitoijera, D.W., Ngonyani, H., Furer, B., Russell, T.L., Mukabana, W.R., Killeen, G.F. and Moore, S.J. (2010) Screening mosquito house entry points as a potential method for integrated control of endophagic filariasis, arbovirus and malaria vectors. *PLoS Neglected Tropical Diseases* 4, article ID e773.

Omumbo, J.A., Lyon, B., Waweru, S.M., Connor, S.J. and Thomson, M.C. (2011) Raised temperatures over the Kericho tea estates: revisiting the climate in the East African highlands malaria debate. *Malaria Journal* 10, 12.

Oyewole, I. and Awolola, T. (2006) Impact of urbanisation on bionomics and distribution of malaria vectors in Lagos, southwestern Nigeria. *Journal of Vector Borne Diseases* 43, 173.

Robert, V., Macintyre, K., Keating, J., Trape, J.-F., Duchemin, J.-B., Warren, M. and Beier, J.C. (2003) Malaria transmission in urban sub-Saharan Africa. *American Journal of Tropical Medicine and Hygiene* 68, 169–176.

Rueda, L., Patel, K., Axtell, R. and Stinner, R. (1990) Temperature-dependent development and survival rates of *Culex quinquefasciatus* and *Aedes aegypti* (Diptera: Culicidae). *Journal of Medical Entomology* 27, 892–898.

Russell, T., Govella, N., Azizi, S., Drakeley, C., Kachur, S.P. and Killeen, G. (2011) Increased proportions of outdoor feeding among residual malaria vector populations following increased use of insecticide-treated nets in rural Tanzania. *Malaria Journal* 10, 80.

Sachs, J.D. (2002) A new global effort to control malaria. *Science* 298, 122–4.

Sachs, J. and Malaney, P. (2002) The economic and social burden of malaria. *Nature* 415, 680–685.

Sattler, M., Mtasiwa, D., Kiama, M., Premji, Z., Tanner, M., Killeen, G. and Lengeler, C. (2005) Habitat characterization and spatial distribution of *Anopheles* sp. mosquito larvae in Dar es Salaam (Tanzania) during an extended dry period. *Malaria Journal* 4, 4.

Shanks, G.D., Biomndo, K., Guyatt, H.L. and Snow, R.W. (2005) Travel as a risk factor for uncomplicated *Plasmodium falciparum* malaria in the highlands of western Kenya. *Transactions of the Royal Society of Tropical Medicine and Hygiene* 99, 71–74.

Siri, J.G., Wilson, M.L., Murray, S., Rosen, D.H., Vulule, J.M., Slutsker, L. and Lindblade, K.A. (2010) Significance of travel to rural areas as a risk factor for malarial anemia in an urban setting. *American Journal of Tropical Medicine and Hygiene* 82, 391–397.

Snow, R., Craig, M., Deichmann, U. and Marsh, K. (1999) Estimating mortality, morbidity and disability due to malaria among Africa's nonpregnant population. *Bulletin of the World Health Organization* 77, 624–640.

Staedke, S.G., Nottingham, E.W., Cox, J., Kamya, M.R., Rosenthal, P.J. and Dorsey, G. (2003) Short report: proximity to mosquito breeding sites as a risk factor for clinical malaria episodes in an urban cohort of Ugandan children. *American Journal of Tropical Medicine and Hygiene* 69, 244–246.

Stoler, J., Weeks, J.R., Getis, A. and Hill, A.G. (2009) Distance threshold for the effect of urban agriculture on elevated self-reported malaria prevalence in Accra, Ghana. *American Journal of Tropical Medicine and Hygiene* 80, 547–554.

Tatem, A., Guerra, C., Kabaria, C., Noor, A. and Hay, S. (2008) Human population, urban settlement patterns and their impact on *Plasmodium falciparum* malaria endemicity. *Malaria Journal* 7, 218.

Tatem, A., Gething, P., Smith, D. and Hay, S. (2013) Urbanization and the global malaria recession. *Malaria Journal* 12, 133.

Tuno, N., Okeka, W., Minakawa, N., Takagi, M. and Yan, G. (2005) Survivorship of *Anopheles gambiae* sensu stricto (Diptera: Culicidae) larvae in western Kenya highland forest. *Journal of Medical Entomology* 42, 270–7.

UN-DESA (2012) *World Urbanization Prospects: The 2011 Revision.* Department of Economic and Social Affairs, United Nations, New York, USA.

UN-HABITAT (2008) *State of the World's Cities 2008/2009: Harmonious Cities.* United Nations Human Settlements Programme, Nairobi, Kenya.

Utzinger, J. and Keiser, J. (2006) Urbanization and tropical health — then and now. *Annals of Tropical Medicine and Parasitology* 100, 517–533.

Walker, M., Winskill, P., Basanez, M.-G., Mwangangi, J., Mbogo, C., Beier, J. and Midega, J. (2013) Temporal and micro-spatial heterogeneity in the distribution of *Anopheles* vectors of malaria along the Kenyan coast. *Parasites and Vectors* 6, 311.

Wang, S.-J., Lengeler, C., Smith, T., Vounatsou, P., Cisse, G., Diallo, D., Akogbeto, M., Mtasiwa, D., Teklehaimanot, A. and Tanner, M. (2005a) Rapid urban malaria appraisal (RUMA) in sub-Saharan Africa. *Malaria Journal* 4, 40.

Wang, S.-J., Lengeler, C., Smith, T., Vounatsou, P., Diadie, D., Pritroipa, X., Convelbo, N., Kientga, M. and Tanner, M. (2005b) Rapid urban malaria appraisal (RUMA) I: epidemiology of urban malaria in Ouagadougou. *Malaria Journal* 4, 43.

Wang, S.-J., Lengeler, C., Mtasiwa, D., Mshana, T., Manane, L., Maro, G. and Tanner, M. (2006a) Rapid urban malaria appraisal (RUMA) II: epidemiology of urban malaria in Dar es Salaam (Tanzania). *Malaria Journal* 5, 28.

Wang, S.-J., Lengeler, C., Smith, T., Vounatsou, P., Akogbeto, M. and Tanner, M. (2006b) Rapid urban malaria appraisal (RUMA) IV: epidemiology of urban malaria in Cotonou (Benin). *Malaria Journal* 5, 45.

Wang, S.-J., Lengeler, C., Smith, T., Vounatsou, P., Cisse, G. and Tanner, M. (2006c) Rapid urban malaria appraisal (RUMA) III: epidemiology of urban malaria in the municipality of Yopougon (Abidjan). *Malaria Journal* 5, 29.

Watson, R.T., Zinyowera, M.C. and Moss, R.H. (1996) *Climate Change 1995 Impacts, Adaptations and Mitigation of Climate Change: Scientific-Technical Analyses.* Cambridge University Press, Cambridge, UK.

Yadouleton, A., Asidi, A., Djouaka, R., Braima, J., Agossou, C. and Akogbeto, M. (2009) Development of vegetable farming: a cause of the emergence of insecticide resistance in populations of *Anopheles gambiae* in urban areas of Benin. *Malaria Journal* 8, 103.

Yadouleton, A., N'Guessan, R., Allagbe, H., Asidi, A., Boko, M., Osse, R., Padonou, G., Kinde, G. and Akogbeto, M. (2010) The impact of the expansion of urban vegetable farming on malaria transmission in major cities of Benin. *Parasites and Vectors* 3, 118.

Yamamoto, S., Louis, V., Sie, A. and Sauerborn, R. (2009) The effects of zooprophylaxis and other mosquito control measures against malaria in Nouna, Burkina Faso. *Malaria Journal* 8, 283.

Zhou, G., Munga, S., Minakawa, N., Githeko, A.K. and Yan, G. (2007) Spatial relationship between adult malaria vector abundance and environmental factors in western Kenya highlands. *American Journal of Tropical Medicine and Hygiene* 77, 29–35.

11 Climate Change and Vector-borne Diseases in the Urban Ecosystem in India

Ramesh C. Dhiman* and Poonam Singh

National Institute of Malaria Research (ICMR), Sector-8, Dwarka, Delhi, India

11.1 Introduction

Of the various communicable diseases, vector-borne diseases (VBDs) are of considerable public health importance. With changing ecological, climatic, socioeconomic and developmental activities, the spatial and temporal distribution of VBDs are also changing. New areas are reporting re-emergence of VBDs, which were forgotten for decades. With the institution of strengthened intervention measures, well-known VBDs like malaria are coming under control, while diseases like dengue fever and chikungunya are emerging faster due to urbanization and globalization (Gubler, 2011). Climate change is also one of the important emerging public health issues by which VBDs are most likely to be affected (IPCC, 2007, 2014). The fast pace of urbanization warrants an understanding of the dynamics of common VBDs in the urban ecosystem for better preparedness for control.

In India, there are six major VBDs: malaria, dengue, chikungunya, lymphatic filariasis, Japanese encephalitis and kala-azar (NVBDCP, 2005), in addition to scrub typhus, cutaneous leishmaniasis, leptospirosis, plague, etc. In the present paper, only the VBDs predominantly endemic in urban areas are discussed in the context of climate change. The endemicity of malaria is gradually reducing and there are a few hard-core foci in tribal and forested areas (NVBDCP, 2005; Dhiman *et al.*, 2010). As the urban ecosystem warrants a different intervention strategy to rural areas, the problem of malaria control in urban areas having more than 50,000 population has been planned under the Urban Malaria Scheme. Presently the scheme is implemented in 131 towns in 19 states and Union Territories in India.

Dengue, which is associated with urbanization, is spreading fast. Of 35 states, dengue has been reported in 32 states. The incidence has increased from 5534 in 2007 to 75,808 in 2012. Recording of dengue cases in Delhi city has exceeded the previous records and up to October 2015, more than 14,000 cases have been reported. The incidence of chikungunya, which reached a peak of 73,000 in 2009, is showing a declining trend, with around 20,000 reports per annum. Lymphatic filariasis is endemic in 250 districts in 20 states/Union Territories and is under elimination mode. Kala-azar, which is confined mainly to the eastern part of the country, i.e. Bihar, West Bengal and eastern Uttar Pradesh, is also targeted for elimination.

11.2 Climatic Zones in India

There are six climatic zones in India: hot and dry, warm and humid, moderate (also known

*E-mail: r.c.dhiman@gmail.com

as temperate), cold and cloudy, cold and sunny, and composite, based on temperature (T) and relative humidity (RH) (Bansal and Minke, 1988). The hot and dry zone experiences >30°C T and <55% RH; warm and humid is characterized by >30°C T and >55% RH; the moderate zone has 25–30°C T and <75% RH; cold and cloudy <25°C T and >55% RH; cold and sunny experiences <25°C T and <55% RH. The composite zone is the one in which 6 months do not fall under any of the five categories mentioned above (Fig. 11.1).

The high endemicity of malaria is confined to the warm and humid zone. The vulnerable zone due to climate change is the northern part of the composite zone, cold and cloudy and cold and sunny so as to make the lower threshold of transmission suitable for most of the VBDs. The warm and humid zone, particularly the states of Odisha and Andhra Pradesh, which experiences extremes of temperature may show reduction in VBDs due to an increase in the upper threshold of temperature. However, further research is required to deduce the impact

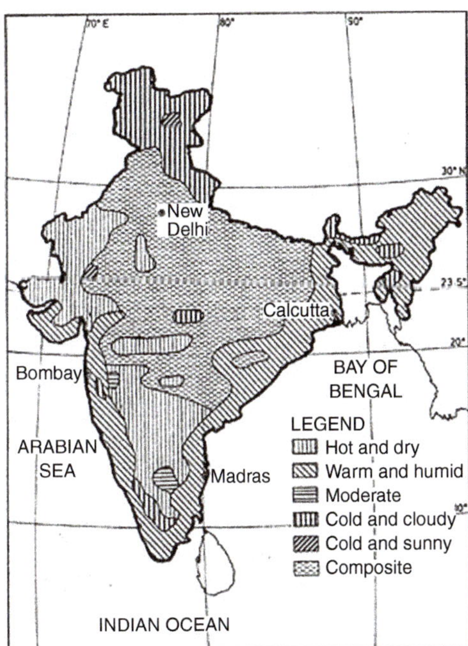

Fig. 11.1. Climatic zones of India. (Source: Bansal and Minke, 1988.)

in conjunction with ecological conditions, which may provide an alternative to the mosquito vectors for adaptation.

11.3 Urban Ecosystem and Common Vector-borne Diseases

Urbanization is increasing at a rapid pace in India. There has also been a shift in emphasis from rural development to urban areas since 2005 with the launch of Jawaharlal Nehru National Urban Renewal Mission. Further, the developmental plan for 100 smart cities has also been launched in India (SMC, 2015). At present, the urban population of India is around 377 million, and this is likely to double by the year 2031 (Nagendra *et al.*, 2013). The urban ecosystem is characterized by the influx of the rural population to cities, change of land use to multi-storeyed buildings, the introduction of a piped water supply, a shift from wood fires to cooking gas, sewage collection, dependence on vehicular transport and various lifestyle changes. Water bodies have been almost eradicated, leading to changes in biodiversity. The piped water supply, which is not reliable, leads to water storage practices. Collection of garbage and delays in disposal leads to the creation of temporary water collections, resulting in breeding grounds for mosquitoes and flies. Wastewater drains provide a suitable breeding habitat for *Culex* mosquitoes. Migration also poses a public health problem in the introduction of disease, as well as dissemination to other areas. In developing countries, owing to lack of complete urbanization, remnants of the rural ecosystem also remain, confined between urbanized areas within a city. Habitats in the urban environment lack diversity, relatively speaking, but are well provided with food and places to live, and free of competitors and predators (Lines *et al.*, 1994). Therefore, the ecosystem of urban areas becomes suitable for vectors of dengue, chikungunya, filariasis and, to a lesser extent, to vectors of malaria. As regards control of VBDs in urban areas, the strategy for vector control usually remains directed towards urban areas, leaving a few foci of transmission in rural areas – akin to the ecosystem within a city.

11.3.1 Malaria

The ecology of vectors as well as human beings in urban areas is different from that in rural areas, warranting a separate paradigm, i.e. urban malaria and the plan for control of malaria in India. Presently 131 cities in 19 states and Union Territories of India are under the Urban Malaria Scheme (UMS). In urban areas, malaria is transmitted by *Anopheles stephensi*, while in rural foci within urban areas, malaria is transmitted by *Anopheles culicifacies*. The peri-urban areas are suitable for the breeding of *An. culicifacies*. Construction activities related to civil engineering work for hospitals, metro-rail, hotels, bridges, etc. lead to the creation of breeding habitats at construction sites, as well as overhead tanks in newly constructed societies, making the area suitable for breeding of *An. stephensi*, even leading to outbreaks (Adak *et al.*, 1994).

Migration, particularly immigration, is a frequent phenomenon in urban centres, in which the poor population from endemic areas like Odisha, Chhattisgarh, Jharkhand and Assam bring a reservoir of infection of malaria. In the outskirts, due to irregular colonization, residential plots with small boundaries serve as suitable habitats for the breeding of *An. culicifacies*.

The data on temperature and relative humidity (RH) for the baseline years 1961–1990 and projected year 2030 were extracted from the PRECIS model (basically derived from HadRM3) and provided by the Indian Institute of Tropical Meteorology in Pune. In the urban centres of Delhi, Chennai and Bangalore, baseline temperature ranges for 1961–1990 are 11.70–37.08°C, 25.19–35.35°C, 22.20–29.01°C, respectively (Fig. 11.1) while in 2030, temperatures are projected to be in the ranges 10.64–41.57°C, 26.64–36.62°C, 24.43–30.52°C, respectively.

Of the five metropolitan cities of India, *An. culicifacies*-transmitted malaria is restricted mainly to Delhi and Bangalore, and breeding is likely to be affected by day-time temperature. *An. stephensi*, which prefers to breed in containers not directly exposed to the sun, is likely to have a faster development time during the winter, in view of the projected rise in temperature. Thus, it is probable that transmission will occur almost throughout the year in Chennai and Bangalore.

In Delhi, projected temperature shows an increase during March, April and June–September, while a reduction in temperature was also observed during January–February and October–December, as compared to baseline. As regards RH, 8 months show an increase in humidity (January–May, October–December) and a decrease in June–September, which are critical months for transmission of malaria. Transmission windows (TWs) based on a minimum required temperature of 18–32°C (Dhiman *et al.*, 2011) are open for 6 months at baseline as well as in the projected scenario. When T and RH (>55%) were considered together for determination of TWs, only 4 months were found to be suitable (Fig. 11.2). This shows that in the projected scenario of climate change, the windows of transmission are not affected. However, as the temperature is projected to increase for 6 months of the year, the intensity of transmission is likely to accelerate.

In Chennai, the projected temperature shows an increase up to a maximum of 4.14°C for the whole year as compared to baseline. There is no difference in TWs based on temperature alone, or in combination with RH, at baseline as well as the projected scenario (Fig. 11.3). But by the projected year 2030, TWS are open for only 6 months as compared to 9 months for baseline data.

In Bangalore also, the projected temperature shows an overall increase in T for all the 12 months as compared to baseline (Fig. 11.4). With baseline temperature, the TWs are open for 12 months, while with T + RH, TWs are reduced by 2 months (10 months open) due to a reduction in required RH. In the projected scenario for 2030, TWs with T + RH show an opening for 11 months. This indicates that in the projected scenario,

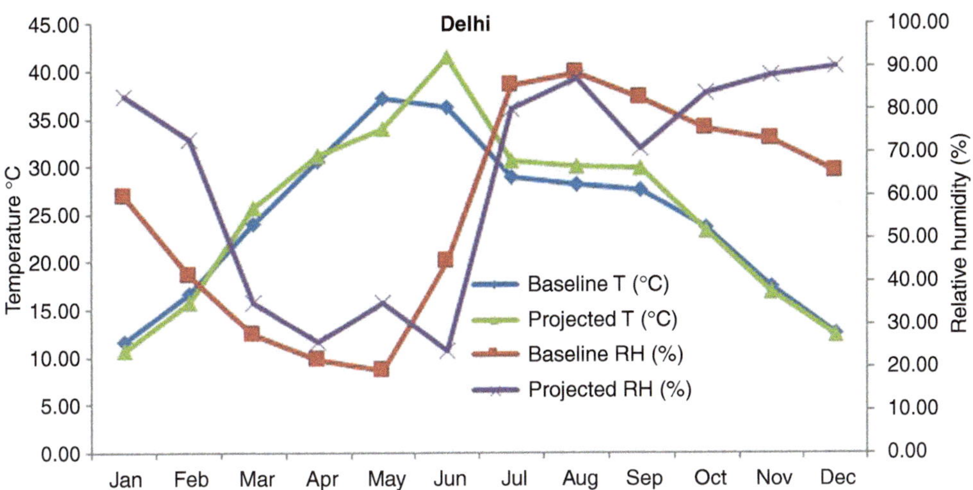

Fig. 11.2.   Baseline (1961–1990) and projected (to 2030) temperature and RH in Delhi.

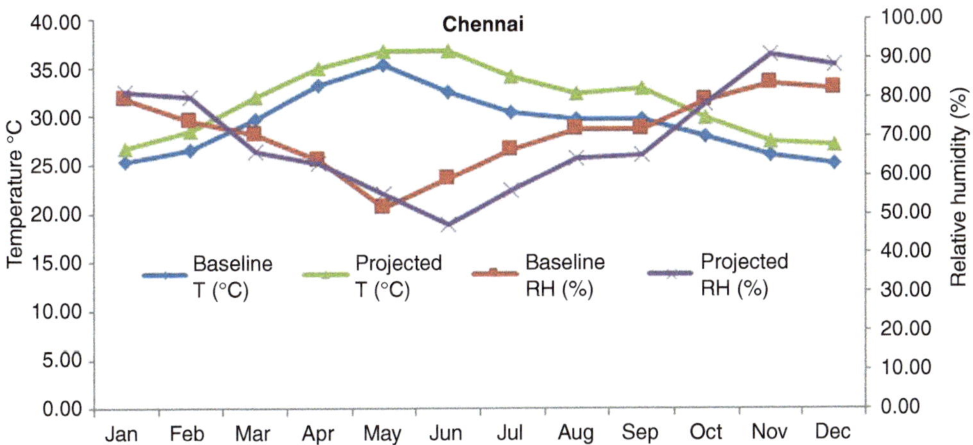

Fig. 11.3.   Baseline (1961–1990) and projected (to 2030) temperature and RH in Chennai.

there is an increase of 1 month. But the overall increase in temperature is likely to exacerbate the intensity of malaria transmission in Bangalore.

11.3.2 Dengue

Similarly, the vectors of dengue breed in container waters, but peri-domestic garbage also serves as suitable breeding areas for *Aedes aegypti* (Arunachalam *et al.*, 2010). The projected temperatures shown above are likely to affect the continuation of a high density of vector species even in colder months. This is evident from the extension of transmission of dengue in Delhi from October to December (personal observation). Under the collateral impacts, water availability is likely to be affected in view of the climate change, leading to increased water-storage practices, which in turn would lead to enhanced breeding grounds of *Ae. aegypti*.

Aedes albopictus, which is known to transmit dengue in peri-urban areas, has also been reported as a vector from Delhi (Kumari *et al.*, 2011) and southern (Tyagi

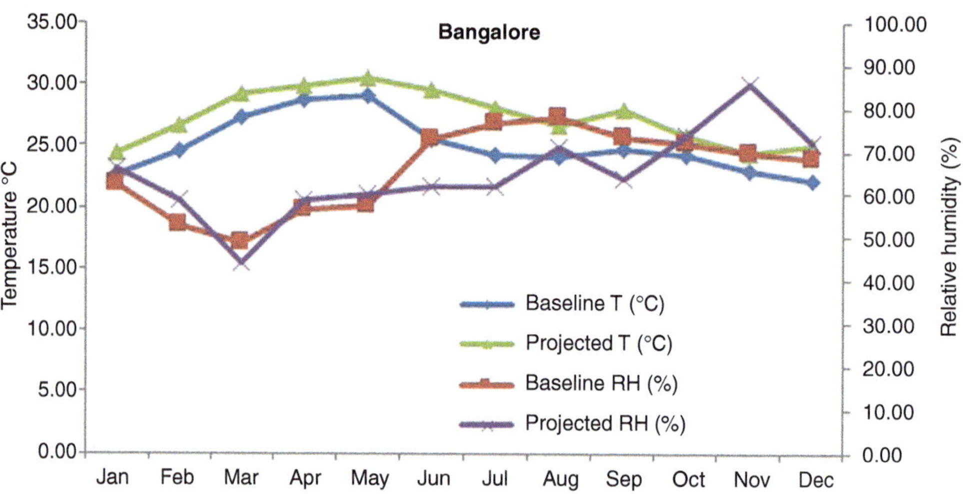

Fig. 11.4.   Baseline (1961–1990) and projected (to 2030) temperature and RH in Bangalore (Karnataka).

et al., 2006) and north-eastern parts of India (Das *et al.*, 2014; Khan *et al.*, 2014). Therefore, breeding and survival of this species is likely to be affected by changes in ecological conditions due to urbanization.

The spread of dengue has also been related to water supply and density of human dwellings (Schmidt *et al.*, 2011; Padmanabha *et al.*, 2012). With growing urbanization and climatic changes dengue is likely to expand spatially and temporally. Schmidt *et al.* (2011), in a study undertaken in Vietnam city, found that rural areas may have similar rates of dengue fever to cities, and that areas with an adequate water supply did not experience severe outbreaks. On the other hand, in a study undertaken in India in urban and rural settings in Maharashtra, Cecilia (2014) found higher seropositivity (58.5%) in urbanized villages as compared to 41.2% in rural areas. Recommendations were to improve the water supply, in addition to vector control for the efficiency of control efforts. In addition to the impact of projected temperature on the development of vector species, the behaviour of communities in emptying and using stored water may also affect production of *Ae. aegypti*, thus requiring targeted intervention considering the behaviour in usage of water and geographic variation in temperature (Padmanabha *et al.*, 2010).

Padmanabha *et al.* (2012) found the epidemiological significance of superproductive containers and human population density in the epidemiology of dengue within a person's home and when visiting others. It was emphasized that variation in population density within a city should be considered in dengue-control policy.

Padmanabha *et al.* (2011) also found a relationship between increased development rate and reduced starvation survival of early instar larvae, particularly in the lower and middle temperatures of 20–30°C, leading to the density regulation of *Ae. aegypti* populations. It indicates that at the projected increase in temperature, the survival rate of *Ae. aegypti* may be adversely affected in areas like Delhi, which experience high temperatures during the summer and monsoon. On the other hand, in areas with moderate temperature, the intensity of transmission is likely to increase due to the increased developmental rate. As the thresholds of temperature required for dengue transmission are not well defined (IPCC, 2001) and other factors like water availability and storage practices and lifestyle of communities also influence dengue transmission, the projected scenario does not seem very helpful for preparedness.

11.3.2.1 Projected scenario of dengue transmission by 2030

Transmission windows for dengue were determined using A1B scenario for baseline (1961–1990) and projected scenario by 2030 as mentioned above. The criterion of determining TWs was 20°C minimum, with 32°C as the upper cut-off point, based on the occurrence of dengue fever cases in Delhi and the published literature (Rueda *et al.*, 1990; Focks *et al.*, 1995).

11.3.2.2 Transmission window based on temperature

In the A1B scenario, 154 pixels in India show unsuitability for dengue transmission. These pixels are confined to the northern part (Fig. 11.5 and Table 11.1). Under the projected scenario, there is a slight opening of TWs in the northern part, while there is an increase in all categories of TWs for 10–12 months of the year, which show a reduction in the southern part.

11.3.2.3 Based on temperature and relative humidity

With combined TWs of dengue based on T and RH (55–90%), there is a reduction in pixels of category I of TWs in the projected scenario from 174 to 162, indicating the opening of a few windows of transmission. There is an increase in TWs in category II from 114 to 171 under the projected scenario (Fig. 11.6 and Table 11.2). The intensity of transmission in categories III to V (4–6, 7–9 and 10–12 months) shows a reduction in the projected scenario by 2030.

The limitations of the projected transmission windows are the lack of knowledge about definite thresholds of dengue transmission (IPCC, 2001), water availability and storage practices, lifestyle and intervention measures for vector and dengue control.

11.3.3 Chikungunya

Chikungunya is also a climate-sensitive disease transmitted mainly by *Ae. aegypti* mosquitoes, and it has been a public health problem since 2006, when it spread to non-endemic areas in India mainly through travellers (Mohan *et al.*, 2010; Mahapatra, 2013). The economic impact during the outbreak of 2006 was estimated to be 25,588 DALYS (Disability Adjusted Life Years) (Mohan *et al.*, 2010). There is a lack of time series retrospective data and thresholds of transmission, and correspondingly scant work related to climate change. Detailed analytical work is warranted to link climatic conditions with the distribution of chikungunya, with a particular focus on the future scenario.

11.3.4 Lymphatic filariasis

Filariasis is caused by the helminthic parasites *Wuchereria bancrofti* and *Brugia malayi*, transmitted by *Culex quiquefasciatus* and *Mansonia annulifera* mosquitoes. The thresholds of transmission are not known for filariasis, therefore the impact of climate change has not been assessed for this disease. It is distributed in almost half of the country and 600 million of the population are at risk (NVBDCP, 2005). Although the disease is in elimination mode in India, it is difficult to control breeding of *Culex* mosquitoes, a nuisance mosquito in urban areas. On the other hand, *M. annulifera* prefers to breed in water bodies with water hyacinth. The removal of weeds from such water bodies can prevent breeding of vectors. Good hygiene and campaigns like 'Swachh Bharat Abhiyan' (Clean India Campaign, launched by the Government of India) can bring about changes to reduce breeding of vectors of filariasis.

11.4 Adaptation in Mosquitoes

In view of the projected rise in temperature, mosquitoes are likely to adapt themselves for survival. In the western and eastern parts of India, Rajasthan and Odisha states experience extremes of heat. Mosquito vectors of malaria prefer to breed in TANKA (large water-storage tanks) and rest indoors

R.C. Dhiman and P. Singh

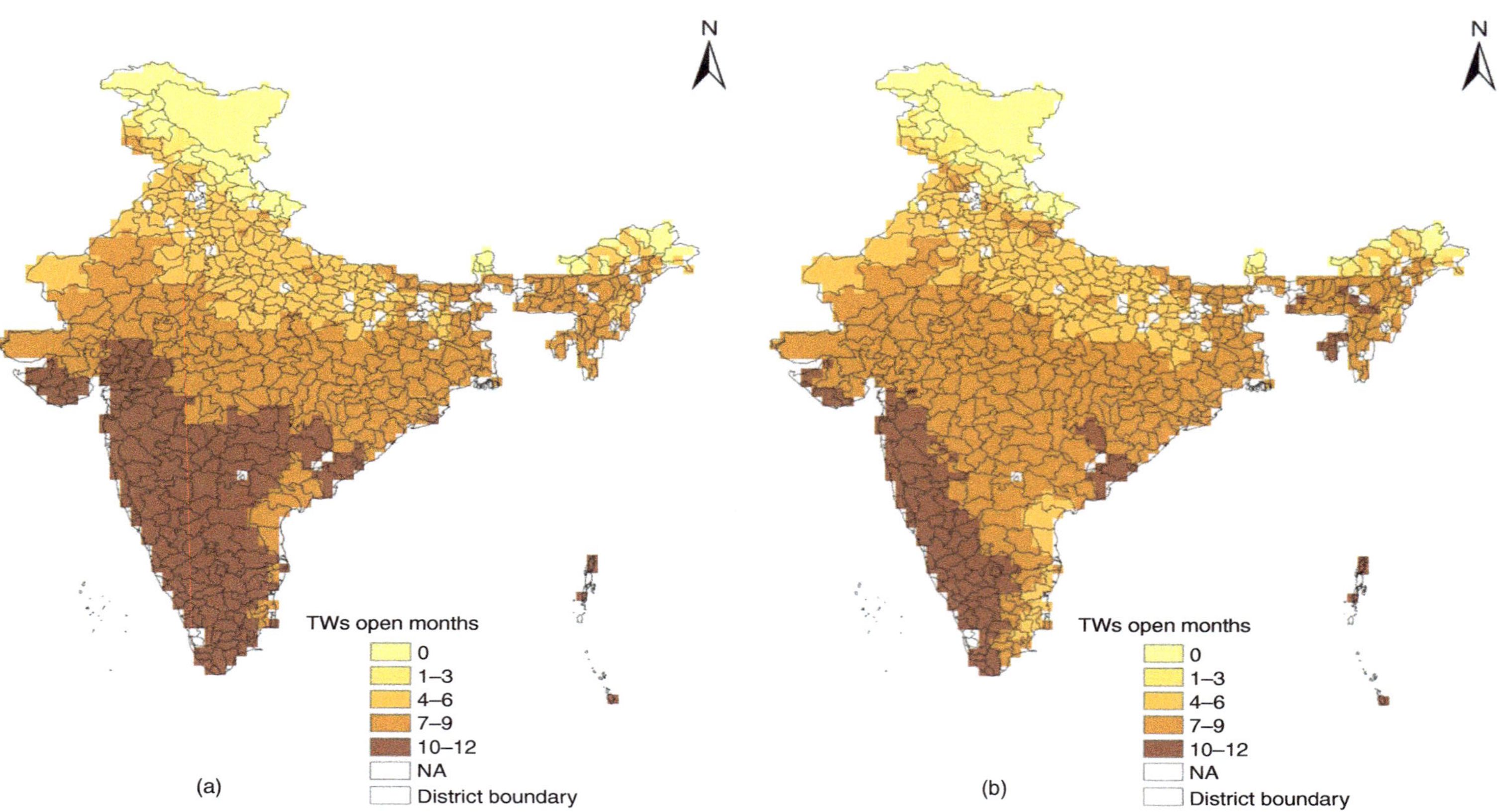

Fig. 11.5. Transmission windows of dengue based on temperature: (a) baseline (1961–1990); (b) projected (to 2030).

Table 11.1. Pixels of transmission windows of different categories of dengue in India, based on temperature; A1B scenario.

Scenario	No. of pixels open in different categories of transmission window					Remarks (PRECIS data not available)
	I (0)	II (1–3)	III (4–6)	IV (7–9)	V (10–12)	
Baseline	154	16	278	589	416	42
Projection (by 2030)	142	34	312	765	200	42

in water-storage containers. In southern India, where moderate temperature ranges from 22°C to 35°C, *An. stephensi* prefers to breed in wells and water harvesting devices in the urban areas of Chennai. Theoretically, areas like Odisha and Andhra Pradesh under a warm and humid climatic zone, may experience a reduction in the transmission months for malaria (Dhiman *et al.*, 2011), as the upper threshold of transmission suitability crosses during the summer. However, the presence of forested and foothill areas in these states and the possible development of heat-shock proteins in mosquitoes as an adaptation for their survival may not result in a reduction of the transmission months. Therefore, the issue of climate change should be viewed in conjunction with ecological change in terms of the future scenario of VBDs, the epidemiology of which is ecology-driven.

11.5 Extreme Weather Events in Urban Areas

During the early 2000s, Mumbai and Chennai witnessed extremes of weather events in terms of severe floods. In Mumbai, there was a devastating flood on 26 July 2005, which brought the city to a standstill (Sharma, 2012), and Mumbai floods have become a common phenomenon. In such situations, the stagnant water and disruption of water supply, leading to poor storage practices or scarcity of water, create congenial conditions for mosquito breeding. In 2015, the city of Chennai broke the previous record of 1088 mm of rainfall for the year 1918: the average November rainfall for Chennai city is 407 mm, while in November 2015 (up to 24 November), 1185 mm of rain

was recorded (Skymet, 2015). About 56.6% of malaria cases from the state of Tamil Nadu are recorded in Chennai city alone (NHMTN, 2013). Such extreme events are likely to alter the transmission patterns of VBDs like malaria, dengue and leptospirosis, which is prevalent in southern India.

11.6 Conclusion

In view of rapid urbanization, there is a growing threat of VBDs, particularly dengue and chikungunya. Because of the changing ecology and climate, the epidemiology of VBDs, particularly dengue, needs to be studied in terms of water supply, population density, and the rural and urban ecosystems, for effective vector control. The risk factors of dengue – dense population, water deficit areas, preferred breeding grounds of vector species – and the role of *Ae. albopictus* in periurban areas need to be determined in order to prepare a plan to avert VBDs. The strategy of vector control in urban areas requires an intersectoral approach, involving public health, sanitation, water supply and urban developments, along with the community. Control strategy needs revisiting: in the urban ecosystem a mixed strategy is needed, in view of the existence of the rural ecosystem within cities in developing countries. In areas vulnerable to climate change, community health education for disease prevention and control is the vital need of the hour.

Acknowledgements

Financial assistance from the Ministry of Environment and Forests, Government of

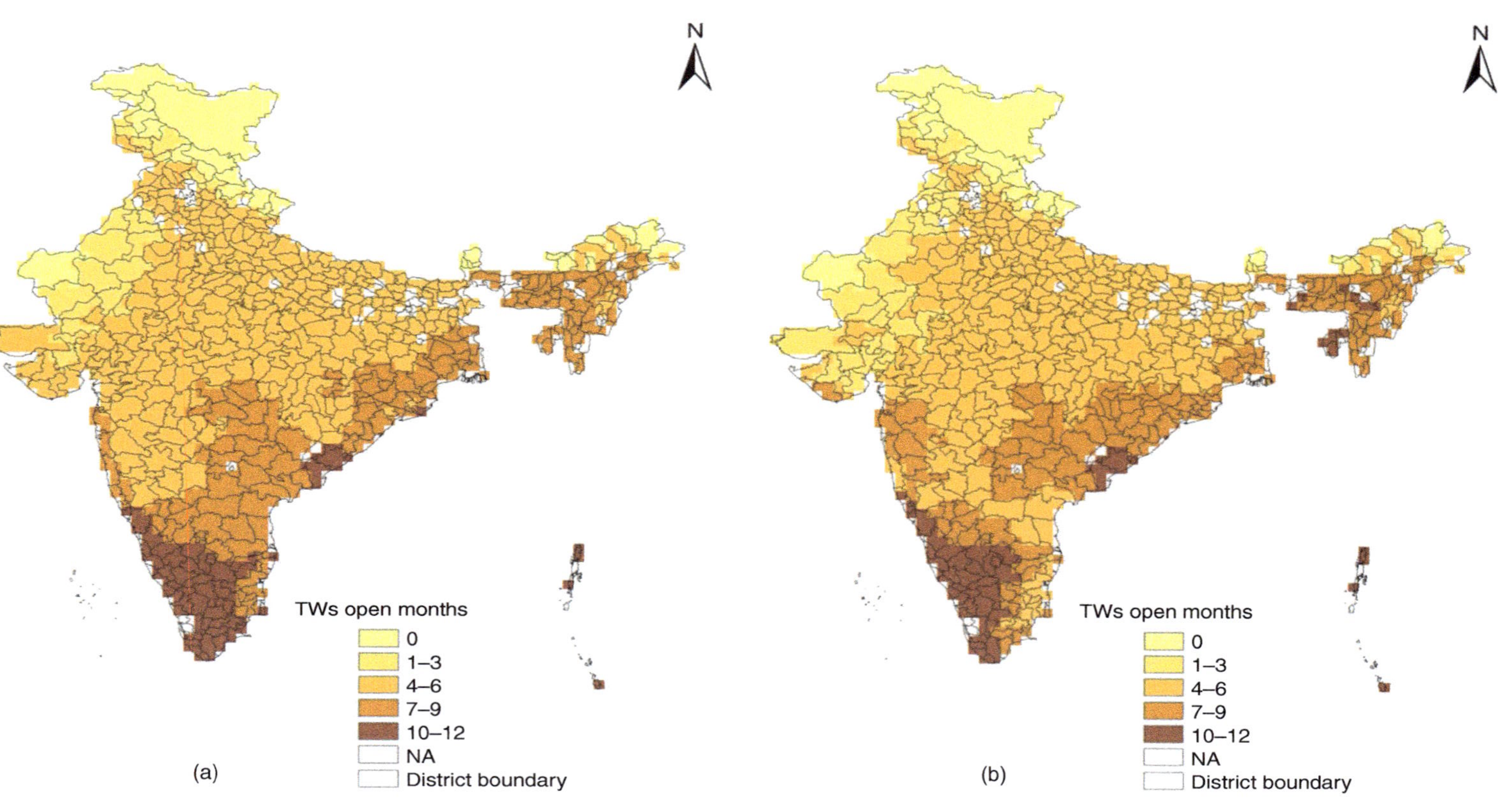

Fig. 11.6.   Transmission windows of dengue based on temperature and relative humidity (A1B scenario): (a) baseline (1961–1990); (b) projected (to 2030).

Table 11.2. Pixels of transmission windows of different categories of dengue in India, based on temperature and relative humidity.

Scenario	No. of pixels open in different categories of transmission window					Remarks (PRECIS data not available)
	I (0)	II (1–3)	III (4–6)	IV (7–9)	V (10–12)	
Baseline	174	114	704	354	107	42
Projection (by 2030)	162	171	688	333	99	42

India under the aegis of the NATCOM II project is duly acknowledged. Thanks are also due to the Indian Institute of Tropical Meteorology, Pune, for providing data for the PRECIS model.

References

Adak, T., Batra, C.P., Mittal, P.K. and Sharma, V.P. (1994) Epidemiological study of malaria outbreak in a hotel construction site of Delhi. *Indian Journal of Malariology* 31(3), 126–131.

Arunachalam, N., Tana, S., Espino, F., Kittayapong P., Abeyewickreme, W., Wai, K.T., Tyagi, B.K., Kroeger, A., Sommerfeld, J. and Petzold, M. (2010) Eco-bio-social determinants of dengue vector breeding: a multi country study in urban and periurban Asia. *Bulletin of the World Health Organization* 88(3), 173–184. DOI: 10.2471/BLT.09.067892.

Bansal, N.K. and Minke, G. (eds) (1988) Climatic Zones and Rural Housing in India. Kernforschungsanlage, Jülich, Germany.

Cecilia, D. (2014) Current status of dengue and chikungunya in India. *WHO South-East Asia Journal of Public Health* 3(1), 22–27.

Das, M., Gopalakrishnan, R. and Kumar, D. (2014) Spatiotemporal distribution of dengue vectors and identification of high risk zones in district Sonitpur, Assam, India. *Indian Journal of Medical Research* 140, 278–284.

Dhiman, R.C., Pahwa, S., Dhillon, G.P.S. and Dash, A.P. (2010) Climate change and threat of vector borne diseases in India: are we prepared? *Parasitology Research* 106, 763–773. DOI: 10.1007/s00436-010-1767-4.

Dhiman, R.C., Chavan, L., Pant, M. and Pahwa, S. (2011) National and regional impacts of climate change on malaria by 2030. *Current Science* 101(3), 372–383.

Focks, D.A., Daniels, E., Haile, D.G. and Keesling, J.E. (1995) A simulation model of the epidemiology of urban dengue fever: literature analysis, model development, preliminary validation, and samples of simulation results. *American Journal of Tropical Medicine and Hygiene* 53, 489–506.

Gubler, D.J. (2011) Dengue, urbanization and globalization: the unholy trinity of the 21st century. *Tropical Medicine and Health* 39(4), 3–11. DOI: 10.2149/tmh.2011-S05.

IPCC (2001) Climate change 2001: impacts, adaptation and vulnerability. In: McCarthy, J.J., Canziani, O.F., Leary, N.A., Dokken, D.J. and White, K.S. (eds) *Contribution of Working Group II to the Third Assessment Report of the Intergovernmental Panel on Climate Change*. Cambridge University Press, Cambridge, UK and New York, USA, pp. 1–1032.

IPCC (2007) Climate change 2007: the physical science basis. In: Solomon, S., Qin, D., Manning, M., Chen, Z., Marquis, M., Averyt, K.B., Tignor, M. and Miller, H.L. (eds) *Working Group I Contribution to the Fourth Assessment Report of the Intergovernmental Panel on Climate Change*. Cambridge University Press, Cambridge, UK and New York, USA, pp. 1–1009.

IPCC (2014) Summary for policymakers. In: Field, C.B., Barros, V.R., Dokken, D.J., Mach, K.J., Mastrandrea, M.D., Bilir, T.E., Chatterjee, M., Ebi, K.L., Estrada, Y.O., Genova, R.C., Girma, B., Kissel, E.S., Levy, A.N., MacCracken, S., Mastrandrea, P.R. and White, L.L. (eds) *Climate Change 2014: Impacts, Adaptation, and Vulnerability. Part A: Global and Sectoral Aspects. Contribution of Working Group II to the Fifth Assessment Report of the Intergovernmental Panel on Climate Change*. Cambridge University Press, Cambridge, UK and New York, USA, pp. 1–32.

Khan, S.A., Dutta, P., Topno, R., Soni, M. and Mahanta, J. (2014) Dengue outbreak in a hilly state of Arunachal Pradesh in Northeast India. *Scientific World Journal* 2014, article ID 584093. DOI: 10.1155/2014/584093. 6 pp.

Kumari, R., Kumar, K. and Chauhan, L.S. (2011) First dengue virus detection in *Aedes albopictus* from Delhi, India: its breeding ecology and role in dengue transmission. *Tropical Medicine and International Health* 16(8), 949–954. DOI: 10.1111/j.1365-3156.2011.02789.x.

Lines, J., Harpham, T., Leake, C. and Schofield, C. (1994) Trends, priorities and policy directions in the control of vector-borne diseases in urban environments. *Health Policy and Planning* 9(2), 113–129. DOI: 10.1093/heapol/9.2.113.

Mahapatra, T. (2013) Recent resurgence of Chikungunya fever in Delhi, India. *Annals of Tropical Medicine and Public Health* 6(2), 149–150.

Mohan, A., Kiran, D.H., Manohar, I.C. and Kumar, D.P. (2010) Epidemiology, clinical manifestations, and diagnosis of chikungunya fever: lessons learned from the re-emerging epidemic. *Indian Journal of Dermatology* 55, 54–63.

Nagendra, H., Sudhira, H.S., Katti, M. and Schewenius, N. (2013) Sub-regional assessment of India: effects of urbanization on land use, biodiversity and ecosystem services. In: Elmqvist, T., Fragkias, M., Goodness, J., Güneralp, B., Marcotullio, P.J., McDonald, R.I., Parnell, S., Schewenius, M., Sendstad, M., Seto, K.C. and Wilkinson, C. (eds) *Urbanization, Biodiversity and Ecosystem Services: Challenges and Opportunities: A Global Assessment.* Springer, Dordrecht, The Netherlands, pp. 66–74. DOI 10.1007/978-94-007-7088-1_6.

NHMTN (2013) *Vector Borne Disease Control Programme (VBDC).* National Health Mission Tamil Nadu, Department of Health and Family Welfare, State Health Society, Tamil Nadu. Available at: http://www.nrhmtn.gov.in/vbdc.html (accessed 2 June 2016).

NVBDCP (2005) Diseases. National Vector Borne Disease Control Programme, Directorate General of Health Services, Ministry of Health & Family Welfare, Government of India. Available at: http://www.nvbdcp.gov.in (accessed 2 June 2016).

Padmanabha, H., Soto, E,, Mosquera, M., Lord, C.C. and Lounibos, L.P. (2010) Ecological links between water storage behaviors and *Aedes aegypti* production: implications for dengue vector control in variable climates. *EcoHealth* 7(1), 78–90.

Padmanabha, H., Lord, C.C. and Lounibos, L.P. (2011) Temperature induces trade-offs between development and starvation resistance in *Aedes aegypti* (L.) larvae. *Medical and Veterinary Entomology* 25(4), 445–453. DOI: 10.1111/j.1365-2915.2011.00950.x.

Padmanabha, H., Durham, D., Correa, F., Diuk-Wasser, M. and Galvani, A. (2012) The interactive roles of *Aedes aegypti* super-production and human density in dengue transmission. *PLoS Neglected Tropical Diseases* 6(8), article ID e1799. DOI: 10.1371/journal.pntd.0001799.

Rueda, L.M., Patel, K.J., Axtell, R.C. and Stinner, R.E. (1990) Temperature-dependent development and survival rates of *Culex quinquefasciatus* and *Aedes aegypti* (Diptera: Culicidae). *Journal of Medical Entomology* 27(5), 892–898.

Schmidt, W.P., Suzuki, M., White, R.G., Tsuzuki A., Yoshida L-M., Yanai, H., Haque, U., Tho, L.H. and Anh, D.D. (2011) Population density, water supply, and the risk of dengue fever in Vietnam: cohort study and spatial analysis. *PLoS Med* 8(8), article ID e1001082. DOI: 10.1371/journal. pmed.1001082.

Sharma, V. (2012) July 26, 2005: When Mumbai was washed away. DNAindia.com, 26 July. Available at: http://www.dnaindia.com/analysis/comment-july-26-2005-when-mumbai-was-washed-away-1719785 (accessed 18 October 2016).

Skymet (2015) 50 mm rain in one hour, Chennai rains exceed all time high. Available at: http://www.skymetweather.com/content/weather-news-and-analysis/chennai-to-cross-november-rain (accessed 24 November 2015).

SMC (2015) Smart Cities Mission. Ministry of Urban Development, Government of India. Available at: http://smartcities.gov.in (accessed 2 June 2016).

Tyagi, B.K., Hiriyan, J., Samuel, P., Tewari, C.S. and Paramasivan, R. (2006) Dengue in Kerala: a critical review. *ICMR Bulletin* 36(4–5), 14–22.

12 Climate Change and Urban Human Health

Martha Macedo de Lima Barata

Lagoa, Rio de Janeiro, Brazil

12.1 Introduction

Current climate extremes and projections for future climate changes have resulted in growing attention being given to the health effects of these events on the urban population (Barata *et al.*, 2016). Considering health as 'a state of complete physical, mental and social well-being and not merely the absence of disease or infirmity' (WHO, 1948), indeed, almost all the effects of climate change have a direct or indirect impact on health (Barata *et al.*, 2016).

The urban population is growing at a fast rate. In 1960, it accounted for 34% of the total global population, which in 2014 grew to 54% and it is estimated that by 2017 the majority of people will be living in urban areas (WHO, 2015). This fast urbanization is threatening environmental quality in cities. There is an expanding field of research exploring how the way in which cities are planned and managed can influence the health of its residents (Rydin, 2012). Additionally, long-term projections and models show that climate change may aggravate the health of city dwellers (Barata *et al.*, 2016).

City residents are especially vulnerable to climate change, due to the high population density and dependency on a complex infrastructure system (Barata *et al.*, 2011b). In this context, building urban health resilience, both in research and practice, will require dealing with the dynamic nature of urban health–social–ecological systems and the incorporation of multiple disciplines into governance approaches for healthier cities, including input from scientists, practitioners, designers and planners (McPhearson *et al.*, 2014).

This chapter focuses on how to use research to build a resilient city facing the hazards posed by climate change. This takes into account all the interrelationships between climate and pests in cities. An attempt is also made to present an overview of the scientific knowledge on urban health outcomes and their climate-related drivers, and to discuss the challenges for implementing adaptation strategies in cities in order to reduce pests and climate change hazards.

12.2 Impact of Climate Change on the Health of City Dwellers

Climate change impacts on the health of the population can occur directly (physical effects on the human body) or indirectly. In the latter case, the weather affects the environment and society (economy, institutions, infrastructure, productivity, recreational activities, housing, health systems, etc.), resulting in adverse effects on health (IPCC, 2007; Confalonieri *et al.*, 2015).

*E-mail: baratamml@gmail.com

The overall burden of illness and the deleterious effects of climate change can be magnified in cities in many ways, such as through a large concentration of human population, the density and impermeability of the surfaces, the urban heat island effect, increased pressure on local ecosystems for water, concentration of waste, dependency on a complex infrastructure system, eventual existence of large areas of underserviced urban slums, and the presence of concentrated pests in direct contact with dwellers (Barata *et al.*, 2011b, 2016).

Barata *et al.* (2016) recognize that storms, flood, heat extremes and landslides are among the most important weather-related events in cities that can harm human health. They can lead to increases in deaths, injuries, endemic infections, gastrointestinal, cardiovascular and respiratory diseases, water-borne and food-borne diseases, nutritional deficiencies, trauma, occupational accidents and mental health problems. The changes in temperature, precipitation, storminess, and other characteristics of climate change will vary significantly among regions with diverse geographical characteristics, weather patterns and demographics. Additionally, the effects of climate change on city dwellers will also depend on the social, cultural, economic, environmental and institutional conditions of each city (Fig. 12.1).

Poverty and poor health are highly correlated. In cities, poor residents usually have limited access to healthy housing and essential public health services (Satterthwaite *et al.*, 2008). Dense populations live in informal settlements, such as slums, where basic infrastructure facilities are lacking. In such places, extreme climate consequences like floods can severely impact the health conditions of the dwellers. Outbreaks of pests is one health concern of particular importance in areas where overcrowding and poor infrastructure and sanitation service are predominant (Freitas *et al.*, 2014). These residents are commonly vulnerable to climate variability and climate change (Frumkin *et al.*, 2008).

Extreme climate events can cause disruptions in critical infrastructure operations and affect health systems in cities. For example, the lack of electricity can make it difficult or impossible to refrigerate food, pump water to upper floors, and operate medical support equipment (Beatty *et al.*, 2006; Barata *et al.*, 2016).

12.3 Climate, Pests and Disease in Cities

Climate change is one of the factors that can contribute to the expansion of vector-borne disease, as a wide range of vector borne-diseases are climate sensitive, expanding their range or living longer in warmer weather (Barata *et al.*, 2011b). Rising temperatures are expected to affect the spread and transmission of diseases carried by vectors such as mosquitoes, ticks and mice (Costello *et al.*, 2009). Temperature affects the rate of pathogen maturation and replication within mosquitoes and the density of insects in a particular area (Costello *et al.*, 2009). Climate change may lead to alterations in the seasonal cycle and spatial distribution of some vector-borne diseases. It is important to note that climate is only one of many drivers of vector-borne disease distribution. Malaria, dengue, yellow fever, West Nile virus, schistosomiasis, leishmaniasis, Lyme disease, tick-borne encephalitis, hantavirus infections, and a number of other vector-borne diseases (see Table 12.1) are projected to increase as a result of climate change (Barata *et al.*, 2011b), contributing significantly to the burden of death and disease in urban areas. There is evidence that the Lyme disease vector, the tick species *Ixodes scapularis*, has expanded its range from the United States northward into Canada over the past several decades, in part due to warming temperatures (Ogden *et al.*, 2008, 2010). Trends in malaria in East Africa have been associated with warming, as observed over multiple decades (Omumbo *et al.*, 2011). While climate change might lead to an expansion in the range of important disease vectors, it is also possible that optimal temperature conditions for certain vector species will be exceeded and thus potentially

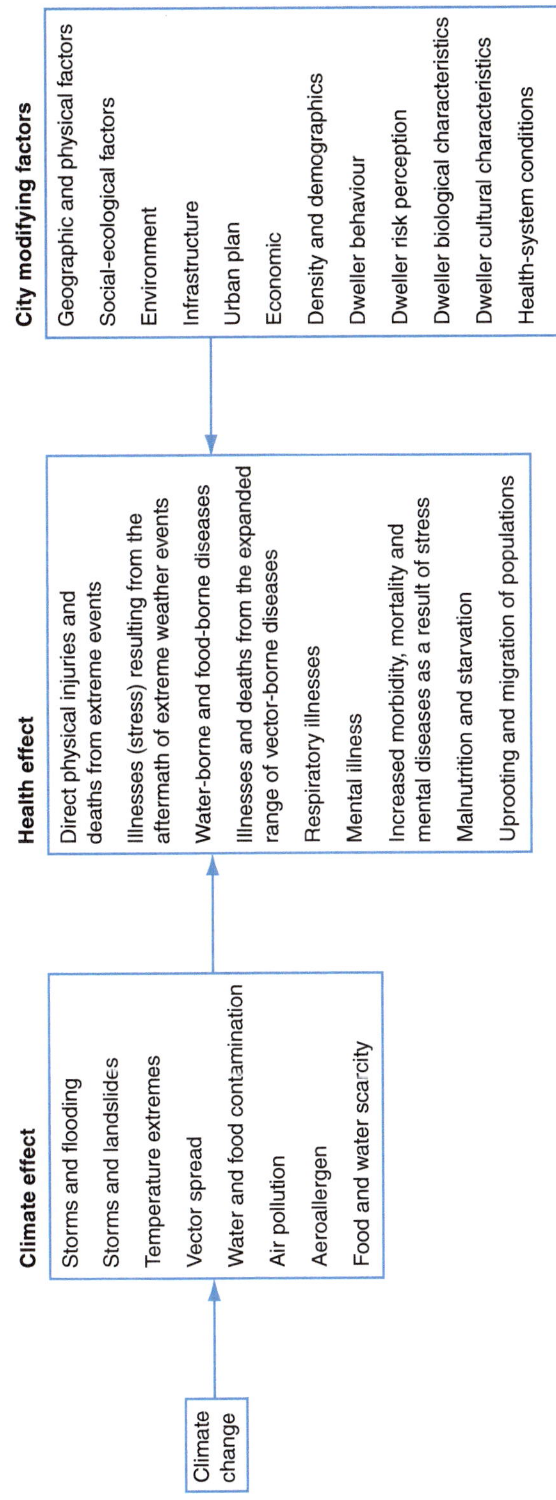

Fig. 12.1. How climate change affects citizens' health. (Adapted from Rosenzweig *et al.*, 2011.)

Table 12.1. Climate-sensitive vector-borne diseases. (Source: Reid and Kovats (2009); Semenza and Menne (2009); Barata *et al.*, 2011b.)

Vector	Diseases
Mosquitoes	Malaria, filariasis, dengue, yellow fever and West Nile virus
Sandflies	Leishmaniasis
Triatomines	Chagas disease
Ixodes ticks	Lyme disease, tick-borne encephalitis
Tsetse flies	African trypanosomiasis
Black flies	Onchocerciasis
Rodents	Leptospirosis, plague, hantavirus

reduce the risk of infection, particularly under high warming scenarios (Smith *et al.*, 2014; Barata *et al.*, 2016).

It is hard to generalize about the extent of the risk that expansion of most mosquito-borne diseases will pose on urban areas, or about the extent to which climate change will contribute to the overall problem (Barata *et al.*, 2011b). Cities do not generally provide the best environments for reproduction of mosquitoes, though standing water in plant pots or urban detritus and waste products such as used car tyres can provide breeding places for these vectors (Hay *et al.*, 2002). Temperature, precipitation levels, humidity and wind speeds, as well as the level of public health services, all affect the proliferation of vectors and the spread of the diseases of concern.

Dwellers can be exposed to water- and food-borne pathogens through a variety of routes, including via the ingestion of polluted drinking water, consumption of contaminated food, inhalation of aerosols containing bacteria, and by direct contact with recreational or flood waters. Many of the pathogens that cause water- and food-borne illnesses in dwellers are sensitive to climate parameters, including increased temperature, changing precipitation patterns, extreme precipitation events, and associated changes in seasonal patterns in the hydrological cycle. Charron *et al.* (2004) chronicled a host of *Escherichia coli, Campylobacter, Cryptosporidium, Toxoplasmosis, Giardia, Leptospirosis,* and non-specific

gastroenteritis outbreaks in North America since 1993 that have been associated with a sustained period of rainfall or an extreme rainfall event which contaminated water and in some cases overwhelmed water-treatment systems. Increased temperatures appear to increase the incidence of common North American diarrhoeal diseases such as campylobacteriosis and salmonellosis (Curriero *et al.*, 2001; ECDPC, 2011; Semenza *et al.*, 2012).

Floods enhance the potential for runoff to carry sediment and chemical pollutants to water supplies (CCSP, 2008), and water-borne illnesses from exposures to pathogens and chemical residues from pesticide run-off (especially from freshly treated properties) in recreational waters have been shown to increase in the hours after extreme rainfall events (Patz *et al.*, 2008). The risk of water-borne illness is greater among the poor, infants, elderly, pregnant women, and immune-compromised individuals (Rose *et al.*, 2001; Barata *et al.*, 2016). The prevalence of food-borne illnesses is undoubtedly high in tropical low-income countries, but the absence of public health reporting of these diseases in most low-income countries makes it difficult to estimate the impact of temperature increases. It is not known whether there is a specific pattern of food-borne illness associated with urban versus rural residency, though higher temperatures in urban environments might be expected to influence the number and extent of outbreaks in cities (Barata *et al.*, 2016).

12.4 Adaptation Strategies

Climate change poses serious challenges to the city public health system, including physical and mental health issues as well as in preventing disease, injury and disability (Ebi and Semenza, 2008). Managerial responsibility towards climate health risk in cities, in the long term, is essential for the sustainable well-being of city dwellers. To achieve this, city managers need to understand the opportunities and weaknesses in both the internal and external environment of the city and then plan the adoption of the best available tools and resources for preventing harm to their citizens. Strategic planning and implementation is necessary in order to achieve the expected outcome.

Proper management of the infrastructure, build construction, land-use plans, utility services, sanitation systems, green space, education and health systems in the cities may result in the reduction of impacts of climate change to the health of their dwellers.

The Urban Climate Change Research Network (UCCRN) has developed a climate risk management framework to support city decision makers in planning strategies to reduce the effect of climate change. Its design and implementation needed knowledge of the hazards, the vulnerability, and the specific adaptive capacity for each city to climate change (Mehrotra *et al.*, 2011).

Identification of how vulnerable a given population is to climate change, namely, what is its propensity or predisposition to be adversely affected by climate change and what are the key determinants to its vulnerability, as well as the location of the most vulnerable population, has immense practical significance for best planning adaptation strategies and prioritizing resource investment. It is noteworthy that effectively responding to climate change is a process, not a one-time assessment of risks (Ebi and Semenza, 2008). It is necessary to apply periodic assessment of the results to achieve the goals.

Surveillance and mapping of vector-borne illnesses as they appear and spread, as well as the conditions that give rise to them, will be an important preparatory step in adaptation (Barata *et al.*, 2011a). Adaptation will also require the application of vector-control strategies, which may include the following.

- Environmental controls: altering breeding sites by draining or filling sites where water collects; ensuring regular disposal of refuse; maintaining clean shelters and personal hygiene.
- Mechanical controls: protective clothing, screens or bed nets, traps, food covers, lids or polystyrene beads in latrines.
- Biological management: using living organisms or products against vector larvae, such as fish that eat larvae (e.g. tilapia), bacteria that produce toxins against larvae, free-floating ferns that prevent breeding, etc.
- Chemical controls: including repellents, insecticides, larvicides and other pesticides (Barata *et al.*, 2011b).

A combination of these methods will be needed, since the control of most vectors will be achieved by applying integrated vector management. This will require the following.

- Knowledge of factors influencing local vector biology, disease transmission and morbidity.
- Collaboration of the health sector with other sectors (such as water, solid waste, sewage disposal, storm water, housing, construction, urban food producers, etc.).
- Health communications, education and engagement with local communities (WHO, 2004; Barata *et al.*, 2011b).

In addition, enhanced or new vaccination programmes may be appropriate for some climate-sensitive diseases (Barata *et al.*, 2011b).

An adaptation strategy for reducing food-borne and water-borne diseases requires knowledge of climate factors that can influence those diseases and the planning and implementation of responses for that (Table 12.2).

12.5 Conclusion

In recent times, despite the uncertainties linking climate change, pests and urban health, the evidence of the potential impacts of climate change on pests in cities and the resulting health effects on its residents is growing. Urgent action is required to ensure the implementation of effective adaptation strategies that are designed to prevent or reduce the potential impact.

Table 12.2. Applying adaptation strategy. (Source: Barata *et al.*, 2011b.)

1. Understanding the relationship between climate variability and food- or water-borne disease
   - Examples of climate factors that can influence food-borne disease:
     - rising temperatures are expected to increase the incidence of food poisoning, as microbes can multiply more quickly
     - warmer weather will support the survival of flies and other pests that can contaminate food
     - hot weather represents a problem when refrigeration failure occurs
     - intense rainfall and floods may also contaminate water sources from which urban residents draw water for washing or cooking food, and contribute to food poisoning.
   - Examples of climate factors that can influence water-borne disease:
     - reduced water supply, leading to reduced sanitation, personal hygiene and effluent dilution
     - extreme rainfall, leading to increased pathogen loading, particularly in areas with inadequate storm-water management, ageing water treatment plants or combined sewer-/storm-water systems
     - direct effects of higher temperatures, favourable to microorganism reproduction.
2. Planning adaptation strategies – examples
   - Managers can reduce food-borne disease risk by implementing:
     - public education, alerting the population to the potential threat of increased contamination and ways to handle food and avoid food-borne diseases
     - timely information to food producers and food handlers
     - public health inspection of places where food is commercially processed or prepared
     - quality-control measures for food storage and handling
     - management of water demand and maintenance of water distribution systems to help avoid critical water shortages
     - watershed protection, such as vegetative buffers to reduce contamination of water from run-off
     - city disposal systems to capture and treat wastes
     - extending and improving urban storm-water management systems, anticipating increases in storm intensity
     - separation of combined sewage and storm-water systems and stronger regulations controlling septic systems
     - well-head alert systems that warn water-system and water-supply managers when rainfall conditions approach levels of concern, similar to predictive forecasts and warning systems
     - public health services such as public alerts, including boil water alerts about the potential threat of increased contamination, and enhanced surveillance and monitoring programmes for water-borne diseases.
3. Implementing planned adaptation strategies
4. Monitoring the results of implemented strategies
5. Verifying the effectiveness of implemented strategies
6. Re-initiating the process of understanding, planning, implementing and monitoring results and verifying the effectiveness of city adaptation strategies

The magnitude and significance of these potential climate-change impacts will vary according to the specific components of each city. Some of them are population density, social, economic, cultural, biological, political, geographic and environmental characteristics, as well as medical and infrastructure services.

Cities are complex systems. They have distinct components with a wide range of potential interactions, and these components self-organize over time and exhibit behavioural patterns that emerge from these diverse and complex interactions.

Recognizing and exploring this complexity can clarify which management strategies may be most effective and may highlight knowledge gaps to be filled through modelling, targeted learning and other strategies to reduce impacts of climate change on urban pests, and consequently on the health of city dwellers.

The costs of health impacts due to pests and climate change (increased illness, injury, disease, death) will further weaken local social systems. So understanding the vulnerabilities and planning the implementation of health-care adaptation measures is essential, and will differ among cities, each of which has its own specific modulating influences, and different social, economic, cultural and political realities.

Successful adaptation strategies need to focus on activities that understand local health vulnerabilities, in order to plan solutions that can reduce or protect the local population from those impacts. Subsequently, it will be necessary to implement solutions, engaging and assuring a cooperative effort between all strategic stakeholders (government, business and financial sectors, as well as scientists and the population) in the city.

Cities provide unique opportunities to build resilience and health protection policies and programmes relating to the potential impacts of pests.

References

Barata M.M.L., Confalonieri U.E.C., Marinho D.P., Luigi G., Silva H.V.O., Oliveira, F.T., Costa Neto C., Oliveira, T.V.S., Marincola, F.V. and Pereira, C.A. (2011a) *Mapa de vulnerabilidade da população do estado do Rio de Janeiro aos impactos das mudanças climáticas nas áreas social, saúde e ambiente*. Rio de Janeiro, Brazil. Available at: http://download.rj.gov.br/documentos/10112/364217/DLFE-40943.pdf/rel_vulbilidado.pdf (accessed 19 April 2012).

Barata, M.M.L., Ligeti, E., De Simone, G., Dickinson, T., Jack, D., Penney, J., Rahman, M. and Zimmerman, R. (2011b) Climate change and human health in cities. In: Rosenzweig, C., Solecki, W., Hammer, S.A. and Mehrotra, S. (eds) *Climate Change and Cities: First Assessment Report of the Urban Climate Change Research Network*. Cambridge University Press, Cambridge, UK, pp. 183–217.

Barata, M.M.L., Kinney, P.L., Dear, K., Ligeti, E., Ebi, K.L., Hess, J., Dickinson, T., Quinn, A.K., Obermaier, M., Sousa, D.S., Jack, D. (2016) Urban health. In: Rosenzweig, C., Solecki, W., Romero-Lankao, P., Mehrotra, S. and Dhakal, S. (eds) *Climate Change and Cities: Second Assessment Report of the Urban Climate Change Research Network*. Cambridge University Press, Cambridge, UK.

Beatty, M.E., Phelps, S., Rohner, C. and Weisfuse, I. (2006) Blackout of 2003: public health effects and emergency response. *Public Health Reports* 121(1), 36–44.

CCSP (2008) Weather and climate extremes in a changing climate: regions of focus: North America, Hawaii, Caribbean and US Pacific Islands. In: Karl, T.R., Meehl, G.A., Miller, C.D., Hassol, S.J., Waple, A.M. and Murray, W.L. (eds) *Synthesis and Assessment Product 3.3, Report by the United States Climate Change Science Program (CCSP) and the Subcommittee on Global Change Research*. Department of Commerce, NOAA's National Climatic Data Center, Washington, DC, USA.

Charron, D., Thomas, M., Waltner-Toews, D., Aramini, J. and Edge, T. (2004) Vulnerability of waterborne diseases to climate change in Canada: a review. *Journal of Toxicology and Environmental Health* 67, 1667–1677.

Confalonieri, U., Barata, M.M.L. and Marinho, D. (2015) Vulnerabilidade climática no Brasil. In: Chang, M., Góes, K., Fernandes, L., Freitas, M.A.V. and Rosa, L.P. (eds) *Metodologias de estudos de vulnerabilidade à mudança do clima*. Coletâneas Mudanças Globais, Vol. 5. Ed. Interciências, Rio de Janeiro, Brazil.

Costello A., Abbas, M., Allen, A., Ball, S., Bell, S., Bellamy, R., Friel, S., Groce, N., Johnson, A., Kett, M., Lee, M., Levy, C., Maslin, M., McCoy, D., McGuire, B., Montgomery, H., Napier, D., Pagel, C., Patel, J., Oliveira, J., Redclift, N., Rees, H., Rogger, D., Scott, J., Stephenson, J., Twigg, J., Wolff, J. and Patterson, C. (2009) Managing the health effects of climate change. *Lancet* 373(9676), 1693–1733. Available at: http://www.thelancet.com/journals/lancet/article/PIIS0140-6736%2809%2960935-1/abstract (accessed 3 June 2016).

Curriero, F.C., Patz, J.A., Rose, J.B. and Lele, S. (2001) The association between extreme precipitation and waterborne disease outbreaks in the United States, 1948–1994. *American Journal of Public Health* 91(8), 1194–1199.

Ebi, K.L. and Semenza, J.C. (2008) Community-based adaptation to the health impacts of climate change. *American Journal of Preventive Medicine* 35(5), 501–507.

ECDPC (2011) *Climate Change and Communicable Diseases in the EU Member States: A Handbook for National Vulnerability, Impact and Adaptation Assessments*. European Centre for Disease Prevention and Control, Stockholm, Sweden.

Freitas, C.M., Silva, M.E. and Osorio-de-Castro, C.G.S. (2014) A redução dos riscos de desastres naturais como desafio para a saúde coletiva. *Ciência e Saúde Coletiva* (Impresso) 5(19), 3628–3628.

Frumkin, H., McMichael, A.J. and Hess, J.J. (2008) Climate change and the health of the public.

American Journal of Preventive Medicine 35(5), 401–402. DOI: 10.1016/j.amepre.2008.08.031.

Hay, S.D., Rogers, S., Randolph, Stern, D., Cox, J., Shanks, G. and Snow, R. (2002) Hot topic or hot air? Climate change and malaria resurgence in East African highlands. *Trends in Parasitology* 18, 530–534.

IPCC (2007) Climate change 2007: 2.3.2 stakeholder involvement in impacts, adaptation and vulnerability. In: Parry, M.L., Canziani, O.F., Palutikof, J.P., van der Linden, P.J. and Hanson, C.E. (eds) *Contribution of Working Group II to the Fourth Assessment Report of the Intergovernmental Panel on Climate Change*. Cambridge University Press, Cambridge, UK.

Mehrotra, S.C., Rosenzweig, W.D., Solecki, C.E., Natenzon, A., Omojola, R., Folorunsho, J. and Gilbride, J. (2011) Cities, disasters and climate risk. In: Rosenzweig, C., Solecki, W.D., Hammer, S.A. and Mehrotra, S. (eds) *Climate Change and Cities: First Assessment Report of the Urban Climate Change Research Network*. Cambridge University Press, Cambridge, UK, pp. 15–42.

McPhearson, T., Andersson, E., Elmqvist, T. and Frantzeskaki, N. (2014) Resilience of and through urban ecosystem services. *Ecosystem Services* 12, 152–156. DOI: 10.1016/j. ecoser.2014.07.012i.

Ogden, N.H., St-Onge, L., Barker, I.K., Brazeau, S., Bigras-Poulin, M., Charron, D.F. and Thompson, R.A. (2008) Risk maps for range expansion of the Lyme disease vector, *Ixodes scapularis*, in Canada now and with climate change. *International Journal of Health Geographics* 7, 24. DOI: 10.1186/1476-072X-7-24.

Ogden, N.H., Bouchard, C., Kurtenbach, K., Margos, G., Lindsay, L.R., Trudel, L. and Milord, F. (2010) Active and passive surveillance and phylogenetic analysis of *Borrelia burgdorferi* elucidate the process of Lyme disease risk emergence in Canada. *Environmental Health Perspectives* 118(7), 909–914. DOI: 10.1289/ ehp.0901766.

Omumbo, J., Lyon, B., Waweru, S., Connor, S. and Thomson, M. (2011) Raised temperatures over the Kericho tea estates: revisiting the climate in the East African highlands malaria debate. *Malaria Journal* 10(12). Available at: http://www. malariajournal.com/content/10/1/12.

Patz, J.A., Vavrus, S.J., Uejio, C.K. and McLellan, S.L. (2008) Climate change and waterborne disease risk in the Great Lakes region of the U.S. *American Journal of Preventive Medicine* 35(5), 451–458. DOI: 10.1016/j. amepre.2008.08.026.

Reid, H. and Kovats, S. (2009) Special Issue on health and climate change of TIEMPO. *Bulletin on Climate and Development* 71.

Rose, J.B., Epstein, P.R., Lipp, E.K., Sherman, B.H., Bernard, S.M. and Patz, J.A. (2001) Climate variability and change in the United States: potential impacts on water- and food-borne diseases caused by microbiologic agents. *Environmental Health Perspectives* 109, Suppl. 2, 211–221.

Rosenzweig, C., Solecki, W.D., Hammer, S.A. and Mehrotra, S. (eds) (2011) *Climate Change and Cities: First Assessment Report of the Urban Climate Change Research Network*. Cambridge University Press, Cambridge, UK.

Rydin, Y., Bleahu, A., Davies, M., Dávila, J.D., Friel, S., De Grandis, G., Groce, N., Hallal, P.C., Hamilton, I., Howden-Chapman, P., Lai, K.-M., Lim, C., Martins, J., Osrin, D., Ridley, I., Scott, I., Taylor, M., Wilkinson, P. and Wilson, J. (2012) Shaping cities for health: complexity and the planning of urban environments in the 21st century. *Lancet* 379, 2079–2108.

Satterthwaite, D., Huq, S., Reid, H., Pelling, M. and Romero Lankao, P. (2008) Adapting to climate change in urban areas: the possibilities and constraints in low- and middle-income nations. *Human Settlements Discussion Paper Series: Climate Change and Cities, No. 1*. International Institute for Environment and Development, London, UK. Available at: http://pubs.iied. org/10549IIED/ (accessed 18 October 2016).

Semenza, J. and Menne, B. (2009) Climate change and infectious diseases in Europe. *Lancet* 9, 365–375.

Semenza, J.C., Herbst, S., Rechenburg, A., Suk, J.E., Höser, C., Schreiber, C. and Kistemann, T. (2012) Climate change impact assessment of food- and waterborne diseases. *Critical Reviews in Environmental Science and Technology* 42(8), 857–890. DOI: 10.1080/10643389.2010.534706.

Smith, K.R., Woodward, A., Campbell-Lendrum, D., Chadee, D.D., Honda, Y., Liu, Q., Olwoch, J.M., Revich, B. and Sauerborn, R. (2014) Human health: impacts, adaptation, and co-benefits. In: Field, C.B., Barros, V.R., Dokken, D.J., Mach, K.J., Mastrandrea, M.D., Bilir, T.E., Chatterjee, M., Ebi, K.L., Estrada, Y.O., Genova, R.C., Girma, B., Kissel, E.S., Levy, A.N., MacCracken, S., Mastrandrea, P.R. and White, L.L. (eds) *Climate Change 2014: Impacts, Adaptation, and Vulnerability. Part A: Global and Sectoral Aspects. Contribution of Working Group II to the Fifth Assessment Report of the Intergovernmental Panel on Climate Change*. Cambridge

University Press, Cambridge, UK and New York, USA, pp. 709–754.

WHO (1948) Constitution of WHO: principles. World Health Organization. Available at: http://www.who.int/about/mission/en (accessed 2 June 2016).

WHO (2004) Global Strategic Framework for Integrated Vector Management. World Health Organization, Geneva, Switzerland. Available at: http://whqlibdoc.who.int/hq/2004/WHO_CDS_CPE_PVC_2004_10.pdf.

WHO (2015) Global Health Observatory (GHO) data. World Health Organization. Available at: http://www.who.int/gho/urban_health/situation_trends/urban_population_growth_text/en (accessed 6 June 2016).

13 Innovative Formulations Useful for Area-wide Application Suitable for Climate Change

David Liszka* and Pawel Swietoslawski

ICB Pharma, Jaworzno, Poland

13.1 Introduction

Significant acceleration of climate change since the mid-20th century, involving so-called global warming, with increase in temperature in the troposphere, is notable. Until this time, the climate changes have been mainly conditioned by natural factors, such as changes in solar activity, changes in Earth's orbit, volcanic eruptions, or the impact of the El-Niño phenomenon. In the last 50 years, anthropogenic factors, i.e. those which are caused by people, until then considered as marginal, came to the fore. One of the major determinants of climate change is greenhouse gases and aerosol emissions (primarily sulfur dioxide). The immediate consequence is an increase in the average temperature of the Earth's surface and oceans. The temperature increase helps to increase the rate of water evaporation, which in turn causes changes in the atmospheric circulation and precipitation. The ultimate consequence of these changes is an increase in the incidence of extreme droughts, storms and floods, which facilitate migration and propagation of disease-carrying urban insects, such as mosquitoes and flies.

The changes in climatic variables can alter parasite ecology by affecting host and geographic distribution, infection pressure, prevalence and intensity of parasites, and can do so directly (via free-living stages) or indirectly (by affecting hosts). Shifts or expansion in distribution, prevalence and intensity of parasites may be closely linked with that of their hosts and may depend on numerous factors. The effects of environmentally detrimental changes in local land use and alterations in global climate disrupt the natural ecosystem and can increase the risk of transmission of parasitic diseases in the human population.

It is known that mosquitoes carry many serious diseases and viruses such as malaria, zika, dengue, and West Nile fever, and in the 2000s, due to progressive global warming, this has become an alerting and serious problem. A team of US scientists, led by Dr Nathan Burkett-Cadena from Auburn University in Alabama, have discovered a very interesting relationship. In the colder winter days mosquitoes switch from feeding mainly on bird blood to mainly the blood of mammals. As a result, there is less chance of infection from a mosquito-borne disease during winters. In early spring, the mosquitoes feed primarily on birds, to later pass to mammals, including humans. Therefore, in the temperate climate zone, the number of cases of infections transmitted by mosquitoes increases significantly at a time when the change of host occurs. Mosquitoes become infected with many viruses while feeding on birds and later they move to feed on

*E-mail: david.liszka@icbpharma.pl

mammals that can easily infect the blood of another host, such as humans. Epidemiologically, the most important factor is the time at which the change of the host takes place. The sooner it happens, the longer the period of intensive feeding by the mosquitoes on mammals. So the risk of infection of the diseases carried by these insects is higher.

13.2 Change in Climate and Pest Management

Climate change is expected to influence pests over large areas and not in isolation. Managing pests over a large area in a given time frame will present unique problems to pest managers. However, there are a number of new techniques being developed to address this issue, which offer potential advantages to traditional and more localized approaches. In outbreak situations, suppression over a broad area can reduce re-infestation and can be considered more effective. Incidents such as the recent tsunamis in Indonesia and Japan, and hurricanes in the USA are reminders that climate change can trigger devastation and pest outbreaks on the same scale.

Currently, the control of most urban pests is still carried out by the use of insecticides. Although other control technologies are often incorporated into the practitioner's system, these technologies are less popular than conventional techniques, which, though effective, are usually harmful to surroundings, leave residues and can affect non-target organisms, warranting an additional back-up plan.

In order to decrease the risks of infection delivered by the disease-carrying insects, scientists around the world are working on various chemical formulations of larvicides and adulticides with various insecticidal active ingredients. Important discoveries and the evolution of the chemical industry in the 20th and 21st centuries have provided many new synthetic insecticides with enormous potential. Introduction of the 'wonder' insecticide, dichlorodiphenyltrichloroethane, best known as DDT, through development of organophosphates, carbamates and finally the discovery of photostable pyrethroids, has led to profound changes in the way pests are being managed today. Unfortunately, although this was a great achievement, pesticides have proven not to be a perfect answer to the pest control problem, mainly because of their high toxicity to non-target species. Consequently, alternative methods were explored, such as moult inhibitors, metabolic disrupters and insect behaviour modifiers.

The environmental problems were not the only ones to become significant as synthetic insecticides came into use. While insecticides have greatly improved human health and agricultural production worldwide, their usefulness has been limited by the evolution of resistance in many major pests, including some that became pests only as a result of insecticide use. Therefore, it is of paramount importance that all future pest-control practices should take into account the possibility of evolutionary resistance. It is also felt that pest susceptibility, a valuable natural resource, is being misused and slowly squandered. It will probably never again be possible to achieve chemical control of insects on the scale achieved between 1945 and 1965 (Wood, 1981). Resistance rapidly increased in a number of agriculturally and medically important insect species. The need of high dosages and more frequent applications to combat these problems requires a newer approach for better pest management.

This chapter discusses a number of novel technologies, such as use of 3D-IPNS (3-Dimensional Immobilizing Polymer Network Structure) acting through a physical mechanism, which is a completely new and promising approach to the current problems in pest management, in addition to a number of other formulations. All these technologies are new and have far-reaching consequences when used.

13.3 Using 3D-IPNS: A Physical Action Formulation

The physical method of pest control, using a sprayable formulation using a conventional sprayer, is a new method, and so far, unique. It can be defined as suppression of insects by physical or mechanical means. This strategy may be used alone, but is often seen as a component of an integrated pest-management programme. In general, insects and mites are considered unlikely to develop resistance to physical control techniques.

Although the use of formulations with a physical mode of action was reported as early as the 18th century, it was not until the mid-19th century that petroleum distillate was first used against scale insects on orange trees. During that time, petroleum oil was used as a emulsion.

An important constraint to the use of agricultural mineral oils is their potential to be phytotoxic. Such phytotoxicity is mostly generated by acidic compounds in oils formed largely by the oxidation of unsaturated molecules. They cause acute effects that are generally observed as burns and necrotic lesions shortly after application of oil to plants. These effects are primarily related to cell death, due to disruption of membranes by the acidic compounds, and are influenced by oil quality, dose and ambient conditions, particularly temperature, humidity and ultraviolet light.

Silicone surfactants are also described as compounds for killing arthropods. The first publication on the application of silicone surfactants as pesticides describes one of them as an aphicide. It is now understood that silicone does not produce a substantial effect in low humidity. The authors drew the conclusion that the use of silicone surfactants as insecticides would only be effective on aquatic insects and small insects such as aphids, whiteflies and mites. It has been proven that organomodified siloxanes provide high efficacy against two-spotted spider mites, while in control of silverleaf whitefly on tomato, the same formulation caused severe phytotoxicity even at the lowest rate (Liu and Stansly, 2000). The hypothesis is that because of its low surface tension, silicone-based surfactants spread rapidly over the insect's body, infiltrating the tracheal system, and the insect dies by drowning (Purcell and Schroeder, 1996).

The 3D-IPNS technology when applied acts by forming a highly advanced polycondensation hybrid material with altered physical properties when compared to the individual components alone. On the treated surface, it forms a spatial, three-dimensional network structure that immobilizes the target pests upon application by either encapsulation or trapping the target species. The chemical spreads efficiently, covering the whole insect body. Thanks to this, it precisely proliferates over the surface and creates a cross-linked, three-dimensional network structure, tightly covering all developmental stages of the pest insects and immobilizing them. Unlike mineral oils, it creates a permeable (breathable) network that does not disrupt the physiological processes by filling the tracheoles and blocking them. This technology allows penetration of the body surface of the pest and, as a secondary effect, immobilizes the valve of the spiracle either preventing the spiracles from closing or keeping them closed. This leads to dehydration, anoxic asphyxiation or osmotic stress, resulting in rupture of the gut in haematophagous parasites that have recently ingested a blood meal, depending on the position the valve is blocked in. This is the effect of inhibition of the parasite's ability to excrete water by transpiration through the spiracles.

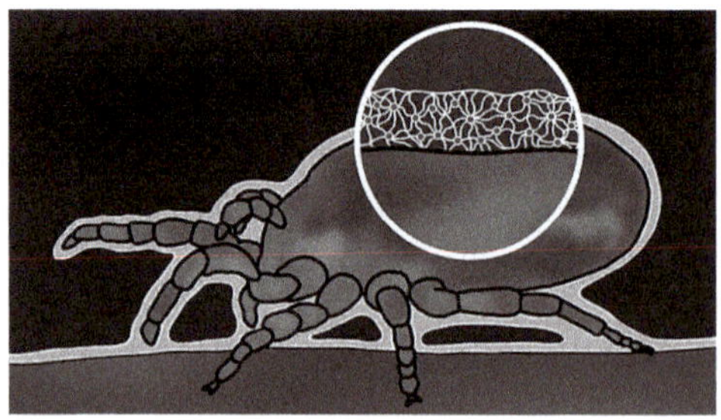

Fig. 13.1. Three-dimensional network structure immobilizing the poultry mite. Source: authors.

13.3.1 Vector insect control

Insecticide resistance of most insects of medical significance, including mosquitoes, has already developed in several countries. Therefore, a physical strategy for the control of adult mosquitoes and flies by entrapping them may circumvent the relatively restricted application of physical control techniques. According to numerous efficacy trials performed against various plant pests, this technology is proving to be safe and will not cause the environmental problems that are so evident in the case of synthetic organic insecticides. This technology can now be used against insect vectors (Fig. 13.2).

13.3.2 Animal health application

A number of insect pests with human significance are expected to be influenced by climate change. Notable among these are pests related to animal and poultry farms. Mites are at the top of this pest list and a number of issues on their control is coming to light. Among them a significant increase in acaricide resistance has been observed, which has resulted in the proliferation of haematophagous arachnids, such as poultry red mite (*Dermanyssus gallinae*) and northern fowl mite (*Ornithonyssus sylviarum*) in endemic regions of the world. These pests are becoming serious ectoparasites in poultry farms, with increased prevalence observed more often in non-intensive systems such as free-range, barns or backyards and also more often in laying hens than in broiler birds. In addition, there is evidence of potential host-expansion and migration of *D. gallinae* infestation to ducks and turkeys. In this context, it is of paramount importance that all future haematophagous ectoparasite intervention strategies should take into account the possibility of resistance development in the pest population.

Poultry red mite (*D. gallinae*) is an important problem in Europe, while northern fowl mite (*O. sylviarum*) is a common external parasite of both domestic fowl and wild birds, mostly in the USA. The current control strategy in poultry production sites typically relies on synthetic acaricides, in spite of advancements in other (immunological and biotechnological) approaches. Repeated long-term use of these constituents has resulted in the development of a 'super-population' of poultry mites, resistant to most neurotoxins. Hence, the availability of effective acaricides has diminished rapidly.

The repeated use of acaricides, sometimes in high concentrations to control infestation of poultry red mites, has meant that the pest has developed resistance. The intensive use of acaricides may also lead to accumulation of the active ingredient in chicken organs, tissues and eggs, causing harm to consumers.

A totally new approach to control these haematophagous arachnids is now possible by the use of 3D-IPNS techniques, a mechanism which is a completely new and promising approach to the current problems in poultry mite management (Figs 13.3 and 13.4). This method results in complete immobilization of the pest, leading to high mortality, thus eliminating the possibility of parasite resistance (Fig. 13.5). This method will effectively tackle the population of such ectoparasites to reduce economical losses, improve welfare, control zoonotic risk for farming workers and provide safety to consumers.

Fig. 13.2. Fly entrapped by three-dimensional network structure. Source: authors.

Fig. 13.3. The deposition of standard acaricidal solution. Source: authors.

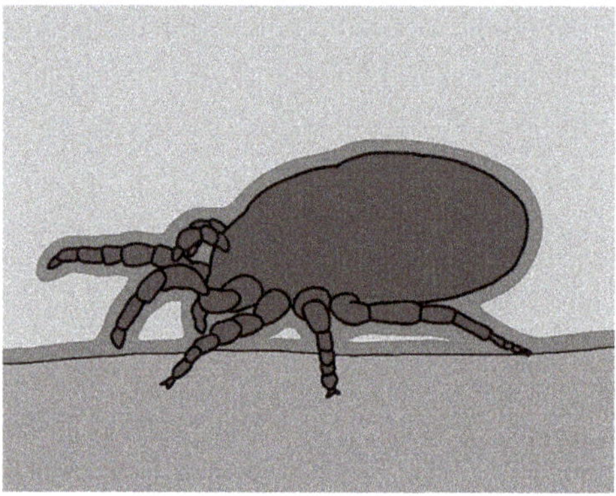

Fig. 13.4. The effect of spreading. Source: authors.

Fig. 13.5. Immobilized poultry mites in cage structure. Source: authors.

13.3.3 3D-IPNS characteristics

The 3D-IPNS technology has proven to be highly effective in combating numerous plant pests. After many years of trials conducted in field conditions, it is considered to be very effective against foliage-feeding pests, including aphids, spider mites, scale insects, psyllid, thrips and whiteflies, infesting cultivated plants, fruit trees and shrubs, ornamentals and certain vegetable and crop species (Fig. 13.6). Taking into account its mode of action, 3D-IPNS technology doesn't affect the population of beneficial Hemiptera (*Anthocoris* and *Orius*) and Hymenoptera, while the beneficial Phytoseiidae are suppressed for a while, but recover completely within about 2 weeks after application. One of the most important characteristics that needs to be emphasized is the fact that this technology is highly potent in controlling resistant populations. Moreover, it does not promote the creation of resistant insect populations through continuous exposure. Lack of pre-harvest intervals and possible accumulation in the environment makes this technology an extremely valuable alternative to synthetic insecticides. It should also be emphasized that apitoxicological assessment of the formulation doesn't pose an unacceptable risk to pollinators. 3D-IPNS can also be provided in the tank mix with insecticides, improving their efficacy by increasing the likelihood that the toxicant is transferred to the insect on contact, enhancing the bioavailability of the active ingredient.

13.4 Using a Physical Action Adjuvant

Adjuvants have been used as long as pesticides. In the early part of this century, animal proteins such as calcium caseinate were used as dispersants for lead arsenate. Animal bone glues were used as stickers (Witt, 2012). Pesticides that were available then were not as effective as now, were difficult to formulate and dispersed inadequately. Only a few natural colloids and surfactants were available as aids or adjuvants. The focus on formulation aids such as adjuvants continued to grow as a means to maximize the effectiveness of a limited number of pesticides available. Today, adjuvants are a well-recognized part of formulation chemistry

(a)

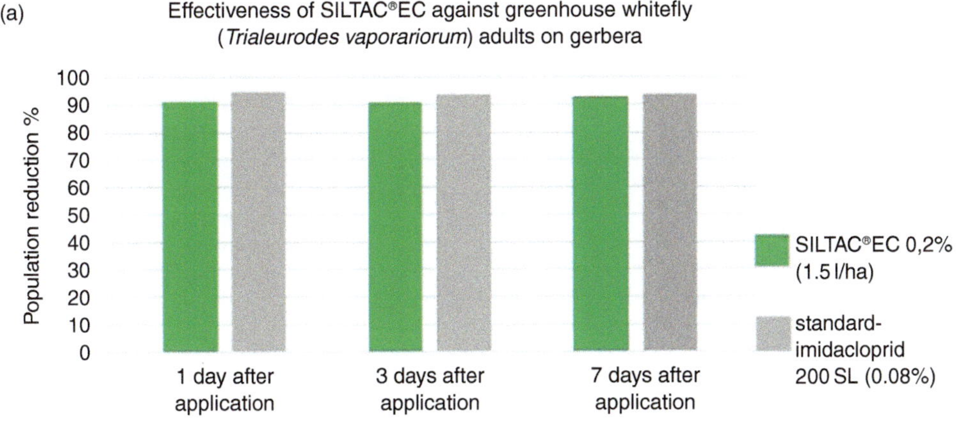

(b)

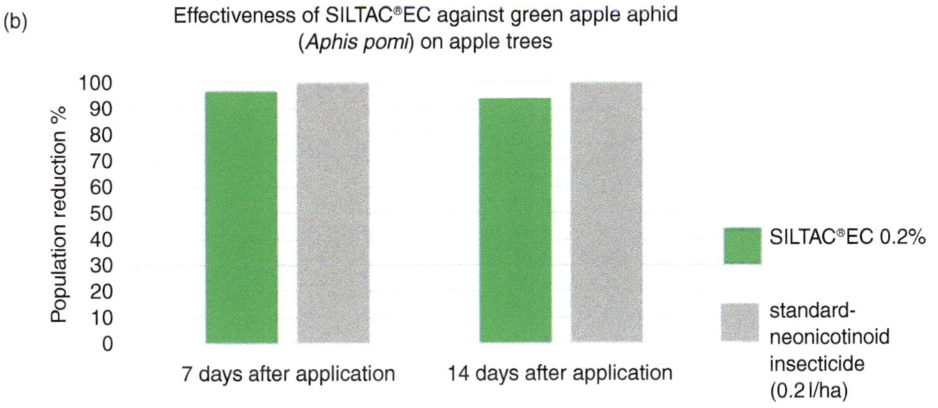

Fig. 13.6. Efficacy of 3D-IPNS technology (SILTAC®EC) against various pest species (a) on ornamentals and (b) in orchards. Source: authors.

and help to increase the efficacy of products. Although these developments were all directed towards the improvement of agriculture formulations, this concept, after modification, can also be utilized in urban pest control. As a consequence, it will allow pest susceptibility to continue and reduce resistance built up in urban pests.

Adjuvants play a significant role in reducing several problems encountered during spray application in spite of advancements made in discovering newer active principles in pesticides and improvement in application techniques. Problems encountered during pesticide application include drift, coverage, adherence, volatilization, penetration, solubility, surface tension, foaming, suspension, evaporation, stability, incompatibility, alkalinity degradation and odour. Adjuvants are formulated to minimize these application problems by buffering, sticking, and also by reducing factors like foaming, spreading, evaporation, emulsifying, drift, volatilization, and odour. Adjuvants also help highlight the area where spray has been applied, increasing compatibility, dispersing and wetting.

This section discusses an adjuvant acting via physical mechanisms, Provecta, which is a new and promising development to address the current problems in the pest control industry. Provecta is a unique mixture of polymeric compounds to be combined in the spray tank with insecticides (toxicants) for more effective treatment. It generates specific actions on contact resulting in dehydration and eventual suffocation through the following mechanisms.

13.4.1 Spreading

When applied, Provecta provides quick and even spreading, helping penetration precisely in the treated surface. It also increases the likelihood that the toxicant is transferred to the insect on contact, enhancing the bioavailability of the active ingredient. This is a crucial consideration when dealing with resistant pest populations harbouring in inaccessible locations, e.g. bed bugs (Fig 13.7).

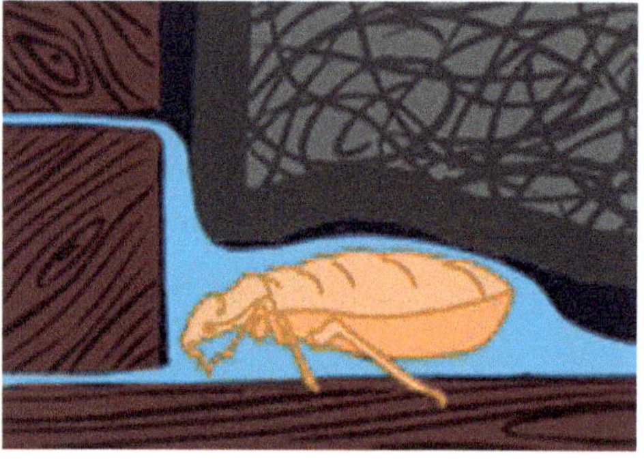

Fig. 13.7. The action of spreading. Source: authors.

13.4.2 Immobilization

After application, Provecta creates a cross-linked, three-dimensional network structure on the treated surface, tightly covering all developmental stages of the insect pests and immobilizing them. This leads to increased exposure to the toxicant. It also improves efficacy through additive effects of the two mechanisms, neurotoxic and physical (Fig. 13.8).

13.4.3 Dehydration and/or suffocation

Once applied, the formulation completely penetrates the body surface of the pest and fills their spiracles (which have hairs at the opening for filtering and a valve for controlling air flow), either preventing them from closing or keeping them closed. This leads to dehydration and suffocation. Both the physical effects eventually kill the insect pest (Fig. 13.9).

All these combined features – enhanced spreading, immobilization (exposure profile modification) and physical effects – significantly improve the efficacy of the formulation and result in quicker mortality. Provecta, an effective adjuvant with physical action against pests, may deal with the increasing problem of resistance to insecticides and environmental concerns with a number of favourable features such as:

Fig. 13.8.   Immobilization through (a) neurotoxic and (b) physical actions. Source: authors.

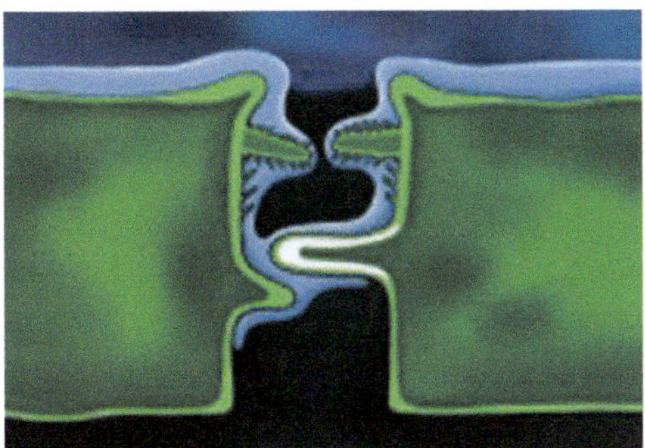

Fig. 13.9. Immobilization of the valve of the respiratory spiracle. Source: authors.

- high efficacy in controlling numerous pest species
- the physical mode of action eliminates the possibility that pests will become resistant
- the synergetic effect results in reduced rate of insecticide or dosages
- perfect at locations where pesticide application predominates, and has resulted in development of resistance
- recommended to be used in a tank mix to control resistant populations in difficult-to-access locations, e.g. against bed bugs
- equivalent to the effects of two different mechanisms added together (neurotoxic and physical).

13.5 Using an Attract-and-kill Formulation

Crawling insect pests in the urban environment, such as cockroaches, could be a challenge in a changing climate. The major reason for failure in controlling such pests is noticeably not due to poor product or application method, but because of access. Most crawling insect pests hide under household goods that are difficult to move or relocate for insecticide treatments.

To address these pressing issues, Attracide, a novel insecticide formulation containing a toxicant and an attractant as active components was developed. The formulation specifically draws out pests such as cockroaches and ants from their hideouts into specific sprayed zones by the use of a patented attractant. The formulation is generally sprayed using a handheld compressed-air sprayer, the conventional tool for pest-control practitioners. The efficacy and usefulness of this product has far-reaching consequences on future methods for practitioners in controlling cockroach and other crawling insect pests.

13.5.1 Product characteristics of an attract-and-kill formulation

One of the most advanced technologies in pest control is the use of microencapsulation to 'attract and kill' the pest. This technique is a novelty in controlling cryptic pests. It is based on a sprayable encapsulated film (Fig. 13.10). It uses solid-core microcapsule technology to hold the active ingredient on both porous and non-porous surfaces for a much longer time than conventional residual formulation. The effective and prolonged control is ensured by mixture of two types of capsules.

1. The attractant is carried in a small microcapsule (d_{90} ~ 2.0 µm). The food grade attractant diffuses out from the capsule, attracting the insects to a treated site.
2. The toxicant is carried within large liquid-core microcapsules (d_{90} ~ 35–55 µm) with a thin shell: when insects walk over the formulation, the microcapsules burst to release the formulation. If they're not trampled, the active ingredient slowly diffuses out from the capsule in a controlled concentration. Large capsules also provide much better efficacy on porous surfaces (contact exposure). Between the shell and core of the capsule, there are trapped carbon dioxide bubbles that increase the efficacy of the trampling mechanism.

13.5.2 Mechanism of action – trampling technology

The attractant, carried in small microcapsules, is delivered prior to the toxicant and lures the insect to a treated site. By passing and crawling the insects pick up the product and carry the capsules loaded with both toxicant and attractants to their nest and unknowingly pass them to the other members of the colony (Fig. 13.10). These capsules are too small to be grabbed intentionally and cockroaches, when grooming their legs or antennae to which capsules have adhered, ingest them, increasing the chances of effectiveness as well as against fipronil-resistant species. The microcapsulated attractant significantly stimulates the grooming behaviour of cockroaches and hence changes the mode of action from contact to stomach.

Further, the efficacy of this formulation is based on the triple mechanism: 1) contact (absorption through cuticle); 2) absorption from stomach; and 3) trampling mechanisms. In case of trampling, the biological efficacy is dependent on the strength of the capsules.

13.6 Using a Controlled-release Capsulated Larvicide

Changes in vector density and distribution are resulting from anthropogenic environmental changes and from widespread disruptions to ecosystem services. These promote the migration of mosquitoes to populate new territories. The ongoing climate change will not only destroy many natural mosquito-breeding sites, but may create new ones. This may force many species to change and modify their feeding preferences by being anthropophilic.

To develop a suitable area-wide application system for controlling mosquito surge, a strategy of combining a microencapsulated *Bti* (*Bacillus thuringiensis* var. *israelensis*) with a microencapsulated formulation of an insect growth regulator (IGR) (pyriproxyfen, (S)-methoprene) was attempted for long-term mosquito management. This will also make it possible to reduce the concentration of the IGR and the dosage of *Bti* provided to the environment without a negative impact on efficacy and reducing environmental concerns to a great extent. The bacterium (*Bti*) produces protein toxins during sporulation that are concentrated in a parasporal body. These proteins are highly toxic to mosquito and blackfly larvae.

However, the efficacy of *Bti* is highly sensitive to the environmental conditions, such as when exposed to UV light. To overcome this, a new formulation was developed with the following characteristics.

1. Protection from photooxidation, providing long-term residual efficacy.
2. The formulation prevented sinking and adsorption of the *Bti* in the top layer of the water body.
3. The protoxin was made to be released and solubilized in the alkaline midgut milieu of the target species.

The above were achieved by the use of a special solid-core encapsulation technology

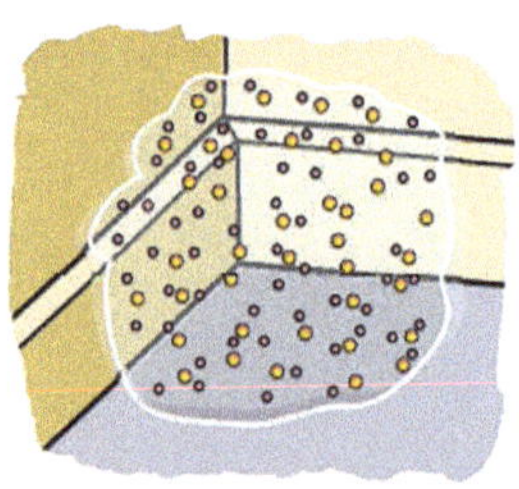
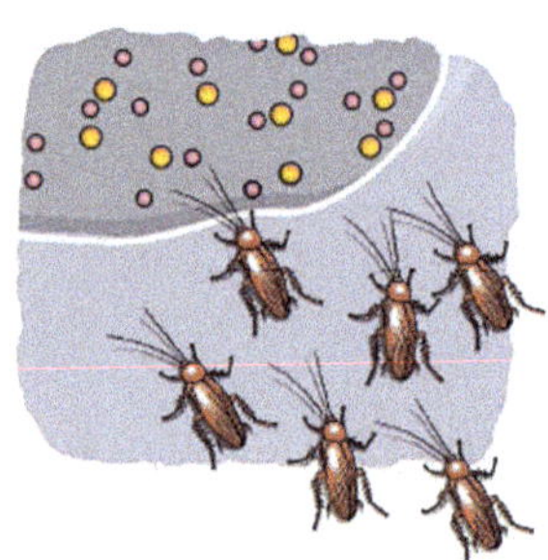
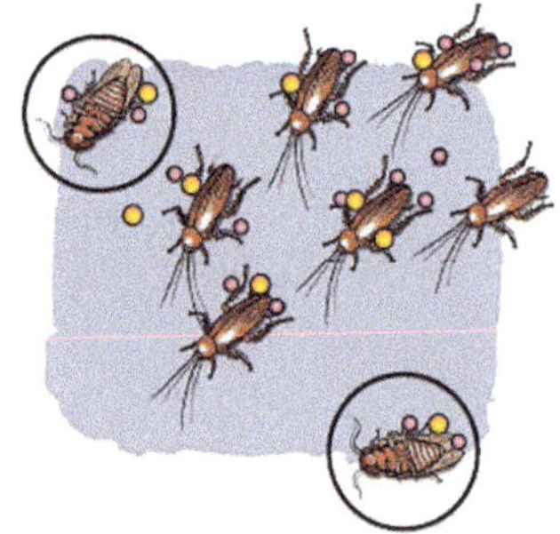

Fig. 13.10. (a) The application site carries encapsulated attractants and the insecticide together. (b) The insects pass over and (c) carry the insecticides with them to other members in the colony. This reduces the population significantly more than a conventional formulation. Source: authors.

(Pyrilarv®), which made it possible to effectively protect *Bti* from rapid photo-degradation. These unique microcapsules are bounded by a layer of gypsum, which is specially modified with polymers and carbon black, enclosing a grain of sand. This form of formulation when applied to mosquito-breeding sites provides additional advantages as listed below.

1. Water activation (Fig. 13.11) – remains stable even in strong sunlight, while water elutes the capsules from the granular carrier. Hence, it can be applied as a mosquito development preventative tool on potential flood areas before flooding. An important characteristic, especially in locations at high risk of mosquito-borne diseases.

2. Long-term residual activity (Fig. 13.12) – water slowly dissolves the gypsum from the granules lying on the bottom of the reservoir, which eventually allows the microcapsules to be released. Microcapsules appear on the surface of the water.

3. Controlled release – long-term residual activity (Fig. 13.13), therefore lower cost per treatment. The active substance is contained in capsules that rupture in the larval gut only (pH-regulated shell degradation). Capsules are delivered by the elution from the gypsum layer of the granules lying on the bottom of the reservoir. This double trigger-release mechanism provides effective control over months. Non-encapsulated *Bti* formulations provide a high level of efficacy; however, the population will begin to recover between 5 and 7 days after each treatment.

4. Good distribution in water – the structure of the product and unique encapsulation technology provides good distribution of the active ingredient in the water. Therefore, it can be used in locations that are not easily accessible.

5. Easy application – conventional ground or aerial application may be used without any special equipment. The high mass core of the granules ensures targeted application,

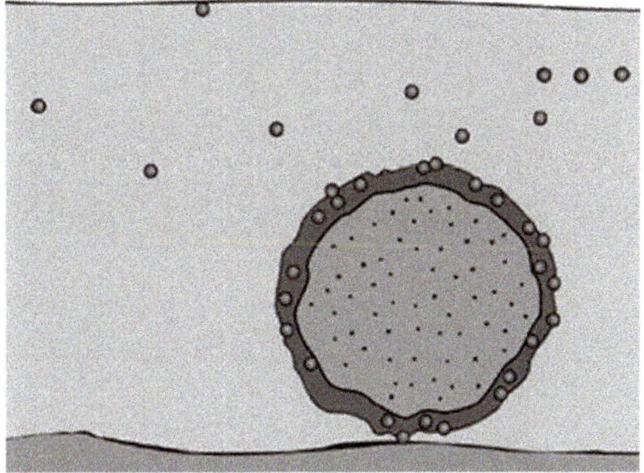

Fig. 13.12. Dissolving of outer capsule and release of microcapsules. Source: authors.

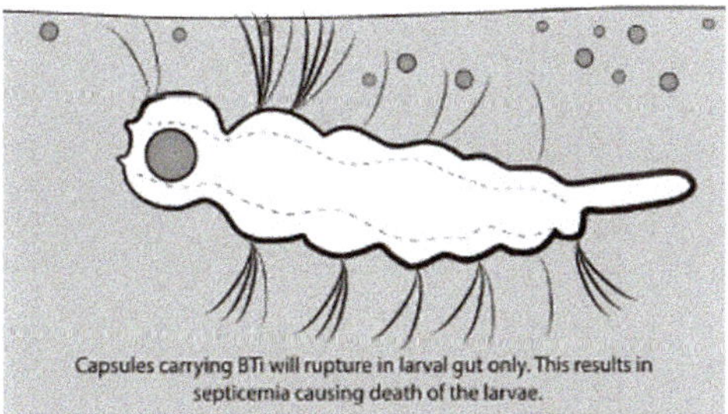

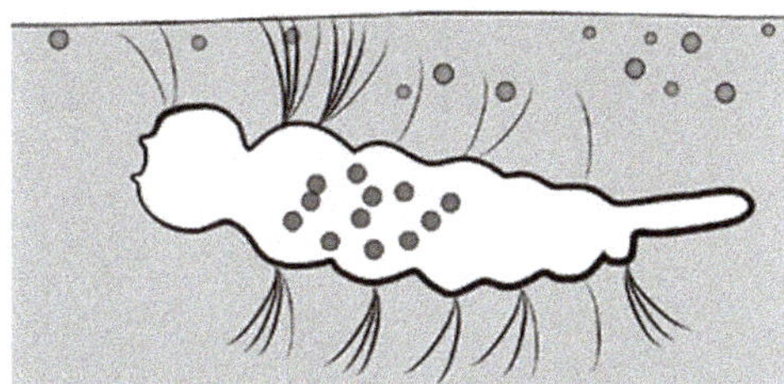

Fig. 13.13. (a) Entry of the capsules into the gut of the larvae. (b) Capsules carrying *Bti* will rupture in the larval gut only, resulting in septicaemia and causing death of the larvae. Source: authors.

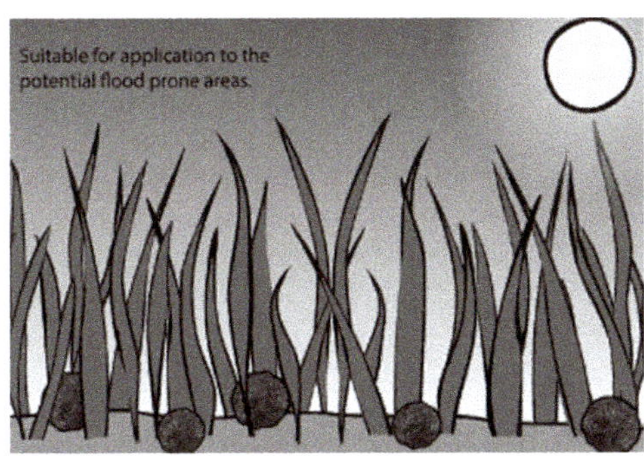

Fig. 13.11. Suitable for application to the potential flood-prone areas. Source: authors.

even in adverse weather conditions when most liquid formulations cannot be applied.

6. Safety of non-target species. The shell of the capsule is not digestible by mammals. *Bti* is not harmful to human beings, so this provides additional application safety. Therefore, this delivery system can also be applied with other substances with a less attractive toxicological profile.

7. Targeted release – microcapsules carrying the *Bti* are released in the area where mosquito breeding takes place (shore area). The capsules will be removed by the waves or wind from the mosquito-breeding site, and the next portion of capsules will quickly be delivered.

13.7 Discussion

In recent times, we have seen the development of a number of high-performance formulations for use in the pest-control industry. Formulations such as microemulsion (ME), capsule suspension (CS), dry-flowable (DF), gels (GEL), granules (G) and baits (B) are some examples which are popular among practitioners. These formulations help achieve targeted delivery along with a reduction in the amount of actual insecticide use. Consequently, dramatic changes in pest-control strategies have taken place and are noticeable globally. Conventional sprayers and indiscriminate sprays have been replaced by precise and targeted delivery.

Insecticides used in urban pest control have a strong impact on human health (Dhang, 2011) and the environment. In addition, insecticide usage is constrained by evolving resistance in pest insects, resulting in the use of higher dosages to provide effective treatments. There is a demand for safe pest management and more efficient methods, including protection for the environment and non-target species. With increases in costs and strict regulations, products with a precise delivery method that are targeted, environmentally safe and easy to use could be perfect candidates for large-scale use in the future.

References

Burkett-Cadena, N.D., Hassan, H.K., Eubanks, M.D., Cupp, E.W. and Unnasch, T.R. (2012) Winter severity predicts the timing of host shifts in the mosquito *Culex erraticus. Biology Letters* 8(4), 567–569.

Dhang, P. (2011) Insecticides as urban pollutants. In: Dhang, P. (ed.) *Urban Pest Management: An Environmental Perspective.* CAB International, Wallingford, Oxfordshire, UK, pp. 1–18.

Liu, T.X. and Stansly, P.A. (2000) Insecticidal activity of surfactants and oils against silverleaf whitefly (*Bemisia argentifolii*) nymphs (Homoptera: Aleyrodidae) on collards and tomato. *Pesticide Management Science* 56, 861–866.

Purcell, M.F. and Schroeder, W.J. (1996) Effect of Silwet L-77 and diazinon on three tephridid fruit flies (Diptera: Tephritidae) and associated endoparasites. *Journal of Economic Entomology* 89, 1566–1570.

Witt, J.M. (2012) Agricultural spray adjuvants. Oregon State University Pesticide Safety Education Program (PSEP) fact sheets. Available at: http://psep.cce.cornell.edu/facts-slides-self/facts/gen-peapp-adjuvants.aspx (accessed 17 March 2016).

Wood, R.J. (1981) Insecticide resistance: genes and mechanisms. In: Bishop, I.A. and Cook, L.M. (eds) *Genetic Consequences of Man-made Change.* Academic Press, San Diego, California, USA, pp. 53–96.